S0-BXN-128

INTERNATIONAL UNION OF PURE AND APPLIED CHEMISTRY

ANALYTICAL CHEMISTRY DIVISION
COMMISSION ON SOLUBILITY DATA

---

# SOLUBILITY DATA SERIES

Volume 9

## ETHANE

## SOLUBILITY DATA SERIES

| | |
|---|---|
| Volume 1 | H. L. Clever, *Helium and Neon* |
| Volume 2 | H. L. Clever, *Krypton, Xenon and Radon* |
| Volume 3 | M. Salomon, *Silver Azide, Cyanide, Cyanamides, Cyanate, Selenocyanate and Thiocyanate* |
| Volume 4 | H. L. Clever, *Argon* |
| Volume 5/6 | C. L. Young, *Hydrogen and Deuterium* |
| Volume 7 | R. Battino, *Oxygen and Ozone* |
| Volume 8 | C. L. Young, *Oxides of Nitrogen* |
| Volume 9 | W. Hayduk, *Ethane* |
| Volume 10 | R. Battino, *Nitrogen and Air* |
| Volume 11 | B. Scrosati and C. A. Vincent, *Alkali Metal, Alkaline-Earth Metal and Ammonium Halides. Amide Solvents* |
| Volume 18 | O. Popovych, *Tetraphenylborates* |

### *Volumes in preparation*

C. L. Young, *Oxides of Sulfur*

H. L. Clever and W. Gerrard, *Hydrogen Halides in Non-Aqueous Solvents*

H. Miyamoto and M. Salomon, *Copper and Silver Halates*

S. Siekierski and J. K. Navratil, *Actinide Nitrates*

E. M. Woolley, *Silver Halides*

Z. Galus and C. Guminski, *Metals and Intermetallic Compounds in Mercury*

J. W. Lorimer, *Alkaline-Earth Metal Sulfates*

A. L. Horvath and F. W. Getzen, *Halogenated Benzenes, Toluenes, Xylenes and Phenols with Water*

H. L. Clever, D. M. Mason and C. L. Young, *Carbon Dioxide*

H. L. Clever and C. L. Young, *Methane*

# SOLUBILITY DATA SERIES

Volume 9

## ETHANE


Volume Editor

WALTER HAYDUK
*University of Ottawa*
*Ontario, Canada*

Evaluators

RUBIN BATTINO
*Wright State University*
*Dayton, Ohio, USA*

H. LAWRENCE CLEVER
*Emory University*
*Atlanta, Georgia, USA*

WALTER HAYDUK
*University of Ottawa*
*Ontario, Canada*

COLIN L. YOUNG
*University of Melbourne*
*Parkville, Victoria, Australia*

Compilers

M. ELIZABETH DERRICK
*Valdosta State University*
*Valdosta, Georgia, USA*

J. CHR. GJALDBAEK
*Royal Danish School of Pharmacy*
*Copenhagen, Denmark*

PATRICK L. LONG
*Emory University*
*Atlanta, Georgia, USA*

E. S. THOMSEN
*Royal Danish School of Pharmacy*
*Copenhagen, Denmark*

PERGAMON PRESS

OXFORD · NEW YORK · TORONTO · SYDNEY · PARIS · FRANKFURT

| | |
|---|---|
| U.K. | Pergamon Press Ltd., Headington Hill Hall, Oxford OX3 0BW, England |
| U.S.A. | Pergamon Press Inc., Maxwell House, Fairview Park, Elmsford, New York 10523, U.S.A. |
| CANADA | Pergamon Press Canada Ltd., Suite 104, 150 Consumers Rd., Willowdale, Ontario M2J 1P9, Canada |
| AUSTRALIA | Pergamon Press (Aust.) Pty. Ltd., P.O. Box 544, Potts Point, N.S.W. 2011, Australia |
| FRANCE | Pergamon Press SARL, 24 rue des Ecoles, 75240 Paris, Cedex 05, France |
| FEDERAL REPUBLIC OF GERMANY | Pergamon Press GmbH, 6242 Kronberg-Taunus, Hammerweg 6, Federal Republic of Germany |

---



First edition 1982

**Library of Congress Cataloging in Publication Data**

Main entry under title:
Ethane.
(Solubility date series ; v. 9)
Includes bibliographical references and index.
1. Ethanes—Solubility—Tables. I. Hayduk, Walter.
II. Series.
QD305.H6E74 1982 547'.01 82-15045

**British Library Cataloguing in Publication Data**

Ethane.—(Solubility data series, ISSN 0191-5622;
v. 9)
1. Ethanes—Solubility—Tables
I. Hayduk, Walter II. International Union
of Pure and Applied Chemistry. *Commission on Solubility Data* III. Series
665.7 QD305.H6

ISBN 0-08-026230-9

*In order to make this volume available as economically and as rapidly as possible the authors' typescripts have been reproduced in their original forms. This method unfortunately has its typographical limitations but it is hoped that they in no way distract the reader.*

*Printed in Great Britain by A. Wheaton & Co. Ltd., Exeter*

CONTENTS

*Foreword* vii
*Preface* xi
*Introduction:* The Solubility of Gases in Liquids xv

Water 1
Water at high pressure 16
Mixed ethene-ethane gas in water at high pressure 23
Aqueous electrolyte solutions 27
Aqueous micellular solutions 61
Aqueous organic solutions 64
Aqueous organic solutions at high pressure 74
Alkanes 77
Alkanes at high pressure 110
Non-polar solvents excluding alkanes 138
Alcohols 166
Polar solvents excluding water, aqueous solutions and alcohols 195
Various organic solvents and hydrogen sulfide at high pressure 232

System Index 253
Registry Number Index 259
Author Index 261

SOLUBILITY DATA SERIES



INTERNATIONAL UNION OF PURE AND APPLIED CHEMISTRY

IUPAC Secretariat: Bank Court Chambers, 2-3 Pound Way,
Cowley Centre, Oxford OX4 3YF, UK

# FOREWORD

*If the knowledge is*
*undigested or simply wrong,*
*more is not better.*

How to communicate and disseminate numerical data effectively in chemical science and technology has been a problem of serious and growing concern to IUPAC, the International Union of Pure and Applied Chemistry, for the last two decades. The steadily expanding volume of numerical information, the formulation of new interdisciplinary areas in which chemistry is a partner, and the links between these and existing traditional subdisciplines in chemistry, along with an increasing number of users, have been considered as urgent aspects of the information problem in general, and of the numerical data problem in particular.

Among the several numerical data projects initiated and operated by various IUPAC commissions, the *Solubility Data Project* is probably one of the most ambitious ones. It is concerned with preparing a comprehensive critical compilation of data on solubilities in all physical systems, of gases, liquids and solids. Both the basic and applied branches of almost all scientific disciplines require a knowledge of solubilities as a function of solvent, temperature and pressure. Solubility data are basic to the fundamental understanding of processes relevant to agronomy, biology, chemistry, geology and oceanography, medicine and pharmacology, and metallurgy and materials science. Knowledge of solubility is very frequently of great importance to such diverse practical applications as drug dosage and drug solubility in biological fluids, anesthesiology, corrosion by dissolution of metals, properties of glasses, ceramics, concretes and coatings, phase relations in the formation of minerals and alloys, the deposits of minerals and radioactive fission products from ocean waters, the composition of ground waters, and the requirements of oxygen and other gases in life support systems.

The widespread relevance of solubility data to many branches and disciplines of science, medicine, technology and engineering, and the difficulty of recovering solubility data from the literature, lead to the proliferation of published data in an ever increasing number of scientific and technical primary sources. The sheer volume of data has overcome the capacity of the classical secondary and tertiary services to respond effectively.

While the proportion of secondary services of the review article type is generally increasing due to the rapid growth of all forms of primary literature, the review articles become more limited in scope, more specialized. The disturbing phenomenon is that in some disciplines, certainly in chemistry, authors are reluctant to treat even those limited-in-scope reviews exhaustively. There is a trend to preselect the literature, sometimes under the pretext of reducing it to manageable size. The crucial problem with such preselection - as far as numerical data are concerned - is that there is no indication as to whether the material was excluded by design or by a less than thorough literature search. We are equally concerned that most current secondary sources, critical in character as they may be, give scant attention to numerical data.

On the other hand, tertiary sources - handbooks, reference books, and other tabulated and graphical compilations - as they exist today, are comprehensive but, as a rule, uncritical. They usually attempt to cover whole disciplines, thus obviously are superficial in treatment. Since they command a wide market, we believe that their service to advancement of science is at least questionable. Additionally, the change which is taking place in the generation of new and diversified numerical data, and the rate at which this is done, is not reflected in an increased third-level service. The emergence of new tertiary literature sources does not parallel the shift that has occurred in the primary literature.

With the status of current secondary and tertiary services being as briefly stated above, the innovative approach of the *Solubility Data Project* is that its compilation and critical evaluation work involve consolidation and reprocessing services when both activities are based on intellectual and scholarly reworking of information from primary sources. It comprises compact compilation, rationalization and simplification, and the fitting of isolated numerical data into a critically evaluated general framework.

The *Solubility Data Project* has developed a mechanism which involves a number of innovations in exploiting the literature fully, and which contains new elements of a more imaginative approach for transfer of reliable information from primary to secondary/tertiary sources. *The fundamental trend of the Solubility Data Project is toward integration of secondary and tertiary services with the objective of producing in-depth critical analysis and evaluation which are characteristic to secondary services, in a scope as broad as conventional tertiary services.*

Fundamental to the philosophy of the project is the recognition that the basic element of strength is the active participation of career scientists in it. Consolidating primary data, producing a truly critically-evaluated set of numerical data, and synthesizing data in a meaningful relationship are demands considered worthy of the efforts of top scientists. Career scientists, who themselves contribute to science by their involvement in active scientific research, are the backbone of the project. The scholarly work is commissioned to recognized authorities, involving a process of careful selection in the best tradition of IUPAC. This selection in turn is the key to the quality of the output. These top experts are expected to view their specific topics dispassionately, paying equal attention to their own contributions and to those of their peers. They digest literature data into a coherent story by weeding out what is wrong from what is believed to be right. To fulfill this task, the evaluator must cover *all* relevant open literature. No reference is excluded by design and every effort is made to detect every bit of relevant primary source. Poor quality or wrong data are mentioned and explicitly disqualified as such. In fact, it is only when the reliable data are presented alongside the unreliable data that proper justice can be done. The user is bound to have incomparably more confidence in a succinct evaluative commentary and a comprehensive review with a complete bibliography to both good and poor data.

It is the standard practice that any given solute-solvent system consists of two essential parts: I. Critical Evaluation and Recommended Values, and II. Compiled Data Sheets.

The Critical Evaluation part gives the following information:

(i) a verbal text of evaluation which discusses the numerical solubility information appearing in the primary sources located in the literature. The evaluation text concerns primarily the quality of data after consideration of the purity of the materials and their characterization, the experimental method employed and the uncertainties in control of physical parameters, the reproducibility of the data, the agreement of the worker's results on accepted test systems with standard values, and finally, the fitting of data, with suitable statistical tests, to mathematical functions;

(ii) a set of recommended numerical data. Whenever possible, the set of recommended data includes weighted average and standard deviations, and a set of smoothing equations derived from the experimental data endorsed by the evaluator;

(iii) a graphical plot of recommended data.

The compilation part consists of data sheets of the best experimental data in the primary literature. Generally speaking, such independent data sheets are given only to the best and endorsed data covering the known range of experimental parameters. Data sheets based on primary sources where the data are of a lower precision are given only when no better data are available. Experimental data with a precision poorer than considered acceptable are reproduced in the form of data sheets when they are the only known data for a particular system. Such data are considered to be still suitable for some applications, and their presence in the compilation should alert researchers to areas that need more work.

The typical data sheet carries the following information:

(i) components - definition of the system - their names, formulas and Chemical Abstracts registry numbers;

(ii) reference to the primary source where the numerical information is reported. In cases when the primary source is a less common periodical or a report document, published though of limited availability, abstract references are also given;

(iii) experimental variables;

(iv) identification of the compiler;

(v) experimental values as they appear in the primary source. Whenever available, the data may be given both in tabular and graphical form. If auxiliary information is available, the experimental data are converted also to SI units by the compiler.

Under the general heading of Auxiliary Information, the essential experimental details are summarized:

(vi) experimental method used for the generation of data;
(vii) type of apparatus and procedure employed;
(viii) source and purity of materials;
(ix) estimated error;
(x) references relevant to the generation of experimental data as cited in the primary source.

This new approach to numerical data presentation, developed during our four years of existence, has been strongly influenced by the diversity of background of those whom we are supposed to serve. We thus deemed it right to preface the evaluation/compilation sheets in each volume with a detailed discussion of the principles of the accurate determination of relevant solubility data and related thermodynamic information.

Finally, the role of education is more than corollary to the efforts we are seeking. The scientific standards advocated here are necessary to strengthen science and technology, and should be regarded as a major effort in the training and formation of the next generation of scientists and engineers. Specifically, we believe that there is going to be an impact of our project on scientific-communication practices. The quality of consolidation adopted by this program offers down-to-earth guidelines, concrete examples which are bound to make primary publication services more responsive than ever before to the needs of users. The self-regulatory message to scientists of 15 years ago to refrain from unnecessary publication has not achieved much. The literature is still, in 1982, cluttered with poor-quality articles. The Weinberg report (in "Reader in Science Information", Eds. J. Sherrod and A. Hodina, Microcard Editions Books, Indian Head, Inc., 1973, p.292) states that "admonition to authors to restrain themselves from premature, unnecessary publication can have little effect unless the climate of the entire technical and scholarly community encourages restraint..." We think that projects of this kind translate the climate into operational terms by exerting pressure on authors to avoid submitting low-grade material. The type of our output, we hope, will encourage attention to quality as authors will increasingly realize that their work will not be suited for permanent retrievability unless it meets the standards adopted in this project. It should help to dispel confusion in the minds of many authors of what represents a permanently useful bit of information of an archival value, and what does not.

If we succeed in that aim, even partially, we have then done our share in protecting the scientific community from unwanted and irrelevant, wrong numerical information.

A. S. Kertes

# PREFACE

The data compiled in this volume for the solubility of ethane in liquids represent the result of an exhaustive literature search. The compilations and evaluations were made with great care in the interest of usefulness and accuracy. It is impossible however, to make this type of compilation both complete and fault-free. Readers and users of this volume are therefore kindly requested to bring to the attention of the Editor any errors or omissions they may find.

It is not possible to claim even for a single solvent, and definitely not for any class of solvents, that there are sufficient data of accuracy to a fraction of a percent so that further experimentation is no longer required, except perhaps for the solubility of ethane in water at atmospheric pressure and at low temperature. In all other cases data are too few and of accuracy too low, for many modern applications. Much of the data in this volume has been classified as *tentative* often only because comparable data have been lacking. When comparable data were available, differences frequently exceeded 2%. We may conclude that there is a need for more, as well as more accurate, solubility data.

The accuracy of solubility data is often limited by the ingenuity of the researcher in the design and construction of the solubility apparatus and the care during its operation. Almost as frequently the accuracy is further limited by the choice of basic data used to calculate the solubility, such as the gas molar volume, partial pressure and partial molal volume in solution in some cases, as well as the solvent or solution density and vapor pressure. In some instances compilers and evaluators have had to guess which data were used by the authors in their calculation of solubility. I wish to make a plea for inclusion of all the actual pertinent data used in the calculation of the gas solubility in all future publications.

Ethane behaves essentially as an ideal gas with deviations from ideality diminishing from approximately 1.0% at 273.15 K to 0.3% at 400 K (1,2,3). In most cases in this volume, the mole fraction solubility was calculated on the basis of ethane being a perfect gas. A notable exception is the solubility in water where the real gas molar volumes were used. For essentially all the remaining data, ideal ethane molar volumes were used in the conversion of solubilities expressed in volumetric units to those expressed as mole fractions. For correcting the solubilities for the non-ideality of ethane, molar volumes may be obtained from the following equation utilizing the compressibility factor, $Z$:

$$PV = Z R T$$

Suggested values of $Z$ as a function of temperature (for atmospheric pressure) as calculated from the second virial coefficients obtained from the recent compilation by Dymond and Smith (3) are:

| $T$/K | $Z$ | $T$/K | $Z$ |
|---|---|---|---|
| 260 | 0.9886 | 300 | 0.9926 |
| 273.15 | 0.9901 | 325 | 0.9942 |
| 280 | 0.9908 | 350 | 0.9955 |
| 298.15 | 0.9925 | 400 | 0.9971 |

In general, the mole fraction solubility is increased as a result of a correction for gas phase non-ideality.

Solvent or solution densities were frequently required for the calculation of solubilities but were not usually available from the original solubility papers. These density data were obtained from various literature sources as required (4,5,6,7,8,9) or estimated by comparison with those for homologous compounds or solutions.

The temperature coefficient of solubility at constant pressure for most solvents was expressed as a linear function of the log of the mole fraction solubility and the inverse of the absolute temperature. A regression line was used as a basis for the smoothed values shown in the compiled sheets. Except for the solubility in water for which a more complex function was used, the simple function was sufficient to accurately represent the effect of temperature.

The contribution and assistance of Professors R. Battino, H.L. Clever and C.L. Young as Evaluators, Compilers and collaborators is most gratefully acknowledged. The assistance of several other compilers is also acknowledged. The support of the IUPAC Commission on Solubility Data is most appreciated. Without their initiative, guidance and support this volume would not have been possible. The painstaking assistance of Mr. C. Blais and typist C. Lachaine is also acknowledged and much appreciated.

Walter Hayduk

Ottawa, Canada

May, 1982

REFERENCES

1. Din, F., Ed. *Thermodynamic Functions of Gases* vol. 3 Butterworth, London, 1961, 193-219.

2. Pompe, A.; Spurling, T.H. *CSIRO Aust. Div. Appl. Organic Chem. Tech. Pap. No.1 (Australia)* 1974, 1-42.

3. Dymond, J.H.; Smith, E.B. *The Virial Coefficients of Gases*, Clarendon Press, Oxford, 1980, 74-80.

4. American Petroleum Institute, Research Project 44 Data Publications.

5. Circular 461 of the U.S. National Bureau of Standards.

6. Smow Table, *Pure and Applied Chemistry*, 1976, *45*, 1-9.

7. Washburn, E.W., Ed. *The International Critical Tables*, vol III, McGraw-Hill, New York, 1931.

8. Wilhoit, R.C.; Zwolinski, B.J. *J. Phys. Chem. Ref. Data* 1973, *2*, Supp. no. 1.

9. Riddick, J.A.; Bunger, W.B. *Technique of Chemistry* Weissberger, A., Ed.; vol. II, Wiley-Interscience, New York, 1970, 3rd Ed.

# THE SOLUBILITY OF GASES IN LIQUIDS

## Introductory Information

C. L. Young, R. Battino, and H. L. Clever

### INTRODUCTION

The Solubility Data Project aims to make a comprehensive search of the literature for data on the solubility of gases, liquids and solids in liquids. Data of suitable accuracy are compiled into data sheets set out in a uniform format. The data for each system are evaluated and where data of sufficient accuracy are available values are recommended and in some cases a smoothing equation is given to represent the variation of solubility with pressure and/or temperature. A text giving an evaluation and recommended values and the compiled data sheets are published on consecutive pages. The following paper by E. Wilhelm gives a rigorous thermodynamic treatment on the solubility of gases in liquids.

### DEFINITION OF GAS SOLUBILITY

The distinction between vapor-liquid equilibria and the solubility of gases in liquids is arbitrary. It is generally accepted that the equilibrium set up at 300K between a typical gas such as argon and a liquid such as water is gas-liquid solubility whereas the equilibrium set up between hexane and cyclohexane at 350K is an example of vapor-liquid equilibrium. However, the distinction between gas-liquid solubility and vapor-liquid equilibrium is often not so clear. The equilibria set up between methane and propane above the critical temperature of methane and below the critical temperature of propane may be classed as vapor-liquid equilibrium or as gas-liquid solubility depending on the particular range of pressure considered and the particular worker concerned.

The difficulty partly stems from our inability to rigorously distinguish between a gas, a vapor, and a liquid; a subject which has been discussed in numerous textbooks. We have taken a fairly liberal view in these volumes and have included systems which may be regarded, by some workers, as vapor-liquid equilibria.

### UNITS AND QUANTITIES

The solubility of gases in liquids is of interest to a wide range of scientific and technological disciplines and not solely to chemistry. Therefore a variety of ways for reporting gas solubility have been used in the primary literature. Sometimes, because of insufficient available information, it has been necessary to use several quantities in the compiled tables. Where possible, the gas solubility has been quoted as a mole fraction of the gaseous component in the liquid phase. The units of pressure used are bar, pascal, millimeters of mercury, and atmosphere. Temperatures are reported in Kelvins.

### EVALUATION AND COMPILATION

The solubility of comparatively few systems is known with sufficient accuracy to enable a set of recommended values to be presented. This is true both of the measurements near atmospheric pressure and at high pressures. Although a considerable number of systems have been studied by at least two workers, the range of pressures and/or temperatures is often sufficiently different to make meaningful comparison impossible.

Occasionally, it is not clear why two groups of workers obtained very different sets of results at the same temperature and pressure, although both sets of results were obtained by reliable methods and are internally consistent. In such cases, sometimes an incorrect assessment has been given. There are several examples where two or more sets of data have been classified as tentative although the sets are mutually inconsistent.

Many high pressure solubility data have been published in a smoothed form. Such data are particularly difficult to evaluate, and unless specifically discussed by the authors, the estimated error on such values can only be regarded as an "informed guess".

Many of the high pressure solubility data have been obtained in a more general study of high pressure vapor-liquid equilibrium. In such cases a note is included to indicate that additional vapor-liquid equilibrium data are given in the source. Since the evaluation is for the compiled data, it is possible that the solubility data are given a classification which is better than that which would be given for the complete vapor-liquid data (or vice versa). For example, it is difficult to determine coexisting liquid and vapor compositions near the critical point of a mixture using some widely used experimental techniques which yield accurate high pressure solubility data. As another example, conventional methods of analysis may give results with an expected error which would be regarded as sufficiently small for vapor-liquid equilibrium data but an order of magnitude too large for acceptable high pressure gas-liquid solubility.

It is occasionally possible to evaluate data on mixtures of a given substance with a member of a homologous series by considering all the available data for the given substance with other members of the homologous series. In this study the use of such a technique has been limited.

The estimated error is often omitted in the original article and sometimes the errors quoted do not cover all the variables. In order to increase the usefulness of the compiled tables *estimated* errors have been included even when absent from the original article. If the error on *any* variable has been inserted by the compiler, this has been noted.

## PURITY OF MATERIALS

The purity of materials has been quoted in the compiled tables where given in the original publication. The solubility is usually more sensitive to impurities in the gaseous component than to liquid impurities in the liquid component. However, the most important impurities are traces of a gas dissolved in the liquid. Inadequate degassing of the absorbing liquid is probably the most often overlooked serious source of error in gas solubility measurements.

## APPARATUS AND PROCEDURES

In the compiled tables brief mention is made of the apparatus and procedure. There are several reviews on experimental methods of determining gas solubilities and these are given in References 1-7.

## METHODS OF EXPRESSING GAS SOLUBILITIES

Because gas solubilities are important for many different scientific and engineering problems, they have been expressed in a great many ways:

### The Mole Fraction, x(g)

The mole fraction solubility for a binary system is given by:

$$x(g) = \frac{n(g)}{n(g) + n(l)}$$

$$= \frac{W(g)/M(g)}{\{W(g)/M(g)\} + \{W(l)/M(l)\}}$$

here n is the number of moles of a substance (an *amount* of substance), W is the mass of a substance, and M is the molecular mass. To be unambiguous, the partial pressure of the gas (or the total pressure) and the temperature of measurement must be specified.

### The Weight Per Cent Solubility, wt%

For a binary system this is given by

$$wt\% = 100\, W(g)/\{W(g) + W(l)\}$$

where W is the weight of substance. As in the case of mole fraction, the pressure (partial or total) and the temperature must be specified. The weight per cent solubility is related to the mole fraction solubility by

$$x(g) = \frac{\{wt\%/M(g)\}}{\{wt\%/M(g)\} + \{(100 - wt\%)/M(1)\}}$$

The Weight Solubility, $C_w$

The weight solubility is the number of moles of dissolved gas per gram of solvent when the partial pressure of gas is 1 atmosphere. The weight solubility is related to the mole fraction solubility at one atmosphere partial pressure by

$$x(g) \text{ (partial pressure 1 atm)} = \frac{C_w M(1)}{1 + C_w M(1)}$$

where M(1) is the molecular weight of the solvent.

The Moles Per Unit Volume Solubility, n

Often for multicomponent systems the density of the liquid mixture is not known and the solubility is quoted as moles of gas per unit volume of liquid mixture. This is related to the mole fraction solubility by

$$x(g) = \frac{n\, v^o(1)}{1 + n\, v^o(1)}$$

where $v^o(1)$ is the molar volume of the liquid component.

The Bunsen Coefficient, $\alpha$

The Bunsen coefficient is defined as the volume of gas reduced to 273.15K and 1 atmosphere pressure which is absorbed by unit volume of solvent (at the temperature of measurement) under a partial pressure of 1 atmosphere. If ideal gas behavior and Henry's law are assumed to be obeyed, then

$$\alpha = \frac{V(g)}{V(1)} \; \frac{273.15}{T}$$

where V(g) is the volume of gas absorbed and V(1) is the original (starting) volume of absorbing solvent. The mole fraction solubility x is related to the Bunsen coefficient by

$$x(g, 1 \text{ atm}) = \frac{\alpha}{\alpha + \frac{273.15}{T} \; \frac{v^o(g)}{v^o(1)}}$$

where $v^o(g)$ and $v^o(1)$ are the molar volumes of gas and solvent at a pressure of one atmosphere. If the gas is ideal,

$$x(g) = \frac{\alpha}{\alpha + \frac{273.15R}{v^o(1)}}$$

Real gases do not follow the ideal gas law and it is important to establish the real gas law used for calculating $\alpha$ in the original publication and to make the necessary adjustments when calculating the mole fraction solubility.

The Kuenen Coefficient, S

This is the volume of gas, reduced to 273.15K and 1 atmosphere pressure, dissolved at a partial pressure of gas of 1 atmosphere by 1 gram of solvent.

The Ostwald Coefficient, L

The Ostwald coefficient, L, is defined as the ratio of the volume of gas absorbed to the volume of the absorbing liquid, all measured at the same temperature:

$$L = \frac{V(g)}{V(1)}$$

If the gas is ideal and Henry's Law is applicable, the Ostwald coefficient is independent of the partial pressure of the gas. It is necessary, in practice, to state the temperature and total pressure for which the Ostwald coefficient is measured. The mole fraction solubility, x(g), is related to the Ostwald coefficient by

$$x(g) = \left[ \frac{RT}{P(g)\ L\ v^{o}(1)} + 1 \right]^{-1}$$

where P is the partial pressure of gas. The mole fraction solubility will be at a partial pressure of P(g). (See the following paper by E. Wilhelm for a more igorous definition of the Ostwald coefficient.)

The Absorption Coefficient, β

There are several "absorption coefficients", the most commonly used one being defined as the volume of gas, reduced to 273.15K and 1 atmosphere, absorbed per unit volume of liquid when the total pressure is 1 atmosphere. β is related to the Bunsen coefficient by

$$\beta = \alpha(1 - P(1))$$

where P(1) is the partial pressure of the liquid in atmospheres.

The Henry's Law Constant

A generally used formulation of Henry's Law may be expressed as

$$P(g) = K_H x(g)$$

where $K_H$ is the Henry's Law constant and x(g) the mole fraction solubility. Other formulations are

$$P(g) = K_2 C(1) \qquad \text{or} \qquad C(g) = K_c C(1)$$

where $K_2$ and $K_c$ are constants, C the concentration, and (1) and (g) refer to the liquid and gas phases. Unfortunately, $K_H$, $K_2$ and $K_c$ are all sometimes referred to as Henry's Law constants. Henry's Law is a *limiting law* but can sometimes be used for converting solubility data from the experimental pressure to a partial gas pressure of 1 atmosphere, provided the mole fraction of the gas in the liquid is small, and that the difference in pressures is small. Great caution must be exercised in using Henry's Law.

The Mole Ratio, N

The mole ratio, N, is defined by

$$N = n(g)/n(1)$$

Table 1 contains a presentation of the most commonly used inter-conversions not already discussed.

For gas solubilities greater than about 0.01 mole fraciton at a partial pressure of 1 atmosphere there are several additional factors which must be taken into account to unambiguously report gas solubilities. Solution densities or the partial molar volume of gases must be known. Corrections should be made for the possible non-ideality of the gas or the non-applicability of Henry's Law.

TABLE 1 Interconversion of parameters used for reporting solubility

$$L = \alpha(T/273.15)$$

$$C_w = \alpha/v_o \rho$$

$$K_H = \frac{17.033 \times 10^6 \rho(\text{soln})}{\alpha\ M(1)} + 760$$

$$L = C_w\ v_{t,gas}\ \rho$$

where $v_o$ is the molal volume of the gas in $cm^3 mol^{-1}$ at 0°C, $\rho$ the density of the solvent at the temperature of the measurement, $\rho_{soln}$ the density of the solution at the temperature of the measurement, and $v_{t,gas}$ the molal volume of the gas ($cm^3 mol^{-1}$) at the temperature of the measurement.

SALT EFFECTS

Salt effect studies have been carried out for many years. The results are often reported as Sechenov (Setchenow) salt effect parameters. There appears to be no common agreement on the units of either the gas solubility, or the electrolyte concentration.

Many of the older papers report the salt effect parameter in a form equivalent to

$$k_{scc}/\text{mol dm}^{-3} = (1/(c_2/\text{mol dm}^{-3}))\ \log\ ((c_1^{\circ}/\text{mol dm}^{-3})/(c_1/\text{mol dm}^{-3}))$$

where the molar gas solubility ratio, $c_1^{\circ}/c_1$, is identical to the Bunsen coefficient ratio, $\alpha^{\circ}/\alpha$, or the Ostwald coefficient ratio, $L^{\circ}/L$. One can designate the salt effect parameters calculated from the three gas solubility ratios as $k_{scc}$, $k_{sc\alpha}$, $k_{scL}$, respectively, but they are identical, and $k_{scc}/dm^3\ mol^{-1}$ describes all of them. The superzero refers to the solubility in the pure solvent.

Recent statistical mechanical theories favor a molal measure of the electrolyte and gas solubility. Some of the more recent salt effects are reported in the form

$$k_{smm}/\text{kg mol}^{-1} = (1/(m_2/\text{mol kg}^{-1})\ \log\ ((m_1^{\circ}/\text{mol kg}^{-1})/(m_1/\text{mol kg}^{-1})$$

In this equation the $m_1^{\circ}/m_1$ ratio is identical to the Kuenen coefficient ratio, $s_1^{\circ}/s_1$, or the solvomolality ratio referenced to water, $A_{sm}^{\circ}/A_{sm}$. Thus the salt effect parameters $k_{smm}$, $k_{sms}$, and $k_{smA_{sm}}$ are well represented by the $k_{smm}/kg\ mol^{-1}$.

Some experimentalists and theoreticians prefer the gas solubility ratio as a mole fraction ratio, $x_1^{\circ}/x_1$. It appears that most calculate the mole fraction on the basis of the total number of ions. The salt effect parameters

$$k_{scx}/\text{dm}^3\ \text{mol}^{-1} = (1/(c_2/\text{mol dm}^{-3}))\ \log\ (x_1^{\circ}/x_1)$$

and

$$k_{smx}/\text{kg mol}^{-1} = (1/(m_2/\text{mol kg}^{-1}))\ \log\ (x_1^{\circ}/x_1)$$

are both in the literature, but $k_{scx}$ appears to be the more common.

The following conversions were worked out among the various forms of the salt effect parameter from standard definitions of molarity, molality, and mole fraction assuming the gas solubilities are small.

$$k_{smc} = (c_2/m_2)\ k_{scc} = (c_2/m_2)\ k_{scm} + F_{1m}$$

$$k_{scm} = k_{scc} - F_{1c} = (m_2/c_2)\ k_{smc} - F_{1c} = (m_2/c_2)\ k_{smm}$$

$$k_{scx} = (m_2/c_2)\ k_{smx} = (m_2/c_2)\ ksmm + F_{2c}$$

$$k_{smm} = k_{smx} - F_{2m} = (c_2/m_2)\ k_{scx} - F_{2m}$$

$$k_{smx} = (c_2/m_2)\ k_{scx} = (c_2/m_2)\ k_{scc} + F_{3m}$$

$$k_{scc} = k_{scx} - F_{3c} = (m_2/c_2)\ k_{smx} - F_{3c}$$

where

$$F_{1m} = (1/m_2)\ \log\ [(\rho°/\rho)\ (1000 + m_2M_2)/1000]$$

$$F_{1c} = (m_2/c_2)\ F_{1m}$$

$$F_{2m} = (1/m_2)\ \log\ [(1000 + \nu m_3\ M_3)/1000]$$

$$F_{2c} = (m_2/c_2)\ F_{2m}$$

$$F_{3m} = (1/m_2)\ \log\ [(1000\rho + (\nu M_3 - M_2)\ c_2)/1000\rho°)$$

$$F_{3c} = (m_2/c_2)\ F_{3m}$$

The factors $F_{1m}$, $F_{1c}$, $F_{2m}$, $F_{2c}$, $F_{3m}$, and $F_{3c}$ can easily be calculated from aqueous electrolyte data such as weight per cent and density as found in Volume III of the International Critical Tables. The values are small and change nearly linearly with both temperature and molality. The factors normally amount to no more than 10 to 20 per cent of the value of the salt effect parameter.

The symbols in the equations above are defined below:

| Component | Molar Concentration $c/\text{mol dm}^{-3}$ | Molal Concentration $m/\text{mol kg}^{-1}$ | Mole Fraction $x$ | Molecular Weight $M/\text{g mol}^{-1}$ |
|---|---|---|---|---|
| Nonelectrolyte | $c_1^o$, $c_1$ | $m_1^o$, $m_1$ | $x_1^o$, $x_1$ | $M_1$ |
| Electrolyte | $c_2$ | $m_2$ | $x_2$ | $M_2$ |
| Solvent | $c_3$ | $m_3$ | $x_3$ | $M_3$ |

The superscript "°" refers to the nonelectrolyte solubility in the pure solvent. The pure solvent and solution densities are $\rho°/\text{g cm}^{-3}$ and $\rho/\text{g cm}^{-3}$, respectively. They should be the densities of gas saturated solvent (water) and salt solution, but the gas free densities will differ negligibly in the $\rho°/\rho$ ratio. The number of ions per formula of electrolyte is symbolized by $\nu$.

The following table gives estimated errors in $k_{scc}$ for various salt concentrations and a range of random errors in the gas solubility measurement

Error in $k_{scc}/\text{dm}^3\ \text{mol}^{-1}$ [a]

| $c_2/\text{mol dm}^{-3}$ | Random Error in gas solubility Measurement ±2% | ±1% | ±0.5% | ±0.1% | ±0.05% |
|---|---|---|---|---|---|
| 1 | ±18% | ±9% | ±5% | ±1.5% | ±1% |
| 0.1 | ±175% | ±87% | ±43% | ±9% | ±4% |
| 0.05 | ±350% | ±174% | ±87% | ±17% | ±9% |
| 0.01 | ±1750% | ±870% | ±435% | ±87% | ±43% |

[a] Based on a $k_{scc}$ value of 0.100.

AQUAMOLAL OR SOLVOMOLAL, $A_{sm}$ or $m_i^{(s)}$

The term aquamolal was suggested by R. E. Kerwin (9). The unit was first used in connection with $D_2O$ and $H_2O + D_2O$ mixtures. It has since been extended in use to other solvents. The unit represents the numbers of moles of solute per 55.51 moles of solvent. It is represented by

$m_i^{(s)}/\text{mol kg}^{-1} = (n_1 M_2/w_2)(w_2/M_0) = m_i(M_2/M_0)$ where an amount of $n_i$ of solute $i$ is dissolved in a mass $w_2$ of solvent of molar mass $M_2$; $M_0$ is the molar mass of a reference solvent and $m_i/\text{mol kg}^{-1}$ is the conventional molality in the reference solvent. The reference solvent is normally water.

TEMPERATURE DEPENDENCE OF GAS SOLUBILITY

In a few cases it has been found possible to fit the mole fraction solubility at various temperatures using an equation of the form

$$\ln x = A + B / (T/100K) + C \ln (T/100K) + DT/100K$$

It is then possible to write the thermodynamic functions $\Delta\overline{G}_1^\circ$, $\Delta\overline{H}_1^\circ$, $\Delta\overline{S}_1^\circ$ and $\Delta\overline{C}_{P_1}$ for the transfer of the gas from the vapor phase at 101,325 Pa partial pressure to the (hypothetical) solution phase of unit mole fraction as:

$$\Delta\overline{G}_1^\circ = -RAT - 100\,RB - RCT \ln (T/100) - RDT^2/100$$

$$\Delta\overline{S}_1^\circ = RA + RC \ln (T/100) + RC + 2\,RDT/100$$

$$\Delta\overline{H}_1^\circ = -100\,RB + RCT + RDT^2/100$$

$$\Delta\overline{C}_{P_1}^\circ = RC + 2\,RDT/100$$

In cases where there are solubilities at only a few temperatures it is convenient to use the simpler equations

$$\Delta\overline{G}_1^\circ = -RT \ln x = A + BT$$

in which case $A = \Delta\overline{H}_1^\circ$ and $-B = \Delta\overline{S}_1^\circ$

REFERENCES


1. Battino, R.; Clever, H. L. *Chem. Rev.* 1966, *66*, 395.
2. Clever, H. L.; Battino, R. in *Solutions and Solubilities*, Ed. M. R. J. Dack, J. Wiley & Sons, New York, 1975, Chapter 7.
3. Hildebrand, J. H.; Prausnitz, J. M.; Scott, R. L. *Regular and related Solutions*, Van Nostrand Reinhold, New York, 1970, Chapter 8.
4. Markham, A. E.; Kobe, K. A. *Chem. Rev.* 1941, *63*, 449.
5. Wilhelm, E.; Battino, R. *Chem. Rev.* 1973, *73*, 1.
6. Wilhelm, E.; Battino, R.; Wilcock, R. J. *Chem. Rev.* 1977, *77*, 219.
7. Kertes, A. S.; Levy, O.; Markovits, G. Y. in *Experimental Thermochemistry* Vol. II, Ed. B. Vodar and B. LeNaindre, Butterworth, London, 1974, Chapter 15.
8. Long, F. A.; McDevit, W. F. *Chem. Rev.* 1952, *51*, 119.
9. Kerwin, K. E., Ph.D. Thesis, University of Pittsburgh 1964.


Revised: April 1982 (R.B., H.L.C.)

COMPONENTS:

(1) Ethane; $C_2H_6$; [74-84-0]

(2) Water; $H_2O$; [7732-18-5]

EVALUATOR:

Rubin Battino
Department of Chemistry
Wright State University
Dayton, Ohio 45435 U.S.A.

CRITICAL EVALUATION:

The recent measurements of the solubility of ethane in water at about atmospheric pressure and in the range 275 to 323 K by Rettich, et al. (1) sets the standard for this system. Real gas corrections were applied to the data and the precision corresponding to one standard diviation at the middle of the temperature range was ±0.13 per cent. The modern measurements of Wen and Hung (2) when combined with those of Ben-Naim and co-workers (3) gave a precision of ±0.41 per cent under the same conditions. However, the smoothed data of this latter group of measurements is on the average 0.66 per cent lower than Rettich, et al.'s results. The difference may be attributed to the fact that the earlier workers did not apply real gas corrections. Also, Wen and Hung's results are systematically lower than those of Ben-Naim and co-workers.

The ethane solubilities reported by four other groups (6-9) are reasonable but not of the precision of those cited above. Winkler's measurements (6) do go to a higher temperature (353 K) but at the higher temperature are about ten per cent below Rettich, et al.'s extrapolated values. The early solubilities of Bunsen (10-13) are amazingly low, being about one-half the recommended values. Schickendantz (14), Schorlemmer (15), and Henrich's (16) nineteenth century values are only of historic interest. Czerski and Czaplinski's single value (17) at 273 K is about twenty per cent low. McAuliffe's work (18,19) is about six per cent high and his hydrocarbon solubilities in water are little better than qualitative. Wetlaufer, et al.'s (20) three values range from good agreement to two per cent high. Both of Rudakow and Lutsyk's measurements (21) are quite low.

Table of recommended smoothed values:

| $T$/K | Mole Fraction $/10^5 x_1$ | $\Delta\bar{G}_1°$ /kJ mol$^{-1}$ | $\Delta\bar{H}_1°$ /kJ mol$^{-1}$ | $\Delta\bar{S}_1°$ /J mol$^{-1}$K$^{-1}$ |
|---|---|---|---|---|
| 273.15 | 7.994 | 21.43 | -26.66 | -176.0 |
| 278.15 | 6.510 | 22.29 | -25.21 | -170.8 |
| 283.15 | 5.400 | 23.13 | -23.77 | -165.6 |
| 288.15 | 4.556 | 23.95 | -22.32 | -160.6 |
| 293.15 | 3.907 | 24.74 | -20.88 | -155.6 |
| 298.15 | 3.401 | 25.51 | -19.43 | -150.7 |
| 303.15 | 3.002 | 26.25 | -17.99 | -145.9 |
| 308.15 | 2.686 | 26.96 | -16.55 | -141.2 |
| 313.15 | 2.434 | 27.66 | -15.10 | -136.6 |
| 318.15 | 2.232 | 28.33 | -13.66 | -132.0 |
| 323.15 | 2.069 | 28.98 | -12.21 | -127.5 |

| COMPONENTS: | EVALUATOR: |
|---|---|
| (1) Ethane; $C_2H_6$; [74-84-0]<br>(2) Water; $H_2O$; [7732-18-5] | Rubin Battino<br>Department of Chemistry<br>Wright State University<br>Dayton, Ohio 45435 U.S.A. |

CRITICAL EVALUATION: continued

Rettich, et al.'s solubilities (1) were smoothed by least squares analysis to give the following equation:

$$\ln x_1 = -90.82250 + 126.9559/(T/100\ \text{K}) + 34.74128 \ln (T/100\ \text{K})$$

In the above equation $x_1$ is the mole fraction solubility at unit fugacity of 101.325 kPa (1 atm). The recommended values for this system are given in the table as smoothed mole fractions at 5 K intervals. Changes in the thermodynamic functions on solutions are also given in the table. The change in heat capacity on solution was constant at 298 J $mol^{-1}$ $K^{-1}$.

References


1. Rettich, T.R.; Handa, Y.P.; Battino, R.; Wihelm, E. *J. Phys. Chem.* 1981, *85*, 3230-3237.
2. Wen, W-Y.; Hung, J.H. *J. Phys. Chem.* 1970, *74*, 170-180.
3. Ben-Naim, A.; Wilf, J.; Yaacobi, M. *J. Phys. Chem.* 1973, *77*, 95-102.
4. Yaacobi, M.; Ben-Naim, A. *J. Soln. Chem.* 1973, *2*, 425-443.
5. Yaacobi, M.; Ben-Naim, A. *J. Phys. Chem.* 1974, *78*, 175-179.
6. Winkler, L.W. *Ber.* 1901, *34*, 1408-1422.
7. Eucken, A.; Hertzberg, G. *Z. Physik. Chem.* 1950, *195*, 1-23.
8. Morrison, T.J.; Billet, F. *J. Chem. Soc.* 1952, 3819-3822.
9. Claussen, W.F.; Polglase, M.F. *J. Am. Chem. Soc.* 1952,*74*,4817-4819.
10. Bunsen, R. *Ann.* 1855, *93*, 1-50.
11. Bunsen, R. *Phil. Mag.* 1855, *9*, 116-130, 181-201.
12. Bunsen, R. *Ann. Chem. Phys.* 1855, *43*, 496-508.
13. Bunsen, R. "*Gasometr. Methoden*"*2nd ed., Braunschweig,* 1858 .
14. Schickendantz, F. *Ann.* 1859, *109*, 116; *Ann. Chim. Phys.* 1860, *59*, 123.
15. Schorlemmer, C. *Ann.* 1864, *132*, 234.
16. Henrich, F. *Z. Physik. Chem.* 1892, *9*, 435.
17. Czerski, L; Czaplinski, A. *Ann. Soc. Chim. Polonorum, (Poland)* 1962, *36*, 1827-1834.
18. McAuliffe, C. *Nature* 1963, *200*, 1092-1093.
19. McAuliffe, C. *J. Phys. Chem.* 1966, *70*, 1267-1275.
20. Wetlaufer, D.B.; Malik, S.K.; Stoller, L.; Coffin, R.L. *J. Am. Chem. Soc.* 1964, *86*, 508-514.
21. Rudakov, E.S.; Lutsyk, A.I. *Zh. Fiz. Kihm.* 1979, *53*, 1298-1300; *Russ. J. Phys. Chem.* 1979, *53*, 731-733.

**COMPONENTS:**

(1) Ethane; $C_2H_6$; [74-84-0]

(2) Water; $H_2O$; [7732-18-5]

**ORIGINAL MEASUREMENTS:**

Rettich, T.R.; Handa, Y.P.;

Battino, R.; Wilhelm, E.

*J. Phys. Chem.* 1981, *85*, 3230-3237.

**VARIABLES:**

$T$/K: 275.44-323.15

$P$/kPa: 50.74-110.58

**PREPARED BY:**

R. Battino

**EXPERIMENTAL VALUES:**

| $T$/K | $P$/atm | $P^1$/kPa | Henry's Constant[2] $H$/atm | Henry's Constant[2] $H$/kPa | Mole Fraction[3] $/10^5 x_1$ |
|---|---|---|---|---|---|
| 275.45 | 0.6964 | 70.56 | 13775 | 1.3957 | 7.2595 |
| 275.44 | 0.7195 | 72.90 | 13779 | 1.3962 | 7.2574 |
| 278.15 | 0.8403 | 85.14 | 15382 | 1.5586 | 6.5011 |
| 283.16 | 0.6890 | 69.81 | 18513 | 1.8758 | 5.4016 |
| 283.14 | 0.5511 | 55.84 | 18512 | 1.8758 | 5.4019 |
| 283.77 | 0.8522 | 86.35 | 18920 | 1.9171 | 5.2854 |
| 288.15 | 0.7568 | 76.68 | 21959 | 2.2250 | 4.5539 |
| 293.15 | 0.8863 | 89.80 | 25612 | 2.5951 | 3.9044 |
| 298.15 | 0.5733 | 58.09 | 29356 | 2.9745 | 3.4065 |
| 298.15 | 0.9443 | 95.68 | 29387 | 2.9776 | 3.4029 |
| 298.14 | 0.7141 | 72.36 | 29384 | 2.9773 | 3.4032 |
| 298.16 | 0.5008 | 50.74 | 29431 | 2.9821 | 3.3978 |

[1] Calculated by compiler.

[2] Henry's law constant evaluated at saturation pressure of solvent from:

$$H = \lim_{x_1 \to 0} (f_1/x_1) \text{ where } f_1 \text{ is the fugacity.}$$

[3] Mole fraction determined at unit fugacity of 101.325 kPa (1 atm).

continued...

**AUXILIARY INFORMATION**

**METHOD/APPARATUS/PROCEDURE:**

The apparatus used was modelled after that of Benson, Krause and Peterson (1). Degassed water is flowed in a thin film over the surface of a 1 $dm^3$ sphere to contact the gas. After equilibrium is attained the solution is sealed in chamber of calibrated volume. The dissolved gas is extracted and its amount determined by a direct PVT measurement. A sample of the gas phase is analyzed in an identical manner. From the results, the saturation pressure of the solvent and Henry's constant are calculated in a thermodynamically rigorous manner, applying all non-ideal corrections.

**SOURCE AND PURITY OF MATERIALS:**

1. Matheson CP grade; purity 99.0 mole per cent minimum. Also Matheson ultra-high purity, 99.96 mole per cent minimum.

2. Resistivity better than $5 \times 10^4 \Omega m$.

**ESTIMATED ERROR:**

$\delta H/H = 0.0008$

$\delta T/K = 0.01$

**REFERENCES:**

1. Benson, B.B.; Krause, D.; Peterson, M.A.

*J. Soln. Chem.* 1979, *8*, 655-690.

| COMPONENTS: | ORIGINAL MEASUREMENTS: |
|---|---|
| (1) Ethane; $C_2H_6$; [74-84-0] | Rettich, T.R.; Handa, Y.P.; |
| (2) Water; $H_2O$; [7732-18-5] | Battino, R.; Wilhelm, E. |
| | *J. Phys. Chem.* 1981, *85*, 3230-3237. |

| VARIABLES: | PREPARED BY: |
|---|---|
| $T$/K: 275.44-323.15 | R. Battino |
| $P$/kPa: 50.74-110.58 | |

EXPERIMENTAL VALUES: continued

| $T$/K | $P$/atm | $P^1$/kPa | Henry's Constant[2] $H$/atm | Henry's Constant[2] $H$/kPa | Mole Fraction[3] $/10^5 x_1$ |
|---|---|---|---|---|---|
| 303.15 | 0.9114 | 92.35 | 33259 | 3.3700 | 3.0067 |
| 308.16 | 0.8189 | 82.98 | 37220 | 3.7713 | 2.6867 |
| 313.14 | 1.0613 | 107.54 | 41211 | 4.1757 | 2.4265 |
| 318.16 | 1.0675 | 108.16 | 44840 | 4.5434 | 2.2302 |
| 318.14 | 1.0854 | 109.98 | 44859 | 4.5453 | 2.2292 |
| 318.16 | 1.0913 | 110.58 | 44793 | 4.5386 | 2.2325 |
| 318.16 | 0.6694 | 67.83 | 44901 | 4.5496 | 2.2271 |
| 318.16 | 0.9065 | 91.85 | 44884 | 4.5479 | 2.2280 |
| 318.16 | 0.5989 | 60.68 | 44824 | 4.5418 | 2.2309 |
| 323.14 | 0.7310 | 74.07 | 48189 | 4.8828 | 2.0752 |
| 323.15 | 0.9583 | 97.10 | 48233 | 4.8872 | 2.0733 |

[1] Calculated by compiler.

[2] Henry's law constant evaluated at saturation pressure of solvent from:

$$H = \lim_{x_1 \to 0} (f_1/x_1)$$ where $f_1$ is the fugacity.

[3] Mole fraction determined at unit fugacity of 101.325 kPa (1 atm).

The authors give the following smoothing equation which fits their data over the experimental temperature range to 0.08 per cent:

$$\ln H = 1340.027 - 2216.171\ \mathrm{T}^{-1} - 2158.422 \ln \mathrm{T} + 718.779\ \mathrm{T} - 40.50119\ \mathrm{T}^2$$

where: $\mathrm{T} = 10^{-2} T/\mathrm{K}$; $H$/Pa.

**COMPONENTS:**

(1) Ethane; $C_2H_6$; [74-84-0]

(2) Water; $H_2O$; [7732-18-5]

**ORIGINAL MEASUREMENTS:**

Wen, W-Y.; Hung, J.H.

*J. Phys. Chem.* 1970, *74*, 170-180.

**VARIABLES:**

$T$/K: 278.15-308.15

$P$/kPa: 101.325 (1 atm)

**PREPARED BY:**

R. Battino

**EXPERIMENTAL VALUES:**

| $t$/°C | $T$/K | Mole Fraction[1] $/10^5 x_1$ | Ethane Solubility $S$/cm$^3$(STP)/kg $H_2O$ |
|---|---|---|---|
| 5 | 278.15 | 6.513 | 80.19 |
| 15 | 288.15 | 4.512 | 55.55 |
| 25 | 298.15 | 3.347 | 41.20 |
| 35 | 308.15 | 2.621 | 32.27 |

[1] Mole fraction solubility calculated by compiler for a gas partial pressure of 101.325 kPa using a gas molar volume of 22,178.6 cm$^3$(STP)mol$^{-1}$.

**AUXILIARY INFORMATION**

**METHOD/APPARATUS/PROCEDURE:**

Used the method of Ben-Naim and Baer (1) except for addition of teflon stopcocks. Degassed liquid in a volumetric container is forced by a stirrer-created vortex up side-arms and through tubes containing gas saturated with liquid. Gas uptake is determined on buret at constant gas pressure.

**SOURCE AND PURITY OF MATERIALS:**

1. Matheson Co; purity 99.9 per cent.

2. Distilled. Specific conductivity of $1.5 \times 10^{-6}$ mho cm$^{-1}$.

**ESTIMATED ERROR:**

$\delta S/S = 0.003$

$\delta T/K = 0.005$

**REFERENCES:**


1. Ben-Naim, A.; Baer, S. *Trans. Faraday Soc.* 1963, *59*, 2735-2741.

**COMPONENTS:**

(1) Ethane; $C_2H_6$; [74-84-0]

(2) Water; $H_2O$; [7732-18-5]

**ORIGINAL MEASUREMENTS:**

Ben-Naim, A.; Wilf, J.; Yaacobi, M.

*J. Phys. Chem.* 1973, *77*, 95-102.

**VARIABLES:**

$T$/K: 278.15-298.15
$P$/kPa: 101.325 (1 atm)

**PREPARED BY:**

R. Battino

**EXPERIMENTAL VALUES:**

| $t$/°C | $T$/K | Mole fraction[1] $/10^5 x_1$ | Ostwald Coefficient[2] $10^2 L/cm^3 cm^{-3}$ |
|---|---|---|---|
| 5 | 278.15 | 6.53 | 8.20 |
| 10 | 283.15 | 5.41 | 6.91 |
| 15 | 288.15 | 4.55 | 5.91 |
| 20 | 293.15 | 3.89 | 5.14 |
| 25 | 298.15 | 3.37 | 4.53 |

[1] Mole fraction solubility of gas at 101.325 kPa partial pressure of gas calculated with virial correction for ethane. These values are about one per cent higher than those assuming an ideal gas.

[2] Original data.

**AUXILIARY INFORMATION**

**METHOD/APPARATUS/PROCEDURE:**

Used the method of Ben-Naim and Baer (1) except for the addition of teflon stopcocks. Degassed liquid in a volumetric container is forced by a stirrer-created vortex up side-arms and through tubes containing gas saturated with liquid. Gas uptake is determined on a buret at constant pressure.

**SOURCE AND PURITY OF MATERIALS:**

1. Matheson; purity 99.9 per cent.

2. Doubly distilled.

**ESTIMATED ERROR:**

$\delta L/L = 0.005$; compiler's estimate.

**REFERENCES:**

1. Ben-Naim, A.; Baer, S.

*Trans. Faraday Soc.* 1963, *59*, 2735-2741.

**COMPONENTS:**

(1) Ethane; $C_2H_6$; [74-84-0]

(2) Water; $H_2O$; [7732-18-5]

**ORIGINAL MEASUREMENTS:**

Yaacobi, M.; Ben-Naim, A.

*J. Soln. Chem.* 1973, *2*, 425-443.

**VARIABLES:**

$T$/K: 283.15-303.15

$P$/kPa: 101.325 (1 atm)

**PREPARED BY:**

R. Battino, W. Hayduk

**EXPERIMENTAL VALUES:**

| $t$/°C | $T$/K | Mole Fraction[1] $/10^5 x_1$ | Ostwald Coefficient[2] $10^2 L$ /cm$^3$cm$^{-3}$ |
|---|---|---|---|
| 10 | 283.15 | 5.402 | 6.905[3] |
| 15 | 288.15 | 4.546 | 5.912 |
| 20 | 293.15 | 3.886 | 5.139 |
| 25 | 298.15 | 3.373 | 4.533 |
| 30 | 303.15 | 2.970 | 4.054 |

[1] Mole fraction solubility at 101.325 kPa partial pressure of gas calculated by compiler with virial correction for ethane. These values are about one per cent higher than those assuming an ideal gas.

[2] Some of these results appear to be the same as published by Ben-Naim, Wilf and Yaacobi in *J. Phys. Chem.* 1973, *77*, 95 but are quoted here to a higher precision.

[3] Corrected value; a misprint appeared in the paper.

**AUXILIARY INFORMATION**

**METHOD/APPARATUS/PROCEDURE:**

Used the method of Ben-Naim and Baer (1) except for the addition of teflon stopcocks. Degassed liquid in a volumetric container is forced by a stirrer-created vortex up side-arms and through tubes containing gas saturated with liquid. Gas uptake is determined on a buret at constant gas pressure.

**SOURCE AND PURITY OF MATERIALS:**

1. Matheson; purity 99.9 per cent.
2. Doubly distilled.

**ESTIMATED ERROR:**

$\delta L/L$= 0.005; compiler's estimate.

**REFERENCES:**

1. Ben-Naim, A.; Baer, S.

*Trans. Faraday Soc.* 1963, *59*, 2735-2741.

**COMPONENTS:**

(1) Ethane; $C_2H_6$; [74-84-0]

(2) Water; $H_2O$; [7732-18-5]

**ORIGINAL MEASUREMENTS:**

Yaacobi, M.; Ben-Naim, A.

*J. Phys. Chem.* 1974, *78*, 175-178.

**VARIABLES:**

$T$/K: 283.15-303.15

$P$/kPa: 101.325 (1 atm)

**PREPARED BY:**

R. Battino, W. Hayduk

**EXPERIMENTAL VALUES:**

| $t$/°C | $T$/K | Mole Fraction[1] $/10^5 x_1$ | Ostwald Coefficient[2] $10^2 L/\mathrm{cm^3 cm^{-3}}$ |
|---|---|---|---|
| 10 | 283.15 | 5.402 | 6.91 |
| 15 | 288.15 | 4.546 | 5.91 |
| 20 | 293.15 | 3.886 | 5.14 |
| 25 | 298.15 | 3.373 | 4.53 |
| 30 | 303.15 | 2.970 | 4.05 |

[1] Mole fraction solubility at 101.325 kPa partial pressure of gas calculated by compiler with virial correction for ethane. These values are about one per cent higher than those assuming ideal gas.

[2] Original data. Same results published by Yaacobi and Ben-Naim in *J. Solution Chem.* 1973, *2*, 425 and some of these results by Ben-Naim, Wilf and Yaacobi in *J. Phys. Chem.* 1973, *77*, 95.

**AUXILIARY INFORMATION**

**METHOD/APPARATUS/PROCEDURE:**

Used the method of Ben-Naim and Baer (1) except for the addition of teflon stopcocks. Degassed liquid in a volumetric container is forced by a stirrer-created vortex up side-arms and through tubes containing the gas saturated with liquid. Gas uptake is determined on a buret at constant pressure.

**SOURCE AND PURITY OF MATERIALS:**

1. Matheson, 99.9 per cent.

2. Doubly distilled.

**ESTIMATED ERROR:**

$\delta L/L = 0.005$, compiler's estimate.

**REFERENCES:**

1. Ben-Naim, A.; Baer, S.
*Trans. Faraday Soc.* 1963, *59*, 2735-2741.

**COMPONENTS:**

(1) Ethane; $C_2H_6$; [74-84-0]

(2) Water; $H_2O$; [7732-18-5]

**ORIGINAL MEASUREMENTS:**

Winkler, L.W.

*Chem. Ber.* 1901, *34*, 1408-1422.

**VARIABLES:**

$T$/K: 273.51-353.12

$P$/kPa: 101.325 (1 atm)

**PREPARED BY:**

R. Battino

**EXPERIMENTAL VALUES:**

| $t$/°C | $T$/K | Mole Fraction[1] $/10^5 x_1$ | Ostwald Coefficient[2] $10^2 L/cm^3 cm^{-3}$ | Bunsen Coefficient[3] $10^2 \alpha/cm^3 (STP) cm^{-3} atm^{-1}$ |
|---|---|---|---|---|
| 0.36 | 273.51 | 7.819 | 9.741 | 9.728 |
| 10.03 | 283.18 | 5.268 | 6.794 | 6.553 |
| 20.00 | 293.15 | 3.804 | 5.070 | 4.724 |
| 30.00 | 303.15 | 2.925 | 4.022 | 3.624 |
| 39.98 | 313.13 | 2.362 | 3.343 | 2.916 |
| 50.00 | 323.15 | 2.000 | 2.909 | 2.459 |
| 59.98 | 333.13 | 1.780 | 2.655 | 2.177 |
| 70.00 | 343.15 | 1.601 | 2.447 | 1.948 |
| 79.97 | 353.12 | 1.510 | 2.361 | 1.826 |

[1] Mole fraction solubility at 101.325 kPa partial pressure of gas calculated by compiler using a gas molar volumeof 22,178.6 $cm^3 (STP) mol^{-1}$.

[2] Ostwald coefficient calculated by compiler.

[3] Original data. These are averages of up to six values at each temperature. The maximum temperature range was 0.16 K at the lowest temperature measurement.

**AUXILIARY INFORMATION**

**METHOD/APPARATUS/PROCEDURE:**

Basically followed the Bunsen absorption method (1). Method is described in two earlier papers (2,3).

**SOURCE AND PURITY OF MATERIALS:**

1. From the decomposition of diethyl zinc.

2. Distilled.

**ESTIMATED ERROR:**

$\delta\alpha/\alpha$ = 0.01; compiler's estimate

$\delta T$/K = 0.01

**REFERENCES:**

1. Bunsen, R. "*Gasometrische Methoden*" , 2nd edition, Braunschweig, 1858.
2. Winkler, L.W. *Ber.* 1893, *24*, 89-101.
3. ibid. 1893, *24*, 3602-3610.

**COMPONENTS:**

(1) Ethane; $C_2H_6$; [74-84-0]

(2) Water; $H_2O$; [7732-18-5]

**ORIGINAL MEASUREMENTS:**

Eucken, A.; Hertzberg, G.

*Z. Physik. Chem.* 1950, *195*, 1-23.

**VARIABLES:**

$T$/K: 273.2-293.2

$P$/kPa: 101.3

**PREPARED BY:**

R. Battino

**EXPERIMENTAL VALUES:**

| $t$/°C | $T$/K | Mole Fraction[1] $/10^5 x_1$ | Ostwald Coefficient $10^2 L/\text{cm}^3\,\text{cm}^{-3}$ |
|---|---|---|---|
| 0 | 273.2 | 7.93 | 9.87 |
| 20 | 293.2 | 3.82 | 5.09 |

[1] Mole fraction at 101.325 kPa partial pressure of gas calculated by compiler using an ethane molar volume of 22,178.6 $cm^3$(STP)$mol^{-1}$.

**AUXILIARY INFORMATION**

**METHOD/APPARATUS/PROCEDURE:**

A volumetric/manometric method was used and briefly described in the paper.

**SOURCE AND PURITY OF MATERIALS:**

No details given.

**ESTIMATED ERROR:**

$\delta L/L$ = 0.01; compiler's estimate.

**REFERENCES:**

**COMPONENTS:**

(1) Ethane; $C_2H_6$; [74-84-0]

(2) Water; $H_2O$; [7732-18-5]

**ORIGINAL MEASUREMENTS:**

Morrison, T.J.; Billet, F.

*J. Chem. Soc.* 1952, 3819-3822.

**VARIABLES:**

$T$/K: 285.5-345.6
$P$/kPa: 101.325 (1 atm)

**PREPARED BY:**

R. Battino

**EXPERIMENTAL VALUES:**

| $t$/°C | $T$/K | Mole Fraction[1] $/10^5 x_1$ | Solubility[2] $S$/cm$^3$(STP)kg$^{-1}$ |
|---|---|---|---|
| 12.3 | 285.5 | 4.342 | 53.46 |
| 12.6 | 285.8 | 4.273 | 52.60 |
| 16.4 | 289.6 | 3.861 | 47.53 |
| 17.6 | 290.8 | 3.747 | 46.13 |
| 24.6 | 297.8 | 3.131 | 38.55 |
| 30.5 | 303.7 | 2.740 | 33.73 |
| 32.5 | 305.7 | 2.629 | 32.36 |
| 35.3 | 308.5 | 2.504 | 30.83 |
| 40.9 | 314.1 | 2.295 | 28.25 |
| 49.1 | 322.3 | 2.049 | 25.23 |
| 62.4 | 335.6 | 1.777 | 21.88 |
| 69.5 | 342.7 | 1.690 | 20.80 |
| 71.4 | 344.6 | 1.662 | 20.46 |
| 72.4 | 345.6 | 1.662 | 20.46 |

[1] Mole fraction solubility at 101.325 kPa partial pressure of gas calculated by compiler using a molar volume of 22,178.6 cm$^3$(STP)mol$^{-1}$.

[2] Solubility in units of cm$^3$(STP)/kg $H_2O$. Smoothing equation given by authors:

$$\log_{10}S = -87.699 + 4730/(T/K) + 29.67 \log_{10}(T/K).$$

**AUXILIARY INFORMATION**

**METHOD/APPARATUS/PROCEDURE:**

Original apparatus described in references (1,2). Degassed solvent as obtained from a new design of apparatus described in this paper, flows in a thin film through the gas down an absorption spiral. The gas uptake and solvent volumes used are read on burets.

**SOURCE AND PURITY OF MATERIALS:**

1. Prepared from Grignard reagents.
2. Distilled.

**ESTIMATED ERROR:**

$\delta S/S$ = 0.01, compiler's estimate.

$\delta T$/K = 0.01

**REFERENCES:**

1. Morrison, T.J. *J. Chem. Soc.* 1952, 3814-3818.
2. Morrison, T.J.; Billet, F. *J. Chem. Soc.* 1948, 2033-2035.

**COMPONENTS:**

(1) Ethane; $C_2H_6$; [74-84-0]

(2) Water; $H_2O$; [7732-18-5]

**ORIGINAL MEASUREMENTS:**

Claussen, W.F.; Polglase, M.F.

*J. Am. Chem. Soc.* 1952, *74*, 4817-4819.

**VARIABLES:**

$T$/K: 274.7-312.9

$P$/kPa: 101.325 (1 atm)

**PREPARED BY:**

R. Battino

**EXPERIMENTAL VALUES:**

| $t$/°C | $T$/K | Mole Fraction[1] $/10^5x_1$ | Ostwald Coefficient[1] $10^2L$/cm cm$^{-1}$ | Bunsen Coefficient[2] $10^2\alpha$/cm$^3$(STP)cm$^{-3}$atm$^{-1}$ |
|---|---|---|---|---|
| 1.5 | 274.7 | 7.53 | 9.42 | 9.31, 9.46, 9.34 |
| 10.5 | 283.7 | 5.27 | 6.80 | 6.55, 6.57, 6.52 |
| 17.5 | 290.7 | 4.24 | 5.61 | 5.27 |
| 19.8 | 293.0 | 3.99 | 5.32 | 4.90, 4.99, 4.99 |
| 29.8 | 303.0 | 3.03 | 4.16 | 3.74, 3.74, 3.75, 3.76 |
| 39.7 | 312.9 | 2.49 | 3.52 | 3.06, 3.07, 3.08 |

[1] Mole fraction solubility at 101.325 kPa partial pressure of gas and Ostwald coefficient calculated by compiler using average of listed Bunsen coefficients and a gas molar volume of 22,178.6 cm$^3$(STP)mol$^{-1}$.

[2] Original data.

**AUXILIARY INFORMATION**

**METHOD/APPARATUS/PROCEDURE:**

Solubility determined by a micro combustion technique. Ethane was bubbled through the water via a sintered glass disc to saturate the water. The train for analysis was composed of an oxygen tank to sweep out the dissolved gas, pressure regulators, mercury manometer, preheater, absorption U-tube containing ascarite and anhydrone, aerator, combustion tube containing copper oxide at 973 K, weighing tubes containing ascarite and anhydrone, and finally, the Marriotte flask. Details are given in the paper.

**SOURCE AND PURITY OF MATERIALS:**

1. 99.9% from Phillips Petroleum Co.; no detectable impurity by infrared.
2. Doubly distilled.

**ESTIMATED ERROR:**

$\delta\alpha/\alpha = 0.01$

$\delta T/K = 0.1$

**REFERENCES:**

**COMPONENTS:**

(1) Ethane; $C_2H_6$; [74-84-0]

(2) Water; $H_2O$; [7732-18-5]

**ORIGINAL MEASUREMENTS:**

Wetlaufer, D. B.; Malik, S. K.; Stoller, L.; Coffin, R. L.

*J. Am. Chem. Soc.* 1964, *86*, 508-514.

**VARIABLES:**

$T$/K: 278.2-318.2

**PREPARED BY:**

C. L. Young

**EXPERIMENTAL VALUES:**

| $T$/K | $10^3$ Conc. of ethane† in soln./mol dm$^{-3}$ | Mole fraction* of ethane $x_{C_2H_6}$ |
|---|---|---|
| 278.2 | 0.00361 | 0.0000652 |
| 298.2 | 0.00186 | 0.0000336 |
| 318.2 | 0.00125 | 0.0000226 |

† at a partial pressure of 101.3 kPa.

* calculated by compiler.

**AUXILIARY INFORMATION**

**METHOD/APPARATUS/PROCEDURE:**

Modified Van Slyke-Neill apparatus fitted with a magnetic stirrer. Solution was saturated with gas and then sample transferred to the Van Slyke extraction chamber.

**SOURCE AND PURITY OF MATERIALS:**

1. Matheson c.p. grade, purity 99 mole per cent or better.

2. Distilled.

**ESTIMATED ERROR:**

$\delta T/K = \pm 0.05$; $\delta x_{C_2H_6} = \pm 2\%$.

**REFERENCES:**

**COMPONENTS:**

(1) Ethane; $C_2H_6$; [74-84-0]

(2) Water; $H_2O$; [7732-18-5]

**ORIGINAL MEASUREMENTS:**

Rudakov, E.S.; Lutsyk, A.I.

*Zh. Fiz. Khim.*, 1979, *53*, 1298-1300.

*Russ. J. Phys. Chem.*, 1979, *53*, 731-733.

**VARIABLES:**

$T$/K: 298.15, 363.15

**PREPARED BY:**

W. Hayduk

**EXPERIMENTAL VALUES:**

| $t$/°C | $T$/K | Partition coefficient[1] $k/cm_2^3 cm_1^{-3}$ | Ostwald coefficient[2] $L/cm^3 cm^{-3}$ | Bunsen coefficient[2] $\alpha/cm^3 (STP) cm^{-3} atm^{-1}$ | Mole fraction[2] $10^4 x_1$ |
|---|---|---|---|---|---|
| 25.0 | 298.15 | 24.0 | 0.0417 | 0.0382 | 0.308 |
| 90.0 | 363.15 | 130.0 | 0.00769 | 0.00579 | 0.0482 |

[1] Original data.

[2] Ostwald and Bunsen coefficients and mole fraction calculated by compiler on basis that partition coefficient is equivalent to the inverse of the Ostwald coefficient and assuming that Henry's law and ideal gas law apply.

**AUXILIARY INFORMATION**

**METHOD/APPARATUS/PROCEDURE:**

Gas chromatographic method used to evaluate partition coefficients. Reactor containing gas and water mechanically shaken.
After phase separation a measured volume of gas introduced into carrier gas for analysis. An equal volume of solution placed into a gas stripping cell for complete stripping of the ethane by the carrier gas. The ratio of areas under the ethane peaks used to determine the solubility. Actual equilibrium pressure not specified.

**SOURCE AND PURITY OF MATERIALS:**

Sources and purities not specified.

**ESTIMATED ERROR:**

$\delta k/k = 0.10$
(authors)

**REFERENCES:**

**COMPONENTS:**

(1) Ethane; $C_2H_6$; [74-84-0]

(2) Water; $H_2O$; [7732-18-5]

**ORIGINAL MEASUREMENTS:**

Czerski, L.; Czaplinski, A.

*Ann. Soc. Chim. Polonorum, (Poland)* 1962, *36*, 1827-1834.

**VARIABLES:**

$T$/K: 273.15
$P$/kPa: 101.3 - 506.6

**PREPARED BY:**

W. Hayduk

**EXPERIMENTAL VALUES:**

| Pressure[1], $P$/atm | Solubility[1], /cm$^3$(STP)dm$^{-3}$ | Bunsen coefficient[2], Ostwald coefficient, $\alpha = L$/cm$^3$cm$^{-3}$ | Mole fraction[2] / $10^5 x_1$ |
|---|---|---|---|
| 1.0 | 78 | 78.4 | 6.30 |
| 3.0 | 235 | | |
| 5.0 | 392 | | |

[1] Original data reported for temperature 0.0°C.

[2] Calculated by compiler assuming that ideal gas law and Henry's law apply, and obtaining an average value for Henry's constant at a gas partial pressure of 101.325 kPa.

AUXILIARY INFORMATION

**METHOD/APPARATUS/PROCEDURE:**

Apparatus consists of a contact chamber agitated by a rocking device. Gas incrementally added from a second smaller chamber of known volume while observing pressure change with each gas addition. Equilibrium established in 2-3 h.

**SOURCE AND PURITY OF MATERIALS:**

Source, purities not available.

**ESTIMATED ERROR:**

$\delta T$/K: 0.05
$\delta P/P$: 0.02
$\delta L/L$: 0.02
(estimated by compiler)

**REFERENCES:**

COMPONENTS:

Ethane; $C_2H_6$; [74-84-0]

Water; $H_2O$; [7732-18-5]
At elevated pressures

EVALUATOR:

Colin L. Young,
School of Chemistry,
University of Melbourne,
Parkville, Victoria 3052,
Australia.

CRITICAL EVALUATION:

The data of McKetta and coworkers (1,2,3) are classified as tentative. These data appear to be internally consistent but when interpolated to 1 atmosphere pressure, the mole fraction solubilities are about 10-15% lower than the values recommended in this volume by Battino for the low pressure solubility of ethane in water. The data of Danneil *et al.* (4) are at higher temperatures and higher pressures than those of McKetta and coworkers (1,2,3) and the two sets of data cannot be meaningfully compared. The data of Danneil *et al.* (4) are internally consistent and are classified as tentative.

References:

1. Culberson, O. L.; McKetta, J. J.
*Trans. AIME Petr. Div.* 1950, *189*, 319.

2. Culberson, O. L.; Horn, A. B.; McKetta, J. J.
*J. Petr. Technol. Trans. AIME Petr. Div.* 1950, *189*, 1.

3. Anthony, R. G.; McKetta, J. J.
*J. Chem. Eng. Data* 1967, *12*, 17.

4. Danneil, A.; Todheide, K.; Franck, E. U.
*Chem. Ing-Tech.* 1967, *13*, 816.

**COMPONENTS:**

(1) Ethane; $C_2H_6$; [74-84-0]

(2) Water; $H_2O$; [7732-18-5]

**ORIGINAL MEASUREMENTS:**

Culberson, O. L.; Horn, A. B.; McKetta, J. J.

*J. Petr. Technol. Trans AIME Pet. Div.* 1950, *189*, 1-6.

**VARIABLES:**

$T$/K = 310.9-444.3

$P$/MPa = 0.41-8.38

**PREPARED BY:**

C. L. Young

**EXPERIMENTAL VALUES:**

| $T$/K | $P$/MPa | Mole fraction of ethane in liquid, $10^4 x_{C_2H_6}$ | $T$/K | $P$/MPa | Mole fraction of ethane in liquid, $10^4 x_{C_2H_6}$ |
|---|---|---|---|---|---|
| 310.93 | 0.41 | 0.893 | 377.59 | 2.22 | 2.56 |
| | 0.76 | 2.04 | | 3.78 | 3.91 |
| | 1.38 | 3.11 | | 5.18 | 5.60 |
| | 2.34 | 5.21 | | 7.72 | 6.61 |
| | 3.92 | 6.47 | 410.93 | 0.78 | 0.475 |
| | 5.29 | 7.09 | | 1.45 | 1.58 |
| | 7.45 | 8.01 | | 2.21 | 2.72 |
| 344.26 | 0.78 | 0.812 | | 3.84 | 4.64 |
| | 1.35 | 1.54 | | 5.41 | 6.00 |
| | 2.17 | 3.09 | | 8.38 | 8.65 |
| | 3.53 | 4.17 | 444.26 | 1.56 | 1.40 |
| | 5.41 | 5.70 | | 2.30 | 2.61 |
| | 8.38 | 6.79 | | 3.65 | 5.03 |
| 377.59 | 0.85 | 0.698 | | 5.12 | 6.71 |
| | 1.46 | 1.30 | | 7.86 | 9.70 |

AUXILIARY INFORMATION

**METHOD/APPARATUS/PROCEDURE:**

Sample equilibrated in large rocking autoclave. Samples analysed by removing water and estimating gas volumetrically. Temperature measured with thermocouple and pressure with Bourdon gauge. Details in source.

**SOURCE AND PURITY OF MATERIALS:**

1. Phillips Petroleum Co., sample purity 99.9 mole per cent minimum.

2. Distilled and degassed.

**ESTIMATED ERROR:**

$\delta T$/K = ±0.5; $\delta P$/MPa = ±0.02 (below 6.8 MPa); ±0.07 (between 6.8 MPa and 33.0 MPa); ±0.13 (above 33.0 MPa); $\delta x_{C_2H_6}$ = ±5% (estimated by compiler).

**REFERENCES:**

**COMPONENTS:**

(1) Ethane; $C_2H_6$; [74-84-0]

(2) Water; $H_2O$; [7732-18-5]

**ORIGINAL MEASUREMENTS:**

Culberson, O. L.; McKetta, J. J.

*Trans. AIME., Petr. Div.*

1950, *189*, 319-322.

**VARIABLES:**

$T$/K = 310.9-444.3

$P$/MPa = 5.07-68.5

**PREPARED BY:**

C. L. Young

**EXPERIMENTAL VALUES:**

| $T$/K | $P$/psia | $P$/MPa | $10^4$ Mole fraction of ethane in liquid, $10^4\ x_{C_2H_6}$ |
|---|---|---|---|
| 310.93 | 1925 | 13.27 | 8.21 |
| | 3115 | 21.48 | 8.90 |
| | 5035 | 34.72 | 10.18 |
| | 5800 | 39.99 | 10.66 |
| | 6330 | 43.64 | 11.05 |
| | 7605 | 52.43 | 10.60 |
| | 9455 | 65.19 | 11.30 |
| 344.26 | 1985 | 13.69 | 7.88 |
| | 3275 | 22.58 | 8.95 |
| | 4885 | 33.68 | 10.11 |
| | 6485 | 44.71 | 10.78 |
| | 7350 | 50.68 | 11.00 |
| | 8330 | 57.43 | 11.78 |
| | 9650 | 66.53 | 11.66 |
| 377.59 | 1965 | 13.55 | 9.44 |
| | 2030 | 14.00 | 9.64 |
| | 2535 | 17.48 | 10.42 |
| | 3455 | 23.82 | 11.32 |
| | 5320 | 36.68 | 12.49 |
| | 7010 | 48.33 | 13.29 |
| | 8480 | 58.47 | 14.34 |
| | 9935 | 68.50 | 15.14 |

(cont.)

**AUXILIARY INFORMATION**

**METHOD/APPARATUS/PROCEDURE:**

Sample equilibrated in large rocking autoclave. Samples of liquid analysed by removing water and estimating gas volumetrically. Temperature measured with thermocouple and pressure with Bourdon gauge. Details in source and ref. (1). Data given in ref. (1) and repeated in the source ref. are not repeated here.

**SOURCE AND PURITY OF MATERIALS:**

1. Phillips Petroleum Co. sample, purity 99.9 mole per cent minimum.
2. Distilled and degassed.

**ESTIMATED ERROR:**

$\delta T$/K = ±0.5; $\delta P$/MPa = ±0.02 (below 68 MPa); ±0.7 (between 68 and 300 MPa); ±1.3 (above 330 MPa); $\delta x_{C_2H_6}$ = ±5% (estimated by compiler).

**REFERENCES:**

1. Culberson, O. L.; Horn, A. B.; McKetta, J. J. *Trans. AIME., Petrol. Div.* 1950, *189*, 1.

COMPONENTS:

(1) Ethane; $C_2H_6$; [74-84-0]

(2) Water; $H_2O$; [7732-18-5]

ORIGINAL MEASUREMENTS:

Culberson, O. L.; McKetta, J. J.

*Trans. AIME., Petr. Div.*

1950, *189*, 319-322.

EXPERIMENTAL VALUES:

| $T$/K | $P$/psia | $P$/MPa | $10^4$ Mole fraction of ethane in liquid, $10^4\,x_{C_2H_6}$ |
|---|---|---|---|
| 410.93 | 736 | 5.07 | 6.60 |
| | 979 | 6.75 | 8.03 |
| | 1470 | 10.14 | 10.65 |
| | 2105 | 14.51 | 12.08 |
| | 2680 | 18.48 | 13.84 |
| | 3585 | 24.72 | 15.30 |
| | 5045 | 34.78 | 17.03 |
| | 6465 | 44.57 | 18.67 |
| | 8055 | 55.54 | 19.01 |
| | 9775 | 67.40 | 20.05 |
| 444.26 | 737 | 5.08 | 7.70 |
| | 992 | 6.84 | 10.39 |
| | 1370 | 9.45 | 13.11 |
| | 1985 | 13.69 | 16.71 |
| | 2605 | 17.96 | 19.70 |
| | 3640 | 25.10 | 23.25 |
| | 4285 | 29.54 | 24.80 |
| | 5035 | 34.72 | 25.15 |
| | 5250 | 36.20 | 26.35 |
| | 6630 | 45.71 | 27.90 |
| | 8320 | 57.36 | 30.60 |
| | 9335 | 64.36 | 32.00 |
| | 9835 | 67.81 | 33.00 |

COMPONENTS:

(1) Ethane; $C_2H_6$; [74-84-0]

(2) Water; $H_2O$; [7732-18-5]

ORIGINAL MEASUREMENTS:

Anthony, R. G.; McKetta, J. J.

*J. Chem. Eng. Data* 1967, *12*, 17-20.

VARIABLES:

$T$/K = 310.9-377.6

$P$/MPa = 2.57-26.03

PREPARED BY:

C. L. Young

EXPERIMENTAL VALUES:

| $T$/K | $P$/MPa | Mole fraction of ethane in liquid, $x_{C_2H_6}$ | in gas, $y_{C_2H_6}$ |
|---|---|---|---|
| 310.9 | 2.566 | - | 0.99733 |
| 344.3 | 2.994 | - | 0.98866 |
| 377.7 | 10.054 | - | 0.98678 |
| 410.9 | 10.799 | - | 0.9633 |
| 344.3 | 3.480 | 0.004070 | - |
| 344.3 | 20.275 | 0.000837 | - |
| 344.4 | 27.611 | 0.001028 | - |
| 377.6 | 28.170 | 0.001153 | - |
| 377.6 | 26.026 | 0.001180 | - |

AUXILIARY INFORMATION

METHOD/APPARATUS/PROCEDURE:

Recirculating flow apparatus with magnetic pump. Temperature measured with thermocouple and pressure measured with gauge. Cell contents equilibrated. Water content of vapor phase determined with electrolytic hygrometer. Hydrocarbon content of water phase determined by using a gas burette. Details in refs. (1), (2) and (3).

SOURCE AND PURITY OF MATERIALS:

1. Minimum purity 99.1 mole per cent.

2. Distilled.

ESTIMATED ERROR:

$\delta T$/K = 0.05-0.10; $\delta P$/MPa ≃ 0.2%;

$\delta x_{C_2H_6}$, $\delta y_{C_2H_6}$ = 0.0002 (compiler's estimate).

REFERENCES:


1. Anthony, R. B. *PhD Thesis, Univ. of Texas*, 1966.
2. Wehe, A. H.; McKetta, J. J. *J. Chem. Eng. Data* 1961, *6*, 167.
3. Wehe, A. H.; McKetta, J. J. *Anal. Chem.* 1961, *33*, 291.

**COMPONENTS:**

(1) Ethane; $C_2H_6$; [74-84-0]

(2) Water; $H_2O$; [7732-18-5]

**ORIGINAL MEASUREMENTS:**

Danneil, A.; Todheide K., and Franck, E.U.

*Chem. Ing-Tech.* 1967, *13*, 816-821.

**VARIABLES:**

$T/K$ = 473.2-673.2

$P/bar$ = 200-3700

**PREPARED BY:**

C.L. Young

**EXPERIMENTAL VALUES:**

| $T/K$ | $P/bar$ | Mole fraction of ethane in liquid, $x_{C_2H_6}$ | in vapour, $y_{C_2H_6}$ |
|---|---|---|---|
| 473.15 | 200 | 0.005 | 0.880 |
| | 500 | 0.005 | 0.930 |
| | 1000 | 0.005 | 0.930 |
| | 1500 | 0.005 | 0.930 |
| | 2000 | 0.005 | 0.930 |
| | 2500 | 0.005 | 0.930 |
| | 3000 | 0.005 | 0.930 |
| | 3500 | 0.005 | 0.930 |
| 523.15 | 200 | 0.007 | 0.690 |
| | 500 | 0.010 | 0.781 |
| | 1000 | 0.0125 | 0.850 |
| | 1500 | 0.015 | 0.885 |
| | 2000 | 0.0175 | 0.902 |
| | 2500 | 0.020 | 0.902 |
| | 3000 | 0.0225 | 0.902 |
| | 3500 | 0.0250 | 0.902 |
| 573.15 | 200 | 0.010 | 0.454 |
| | 500 | 0.020 | 0.587 |
| | 1000 | 0.024 | 0.734 |
| | 1500 | 0.028 | 0.828 |
| | 2000 | 0.032 | 0.855 |
| | 2500 | 0.035 | 0.855 |
| | 3000 | 0.038 | 0.855 |
| | 3500 | 0.041 | 0.855 |

AUXILIARY INFORMATION

**METHOD/APPARATUS/PROCEDURE:**

Static bomb with magnetically operated stirrer. Pressure measured with Bourdon gauge. Temperature measured with NiCr-Ni thermocouple. Samples of vapour and liquid analysed by stripping out hydrocarbon with carbon dioxide and estimating volumetrically. Water estimated gravimetrically.

**SOURCE AND PURITY OF MATERIALS:**

1. Purity 99.8 mole per cent, 0.07 mole per cent propene, 0.06 mole per cent ethene, and 0.017 mole per cent propane

2. Triply distilled.

**ESTIMATED ERROR:**

$\delta T/K = \pm 0.7$; $\delta P/bar = \pm 1\%$; $\delta x_{C_2H_6} = \pm 0.006$; $\delta y_{C_2H_6} = \pm 0.012$.

**REFERENCES:**

COMPONENTS:

(1) Ethane; $C_2H_6$; [74-84-0]

(2) Water; $H_2O$; [7732-18-5]

ORIGINAL MEASUREMENTS:

Danneil, A.; Todheide, K.; Franck, E.U.

*Chem. Ing-Tech.* 1967, *13*, 816-821.

EXPERIMENTAL VALUES:

| $T$/K | $P$/bar | Mole fraction of ethane in liquid, $x_{C_2H_6}$ | in vapour, $y_{C_2H_6}$ |
|---|---|---|---|
| 623.15 | 200 | 0.009 | 0.150 |
| | 300 | 0.035 | 0.230 |
| | 400 | 0.065 | 0.275 |
| | 500 | 0.099 | 0.302 |
| | 600 | 0.143 | 0.305 |
| | 680 | 0.225 | 0.225 |
| | 760 | 0.240 | 0.240 |
| | 800 | 0.125 | 0.320 |
| | 900 | 0.097 | 0.423 |
| | 1000 | 0.085 | 0.489 |
| | 1500 | 0.075 | 0.678 |
| | 2000 | 0.073 | 0.738 |
| | 2500 | 0.072 | 0.756 |
| | 3000 | 0.071 | 0.760 |
| | 3500 | 0.070 | 0.760 |
| 629.15 | 200 | 0.009 | 0.135 |
| | 300 | 0.037 | 0.210 |
| | 400 | 0.080 | 0.227 |
| | 500 | 0.175 | 0.175 |
| | 1205 | 0.295 | 0.295 |
| | 1300 | 0.117 | 0.542 |
| | 1400 | 0.106 | 0.592 |
| | 1500 | 0.104 | 0.625 |
| | 2000 | 0.097 | 0.704 |
| | 2500 | 0.090 | 0.730 |
| | 3000 | 0.087 | 0.740 |
| | 3500 | 0.083 | 0.740 |
| 643.15 | 1680 | 0.315 | 0.315 |
| | 1700 | 0.211 | 0.367 |
| | 1800 | 0.135 | 0.537 |
| | 1900 | 0.123 | 0.589 |
| | 2000 | 0.114 | 0.619 |
| | 2500 | 0.096 | 0.684 |
| | 3000 | 0.093 | 0.712 |
| | 3500 | 0.090 | 0.722 |
| 651.15 | 1990 | 0.320 | 0.320 |
| | 2000 | 0.214 | 0.361 |
| | 2100 | 0.137 | 0.516 |
| | 2200 | 0.118 | 0.574 |
| | 2500 | 0.102 | 0.653 |
| | 3000 | 0.099 | 0.690 |
| | 3500 | 0.096 | 0.692 |
| 658.15 | 2190 | 0.325 | 0.325 |
| | 2200 | 0.202 | 0.411 |
| | 2300 | 0.167 | 0.532 |
| | 2400 | 0.158 | 0.575 |
| | 2500 | 0.152 | 0.603 |
| | 3000 | 0.133 | 0.654 |
| | 3500 | 0.120 | 0.655 |
| 673.15 | 3215 | 0.340 | 0.340 |
| | 3300 | 0.145 | 0.561 |
| | 3400 | 0.138 | 0.591 |
| | 3500 | 0.135 | 0.595 |
| | 3700 | 0.132 | 0.601 |

COMPONENTS:

(1) Ethane; $C_2H_6$; [74-84-0]

(2) Ethene; $C_2H_4$; [74-85-1]

(3) Water; $H_2O$; [7732-18-5]

At elevated pressures

EVALUATOR:

Walter Hayduk
Department of Chemical Engineering
University of Ottawa
Ottawa, Canada K1N 9B4

CRITICAL EVALUATION:

Anthony and McKetta (1) measured the solubilities of gas mixtures composed of ethane and ethene in water at elevated pressures. For those gas compositions rich in one of the gases, it is possible to estimate the solubility of that gas as if it were pure, at the particular conditions of temperature and pressure using Henry's law. This procedure made it possible to compare the data obtained for gas mixtures with those for pure ethane in water as measured by Culbertson et al. (2) in the same laboratory. The latter data, when extrapolated to atmospheric pressure, were judged to be some 10-15% too low (see Critical Evaluation for water at elevated pressures). The estimated solubilities for ethane based on the mixed gas solubilities of Anthony and McKetta (1) are generally higher (in some cases as much as 20%) than those obtained for pure ethane solvent (2).

It may be observed that at high pressures ethene is more soluble than ethane by a factor ranging from approximately 2.9, to 3.8, at a temperature of 311 K and for pressures ranging from approximately 3.5 MPa to 35 MPa, respectively. In some cases, essentially replicate data show variations of up to 10%.

These data are classified as tentative.

References

1. Anthony, R.G.; McKetta, J.J. *J. Chem. Eng. Data* 1967, *12*, 21-28.

2. Culbertson, O.L.; Horn, A.B.; McKetta, J.J. *J. Petr. Technol. Trans. AIME Pet. Div.* 1950, *189*, 1-6.

COMPONENTS:

(1) Ethane; $C_2H_6$; [74-84-0]

(2) Ethene; $C_2H_4$; [74-85-1]

(3) Water; $H_2O$; [7732-18-5]

ORIGINAL MEASUREMENTS:

Anthony, R. G.; McKetta, J. J.

*J. Chem. Eng. Data*

1967, *12*, 21-28.

VARIABLES:

$T$/K = 310.9-410.9

$P$/MPa = 2.06-34.6

PREPARED BY:

C. L. Young

EXPERIMENTAL VALUES:

| $T$/K | $P$/MPa | Mole fractions in liquid $10^3x_{C_2H_6}$ | $10^3x_{C_2H_4}$ | in vapor $y^{\dagger}_{C_2H_6}$ | $y^{\dagger}_{C_2H_4}$ | $10^3y_{H_2O}$ |
|---|---|---|---|---|---|---|
| 311.0 | 3.469 | 0.1993 | 1.2756 | 0.2741 | 0.7251 | 2.320 |
| 311.0 | 3.469 | 0.2520 | 1.0820 | 0.4033 | 0.5967 | 2.139 |
| 310.9 | 3.476 | 0.2905 | 0.9493 | 0.4517 | 0.5483 | 2.159 |
| 311.0 | 3.473 | 0.5616 | 0.3692 | 0.8154 | 0.1845 | 1.817 |
| 311.0 | 6.865 | 0.589 | 0.8083 | 0.7308 | 0.2692 | 0.7846 |
| 311.0 | 6.900 | 0.761 | 0.2007 | 0.9208 | 0.0792 | 0.6778 |
| 311.0 | 6.900 | 0.238 | 2.162 | 0.2387 | 0.7613 | 1.264 |
| 310.9 | 3.720 | No liquid samples | | 1.00 | - | 1.796 |
| 311.0 | 7.013 | taken | | - | 1.00 | 1.519 |
| 377.5 | 3.474 | 0.9797 | 0.9790 | 0.2201 | 0.7799 | 39.19 |
| 377.4 | 3.469 | 0.9219 | 0.9090 | 0.2115 | 0.7885 | 37.75 |
| 377.6 | 3.452 | 0.1289 | 0.7982 | 0.3284 | 0.6716 | 40.58 |
| 377.6 | 3.459 | 0.2194 | 0.5666 | 0.5122 | 0.4878 | 38.89 |
| 377.4 | 3.480 | 0.3723 | 0.1269 | 0.8921 | 0.1079 | 37.59 |
| 377.5 | 6.910 | 0.425 | 0.9597 | 0.6066 | 0.3934 | 20.49 |
| 377.6 | 6.931 | 0.5783 | 0.3698 | 0.8357 | 0.1643 | 20.32 |
| 377.8 | 10.337 | 0.4396 | 1.4684 | 0.4925 | 0.5075 | 15.15 |
| 377.7 | 6.924 | 0.3593 | 0.9107 | 0.6068 | 0.3932 | 20.17 |
| 377.6 | 10.281 | 0.5653 | 0.9977 | 0.6525 | 0.3475 | 14.77 |
| 377.6 | 6.955 | 0.1122 | 1.852 | 0.1481 | 0.8519 | 22.07 |
| 377.7 | 10.337 | 0.1901 | 2.3379 | 0.1943 | 0.8057 | 16.44 |
| 311.0 | 7.448 | 0.8810 | 0.0030 | 0.9906 | 0.0094 | 0.7017 |
| 310.9 | 10.337 | No liquid samples taken | | 0.9906 | 0.0094 | 0.6713 |
| 310.9 | 10.364 | 0.7531 | 0.2452 | 0.9460 | 0.0540 | 0.6809 |

(cont.)

AUXILIARY INFORMATION

METHOD/APPARATUS/PROCEDURE:

Windowed equilibrium cell contained in air thermostat. Vapor re-circulated by magnetic pump. Water content of vapor determined using an electrolytic hygrometer. Hydrocarbon content of aqueous phase determined by a method of partial extraction. Details in refs. (1) and (2).

SOURCE AND PURITY OF MATERIALS:

1. Minimum purity 99.1 mole per cent.
2. Minimum purity 99.3 mole per cent.
3. Distilled.

ESTIMATED ERROR:

$\delta T$/K = ±0.1; $\delta P$/MPa = ±0.2%;

$\delta x$, $\delta y$ = ±5.0%.

REFERENCES:


1. Wehe, A. H.; McKetta, J. J. *Anal. Chem.* 1961, *33*, 291.
2. Wehe, A. H.; McKetta, J. J. *J. Chem. Eng. Data* 1961, *6*, 167.

| COMPONENTS: | ORIGINAL MEASUREMENTS: |
|---|---|
| (1) Ethane; $C_2H_6$; [74-84-0] | Anthony, R. G.; McKetta, J. J. |
| (2) Ethene; $C_2H_4$; [74-85-1] | *J. Chem. Eng. Data.* |
| (3) Water; $H_2O$; [7732-18-5] | 1967, *12*, 21-28. |

EXPERIMENTAL VALUES:

| | | Mole fractions | | | | |
|---|---|---|---|---|---|---|
| | | in liquid | | in vapor | | |
| $T$/K | $P$/MPa | $10^3x_{C_2H_6}$ | $10^3x_{C_2H_4}$ | $y^\dagger_{C_2H_6}$ | $y^\dagger_{C_2H_4}$ | $10^3y_{H_2O}$ |
| 310.9 | 10.247 | 0.7663 | 0.4087 | 0.8676 | 0.1324 | 0.7886 |
| 310.9 | 10.488 | 0.5472 | 1.1368 | 0.6560 | 0.3440 | 0.9710 |
| 310.9 | 10.378 | 0.2517 | 2.417 | 0.2423 | 0.7577 | 1.383 |
| 310.9 | 13.750 | 0.2550 | 2.4870 | 0.2570 | 0.7430 | 1.201 |
| 310.9 | 10.433 | 0.2463 | 2.417 | 0.2701 | 0.7299 | 1.215 |
| 310.9 | 13.735 | 0.2932 | 2.5018 | 0.2639 | 0.7361 | 1.154 |
| 310.9 | 13.784 | 0.4275 | 1.9589 | 0.4227 | 0.5773 | 0.8641 |
| 310.9 | 13.619 | 0.7694 | 0.5006 | 0.8491 | 0.1509 | 0.6238 |
| 310.9 | 20.696 | 1.099 | 0.6561 | 0.8324 | 0.1676 | 0.644 |
| 344.1 | 13.860 | 0.676 | 0.6556 | 0.7958 | 0.2042 | 3.171 |
| 344.2 | 10.475 | 0.651 | 0.6098 | 0.7860 | 0.2140 | 3.429 |
| 344.3 | 13.860 | 0.5644 | 0.4624 | 0.7830 | 0.2170 | 5.108 |
| 344.3 | 3.463 | 0.3508 | 0.2671 | 0.7740 | 0.2260 | 0.24 |
| 344.3 | 13.894 | 0.468 | 1.4237 | 0.5472 | 0.4528 | 3.820 |
| 344.3 | 10.371 | 0.437 | 1.2906 | 0.5268 | 0.4732 | 4.393 |
| 344.3 | 6.879 | 0.3645 | 1.105 | 0.5350 | 0.4650 | 5.727 |
| 344.3 | 3.463 | 0.2386 | 0.6212 | 0.5281 | 0.4719 | 0.916 |
| 344.1 | 13.784 | 0.271 | 2.1715 | 0.2975 | 0.7025 | 3.832 |
| 344.3 | 10.440 | 0.474 | 2.3429 | 0.3103 | 0.6897 | 4.937 |
| 344.3 | 6.865 | 0.2353 | 1.5887 | 0.3011 | 0.6989 | 6.050 |
| 344.3 | 3.480 | 0.1494 | 0.9686 | 0.6917 | 0.3083 | 10.55 |
| 344.4 | 10.302 | 0.547 | 0.8966 | 0.6781 | 0.3219 | 3.992 |
| 344.3 | 13.722 | 0.788 | 0.2930 | 0.8950 | 0.1050 | 2.641 |
| 344.3 | 10.440 | 0.710 | 0.2809 | 0.8996 | 0.1004 | 3.390 |
| 344.3 | 6.865 | 0.698 | 0.2252 | 0.8933 | 0.1067 | 4.709 |
| 344.3 | 2.056 | 0.332 | 0.1144 | 0.8851 | 0.1149 | 10.13 |
| 310.9 | 13.881 | No liquid samples taken | | 0.9809 | 0.0191 | 0.6005 |
| 310.9 | 6.965 | | | 0.9846 | 0.0154 | 0.6505 |
| 311.0 | 6.986 | | | 0.3166 | 0.6834 | 1.196 |
| 311.0 | 6.896 | | | 0.8001 | 0.1999 | 0.7747 |
| 310.9 | 6.944 | | | 0.4773 | 0.5227 | 1.0110 |
| 310.9 | 3.497 | | | 0.4812 | 0.5188 | 2.006 |
| 310.9 | 3.455 | | | 0.2647 | 0.7353 | 1.988 |
| 310.9 | 3.451 | | | 0.2637 | 0.7363 | 2.137 |
| 310.8 | 3.418 | | | 0.8210 | 0.1790 | 1.858 |
| 410.9 | 6.958 | 0.6166 | 0.6304 | 0.7560 | 0.2440 | 57.76 |
| 410.9 | 3.473 | 0.3485 | 0.3275 | 0.7370 | 0.2630 | 108.01 |
| 410.9 | 10.344 | 0.6923 | 1.4217 | 0.5447 | 0.4553 | 43.44 |
| 410.9 | 6.910 | 0.4530 | 1.0230 | 0.5617 | 0.4383 | 58.66 |
| 410.9 | 3.531 | 0.2718 | 0.5422 | 0.5490 | 0.4510 | 105.50 |
| 410.9 | 10.316 | 0.2889 | 2.3140 | 0.2732 | 0.7268 | 38.19 |
| 410.9 | 6.958 | 0.2405 | 1.7225 | 0.2651 | 0.7349 | 61.45 |
| 410.6 | 3.480 | 0.1274 | 0.9286 | 0.2830 | 0.7170 | 109.4 |
| 410.9 | 10.302 | 0.8060 | 0.8261 | 0.7540 | 0.2460 | 40.75 |
| 410.9 | 13.763 | 0.8862 | 0.9168 | 0.7496 | 0.2504 | 33.14 |
| 311.0 | 34.610 | 0.6492 | 1.9548 | 0.5488 | 0.4512 | 1.131 |
| 311.0 | 27.611 | 0.5823 | 1.7366 | 0.5588 | 0.4412 | 1.180 |
| 311.0 | 20.737 | 0.5748 | 1.5942 | 0.5655 | 0.4345 | 1.175 |
| 310.9 | 34.506 | 0.8257 | 1.0773 | 0.7414 | 0.2586 | 0.778 |
| 311.0 | 27.611 | 0.7635 | 1.0084 | 0.7485 | 0.2515 | 0.7180 |
| 310.9 | 20.695 | 0.7061 | 0.9739 | 0.7441 | 0.2559 | 0.8490 |
| 310.9 | 34.630 | 0.4198 | 2.6492 | 0.3731 | 0.6269 | 1.141 |
| 310.9 | 27.594 | 0.4133 | 2.4916 | 0.3653 | 0.6347 | 1.162 |
| 310.9 | 20.786 | 0.3804 | 2.3585 | 0.3636 | 0.6364 | 1.293 |
| 310.9 | 34.506 | 0.3236 | 3.0824 | 0.2696 | 0.7304 | 1.312 |
| 311.0 | 27.577 | 0.2927 | 3.0453 | 0.2687 | 0.7313 | 1.383 |
| 311.0 | 20.734 | 0.2588 | 2.7411 | 0.2729 | 0.7261 | 1.299 |
| 344.3 | 34.534 | 0.3126 | 2.981 | 0.2711 | 0.7289 | 3.976 |

(cont.)

COMPONENTS:

(1) Ethane; $C_2H_6$; [74-84-0]

(2) Ethene; $C_2H_4$; [74-85-1]

(3) Water; $H_2O$; [7732-18-5]

ORIGINAL MEASUREMENTS:

Anthony, R. G.; McKetta, J. J.

*J. Chem. Eng. Data.*

1967, *12*, 21-28.

EXPERIMENTAL VALUES:

| | | Mole fractions | | | | |
|---|---|---|---|---|---|---|
| | | in liquid | | in vapor | | |
| $T$/K | $P$/MPa | $10^3x_{C_2H_6}$ | $10^3x_{C_2H_4}$ | $y^{\dagger}_{C_2H_6}$ | $y^{\dagger}_{C_2H_4}$ | $10^3y_{H_2O}$ |
| 344.3 | 27.653 | 0.2900 | 2.7889 | 0.2658 | 0.7342 | 4.079 |
| 344.3 | 20.786 | 0.2781 | 2.5599 | 0.2697 | 0.7303 | 4.174 |
| 344.4 | 34.479 | 0.7477 | 1.155 | 0.6779 | 0.3221 | 2.752 |
| 344.3 | 27.611 | 0.7003 | 1.1637 | 0.6922 | 0.3078 | 2.736 |
| 344.4 | 20.744 | 0.6413 | 1.0637 | 0.7011 | 0.2989 | 3.004 |
| 344.2 | 34.465 | 0.8778 | 0.6710 | 0.8162 | 0.1838 | 2.643 |
| 344.2 | 27.598 | 0.8188 | 0.6822 | 0.8144 | 0.1856 | 2.580 |
| 344.2 | 20.737 | 0.7336 | 0.6163 | 0.8185 | 0.1815 | 2.623 |
| 344.3 | 34.474 | 0.9484 | 0.2627 | 0.9169 | 0.0831 | 2.442 |
| 344.3 | 27.646 | 0.9120 | 0.3120 | 0.9186 | 0.0814 | 2.325 |
| 344.3 | 20.717 | 0.8529 | 0.2771 | 0.9225 | 0.0775 | 2.518 |
| 344.3 | 13.860 | No liquid sample taken | | 0.9152 | 0.0848 | 2.742 |
| 377.6 | 27.687 | 0.9076 | 1.2314 | 0.7270 | 0.2730 | 6.937 |
| 377.5 | 20.717 | 0.8025 | 1.0655 | 0.7247 | 0.2753 | 10.16 |
| 377.6 | 13.874 | 0.7052 | 0.8968 | 0.7245 | 0.2755 | 12.90 |
| 377.5 | 27.646 | 0.5425 | 2.4124 | 0.4401 | 0.5599 | 9.377 |
| 377.5 | 20.772 | 0.5067 | 2.1783 | 0.4392 | 0.5608 | 11.65 |
| 377.6 | 13.860 | 0.4499 | 1.856 | 0.4374 | 0.5626 | 13.28 |
| 377.6 | 27.563 | 0.3586 | 3.1954 | 0.2637 | 0.7363 | 11.22 |
| 377.6 | 20.744 | 0.3384 | 2.9466 | 0.2608 | 0.7392 | 12.44 |
| 377.5 | 13.874 | 0.2545 | 2.4448 | 0.2587 | 0.9413 | 14.10 |
| 377.5 | 27.680 | 1.093 | 0.3900 | 0.9053 | 0.0947 | 7.568 |
| 377.5 | 20.758 | 1.012 | 0.3680 | 0.9112 | 0.0888 | 9.297 |
| 377.5 | 13.867 | 0.9072 | 0.2998 | 0.9040 | 0.0960 | 11.10 |
| 410.9 | 20.751 | 1.237 | 0.7372 | 0.8325 | 0.1675 | 22.29 |
| 410.9 | 13.853 | 1.0313 | 0.5967 | 0.8477 | 0.1523 | 27.31 |
| 410.9 | 10.364 | 0.8897 | 0.4923 | 0.8336 | 0.1664 | 34.88 |
| 410.9 | 20.641 | 0.8337 | 2.0783 | 0.5762 | 0.4238 | 27.17 |
| 410.9 | 13.791 | 0.7022 | 1.6548 | 0.5628 | 0.4372 | 32.11 |
| 410.9 | 10.389 | 0.5970 | 1.3830 | 0.5608 | 0.4392 | 41.20 |
| 410.9 | 20.699 | 0.5504 | 2.9936 | 0.3753 | 0.6247 | 29.43 |
| 410.9 | 13.784 | 0.4523 | 2.3747 | 0.3662 | 0.6338 | 35.84 |
| 410.9 | 10.375 | 0.3820 | 1.9190 | 0.3606 | 0.6393 | 44.13 |
| 410.8 | 3.473 | No liquid sample taken | | 0.3612 | 0.6388 | 10.64 |

† Mole fraction after removal of water vapor.

COMPONENTS:

(1) Ethane; $C_2H_6$; [74-84-0]

(2) Electrolyte

(3) Water; $H_2O$; [7732-18-5]

EVALUATOR:

H. Lawrence Clever
Department of Chemistry
Emory University
Atlanta, GA 30322 USA

CRITICAL EVALUATION:

An Evaluation of the Solubility of Ethane in Aqueous Electrolyte Solutions.

Not enough workers have measured the solubility of ethane in any one aqueous electrolyte system over common ranges of concentration and temperature to recommend solubility values. Most of the available data are classed as tentative.

In order to have a common basis for comparison, where possible the solubility data have been converted to Sechenov (Setschenow) salt effect parameters at an ethane partial pressure of 101.325 kPa in the form

$$k_{sc\alpha}/\text{dm}^3\ \text{mol}^{-1} = (1/(c_2/\text{mol dm}^{-3}))\log(\alpha°/\alpha)$$

where $c_2$ is the electrolyte concentration in mol $\text{dm}^{-3}$, and $\alpha°$ and $\alpha$ are the Bunsen coefficients in pure water and electrolyte solution, respectively. The Ostwald coefficient ratio, $L°/L$, will give the same value, but the salt effect parameter is symbolized, $k_{scL}$. Both ratios are equivalent to a molar gas solubility ratio, $c_1°/c_1$, thus

$$k_{sc\alpha} = k_{scL} = k_{scc} = (1/(c_2/\text{mol dm}^{-3}))\ \log\ (c°_{1,\text{sat}}/c_{1,\text{sat}})$$

Other forms of the salt effect parameter will be found on the data sheets that follow this discussion. They include

$$k_{sms}/\text{kg mol}^{-1} = (1/(m_2/\text{mol kg}^{-1}))\log(s°/s)$$

$$k_{scx}/\text{dm}^3\ \text{mol}^{-1} = (1/(c_2/\text{mol dm}^{-3}))\log(x°/x)$$

$$k_{smx}/\text{kg mol}^{-1} = (1/(m_2/\text{mol kg}^{-1}))\log(x°/x)$$

where $m_2$ is the electrolyte molality, s°/s is the Kuenen coefficient ratio, and $x°/x$ is the mole fraction gas solubility ratio usually calculated with respect to all ions in the solution. The Kuenen coefficient, s, is referenced to 1 g of water whether it is for pure water or the electrolyte solution. Thus the s°/s ratio is equal to a gas molality ratio $m_1°/m_1$.

The density data required to convert the salt effect parameter from one form to another were taken from the *International Critical Tables*, McGraw-Hill Co., Inc., 1928, Vol. III.

The activity coefficient of the dissolved ethane, $y_1$, is a function of the concentration of all solute species, which in the present systems are the electrolyte and the nonelectrolyte. At a given temperature log $y_1$ can be represented as a power series in $c_1$ and $c_2$

$$\log y_1 = \sum_{nm} k_{nm}\, c_1^n\, c_2^m \qquad (1)$$

If it is assumed that for low concentrations of both nonelectrolyte, $c_1$, and electrolyte, $c_2$, the only important terms are the linear ones,

$$\log y_1 = k_1 c_1 + k_2 c_2 \qquad (2)$$

The expression has been experimentally verified for moderately dilute solutions in which there is no chemical interaction between the solute species.

The measurements of the nonelectrolyte solubility in pure water, $c^{\circ}_{1,\mathrm{sat}}$, and in a salt solution, $c_{1,\mathrm{sat}}$, give directly the activity coefficient of the nonelectrolyte. Long and McDevit (1) show that

$$\log(y_1/y_1^{\circ}) = \log(c^{\circ}_{1,\mathrm{sat}}/c_{1,\mathrm{sat}}) = k_1(c_{1,\mathrm{sat}}-c^{\circ}_{1,\mathrm{sat}}) + k_2 c_2 \qquad (3)$$

And if the nonelectrolyte solubility values are low, as they generally are for a gas, the term in $k_1$ can be neglected, even though $k_1$ is similar in magnitude to $k_2$. Thus for low nonelectrolyte concentration

$$\log(y_1/y_1^{\circ}) = \log(c^{\circ}_{1,\mathrm{sat}}/c_{1,\mathrm{sat}}) = k_2\, c_2 \qquad (4)$$

The solubility data which are evaluated in this section do not always meet the requirements set forth above. Thus, the experimental Setschenow salt effect parameters, $k_s$, may not be equivalent to the theoretically important salt effect parameter, $k_2$.

Although for a given system the experimental values of $k_{scc}$, $k_{sc\alpha}$, and $k_{scL}$ will have the same magnitude and units, there may be a unit problem when one tries to use equation (3). The constant $k_2$ will have inverse $c_2$ units, $dm^3\ mol^{-1}$, and $k_1$ will have inverse $c_1$ units. If the gas solubility is expressed as $c_1/mol\ dm^{-3}$, $k_1$ units will be $dm^3\ mol^{-1}$, but if the Bunsen or Ostwald coefficients are used for $c_{1sat}$ and $c^{\circ}_{1sat}$ in equation 3, then $k_1$ will have units which are the inverse of the Bunsen or Ostwald coefficients, $cm^3\ atm\ (cm^3(STP))^{-1}$ and $cm^3 cm^{-3}$ respectively.

A plot of log $(\alpha^{\circ}/\alpha)$ *vs.* $c_2$ is usually linear over moderate concentrations of the electrolyte. However, curvature of the plot is often seen at above moderate concentrations, and in extreme cases one observes regions of both salting-out and salting-in over different concentration regions of the same isotherm.

The salt effect parameters, $k_{sc\alpha}/dm^3\ mol^{-1}$, are presented by several methods in the evaluation.

In the first method $k_{sc\alpha}$ values are calculated for each experimental determination, and a graph is prepared of $k_{sc\alpha}$ *vs.* $c_2$. If the plot is linear and of zero slope, $k_{sc\alpha}$ is taken to be independent of concentration. In such cases an average value of $k_{sc\alpha}$ is given. If the plot is linear, but of non-zero slope, $k_{sc\alpha}$ is fitted to a linear equation in $c_2$.

In the second method a graph is prepared of $\log(\alpha^{\circ}/\alpha)$ *vs.* $c_2$. A linear plot shows no concentration dependence of the salt effect parameter and the slope is $k_{sc\alpha}$. Recently some workers have fitted such plots that show curvature to a function

$$\log(\alpha^{\circ}/\alpha) = k^{\circ}_{sc\alpha}\, c_2/(1 + k'_{sc\alpha}\, c_2)$$

In a sense the first method, in which the salt effect parameter is given by a linear function of concentration, $k_{sc\alpha} = a + bc_2$, is equivalent to

$$\log(\alpha^{\circ}/\alpha) = (a + bc_2)c_2 = ac_2 + bc_2^2$$

however, the constants should be redetermined from the $\log(\alpha^{\circ}/\alpha)$ and $c_2$ data.

COMPONENTS:

(1) Ethane; $C_2H_6$; [74-84-0]

(2) Electrolyte

(3) Water; $H_2O$; [7732-18-5]

EVALUATOR:

H. Lawrence Clever
Department of Chemistry
Emory University
Atlanta, GA 30322 USA

CRITICAL EVALUATION:

14(1) Ethane + Sulfuric acid [7664-93-6] + Water

Rudakov and Lutsyk (2) measured the solubility of ethane in 0, 80.0, 93.0, and 97.7 wt per cent sulfuric acid at 298.15 and 363.15 K. Their data have been converted to sulfuric acid molality and ethane Ostwald coefficient in order to calculate values of $k_{sc\alpha}$ ($k_{scL}$). The values are:

| $T$/K | Sulfuric Acid | | $k_{sc\alpha}$/dm$^3$ mol$^{-1}$ |
|---|---|---|---|
| | wt % | $c_2$/mol dm$^{-3}$ | |
| 298.15 | 80.0 | 14.0 | 0.040 |
| 363.15 | 93.0 | 16.8 | -0.038 |
| | 97.7 | 17.7 | -0.053 |

Ethane is salted-out at 298.15 K and relatively strongly salted-in at 363.15 in the concentrated sulfuric acid solutions. The salt effect parameters are classed as tentative.

18(1) Ethane + Ammonium chloride [12125-02-9] + Water
18(2) Ethane + Ammonium bromide [12124-97-9] + Water
18(3) Ethane + Tetramethylammonium bromide [64-20-0] + Water
18(4) Ethane + Tetraethylammonium bromide [71-91-0] + Water
18(5) Ethane + Tetrapropylammonium bromide [1941-30-6] + Water
18(6) Ethane + Tetrabutylammonium bromide [1643-19-2] + Water
18(7) Ethane + Guanidinium chloride [50-01-1] + Water
18(8) Ethane + Tetraethanolammonium bromide [4328-04-5] + Water

Ben-Naim and Yaacobi (3) measured the solubility of ethane in water and aqueous 1 mol dm$^{-3}$ ammonium chloride solution at five degree temperature intervals from 283.15 to 303.15 K. Wetlaufer, Malik, Stoller, and Coffin (4) measured the solubility of ethane in water and 4.87 mol dm$^{-3}$ guanidinium chloride at temperatures of 278.2, 298.2, and 318.2 K. Wen and Hung (5) measured the solubility of ethane in water and aqueous ammonium bromide and five tetralkyl ammonium bromide salts at 10 degree intervals between 278.15 and 308.15 K.

All of the salt effect data are classed as tentative since there are no independent data to confirm the present values. The data of Wen and Hung are believed to be more reliable than the other data because they made measurements at more than one electrolyte concentration. The salt effect parameters reported by Wen and Hung are limiting values for 0.1 mol kg$^{-1}$ electrolyte solutions. Their $k_{sms}$ values are believed to be numerically the same as $k_{sc\alpha}$ values for the dilute solutions.

The salt effect parameters for the various solutions are summarized in the table shown on the next page.

| Electrolyte \ T/K | Salt Effect Parameter, $k_{sc\alpha}/dm^3\ mol^{-1}$ | | | | | | |
|---|---|---|---|---|---|---|---|
| | 278.15 | 283.15 | 288.15 | 293.15 | 298.15 | 303.15 | 308.15 |
| $NH_4Cl$ | | 0.120 | 0.109 | 0.105 | 0.106 | 0.114 | |
| $NH_4Br$[1] | 0.082 | | 0.073 | | 0.065 | | 0.056 |
| $(CH_3)_4NBr$[1] | -0.016 | | -0.028 | | -0.040 | | -0.052 |
| $(C_2H_5)_4NBr$[1] | -0.082 | | -0.095 | | -0.117 | | -0.147 |
| $(C_3H_5)_4NBr$[1] | -0.075 | | -0.105 | | -0.141 | | -0.190 |
| $(C_4H_9)_4NBr$[1] | -0.063 | | -0.101 | | -0.155 | | -0.225 |
| $(C_2H_5O)_4NBr$[1] | 0.014 | | 0.002 | | -0.013 | | -0.038 |
| $CH_6ClN_3$[2] | 0.041 | | | | 0.017 | | 0.002[3] |

[1] Author's values stated to be $k_{sms}$ for 0.1 mol $kg^{-1}$ electrolyte. Values of $k_{sc\alpha}$ for 0.1 mol $dm^{-3}$ electrolyte are assumed not to differ significantly from the values above.

[2] Value of $k_{sc\alpha}$ at 4.87 mol $dm^{-3}$ guanidium chloride ($(NH_2)_2CNH.HCl$).

[3] Value for $T$/K = 318.2.

The values of $k_{sc\alpha}$ decrease (become more negative) as the temperature increases. Ammonium chloride and bromide salt-out ethane, but the tetraalkyl ammonium bromides salt-in ethane. At temperatures of 298.15 and 308.15 K the salting-in increases as the alkyl group increases in size, but at the temperatures of 278.15 and 288.15 K there is no regular pattern to the salting-in effect.

94(1) Ethane + Calcium chloride [10043-52-4] + Water

Czerski and Czaplinski (6) measured the solubility of ethane in aqueous calcium chloride solution of 0.5, 1.0 and 1.5 mol $dm^{-3}$ at 273.15 K and pressures ranging from 1 to 15.8 atm. The salt effect parameters were calculated from their data assuming Henry's law was obeyed. The values of the ethane solubility calculated for an ethane partial pressure of 101.3 kPa (1 atm) were used to calculate the $k_{sc\alpha}$ ($k_{scc}$) values below. The values are classed as tentative.

| $T$/K | $c_2$/mol $dm^{-3}$ | $k_{sc\alpha}$ |
|---|---|---|
| 273.15 | 0.5 | 0.347 |
| | 1.0 | 0.368 |
| | 1.5 | 0.371 |

98(1) Ethane + Lithium chloride [7447-41-8] + Water

Both Morrison and Billett (7) and Ben-Naim and Yaacobi (3) report the solubility of ethane in water and aqueous lithium chloride solution. Both reported the solubility of ethane at only one lithium chloride concentration. Morrison and Billett worked with 1.0 mol $kg^{-1}$ lithium chloride and Ben-Naim and Yaacobi worked with 1.0 mol $dm^{-3}$ lithium chloride. The evaluator has recalculated the Morrison and Billett data as $k_{sc\alpha}$ values. The results are summarized in the following table.

COMPONENTS:

(1) Ethane; $C_2H_6$; [74-84-0]

(2) Electrolyte

(3) Water; $H_2O$; [7732-18-5]

EVALUATOR:
H. Lawrence Clever
Department of Chemistry
Emory University
Atlanta, GA 30322 USA

CRITICAL EVALUATION:

| $T$/K | Morrison, Billett | | | Ben-Naim, Yaacobi |
|---|---|---|---|---|
| | $k_{sms}$/kg mol$^{-1}$ | $k_{smx}$/kg mol$^{-1}$ | $k_{sc\alpha}$/dm$^3$ mol$^{-1}$ | $k_{sc\alpha}$/dm$^3$ mol$^{-1}$ |
| 283.15 | | | | 0.147 |
| 285.75 | 0.155 | 0.170 | 0.166 | |
| 288.15 | | | | |
| 293.15 | | | | 0.140 |
| 298.15 | | | | 0.135 |
| 303.15 | 0.124 | 0.139 | 0.135 | 0.130 |
| 322.55 | 0.110 | 0.125 | 0.122 | 0.126 |
| 344.85 | 0.107 | 0.122 | 0.122 | |

The results from the two laboratories agree to within 7 to 14 per cent, with the agreement improving as the temperature increases. The results are classed as tentative.

99(1) Ethane + Sodium chloride [7647-14-5] + Water

Five laboratories have reported the solubility of ethane in aqueous sodium chloride solution at temperatures between 273.15 and 348.15 K. At temperatures of 293.15 and 298.15 K agreement among $k_{sc\alpha}$ values is 2 to 2.5 per cent. However at other temperatures the values show differences of 15 per cent and more. The values of $k_{sc\alpha}$ are summarized below.

| $T$/K | Euken, Hertzberg | Morrison, Billett | Mishnina, Audeeva, Bozkovakaya | Czerski, Czaplinski | Ben-Naim, Yaacobi | Smoothed Values |
|---|---|---|---|---|---|---|
| 273.15 | 0.205 | | | 0.279 | | 0.203 |
| 283.15 | | | 0.188 | | 0.194 | 0.190 |
| 285.75 | | 0.195 | | | | 0.187 |
| 288.15 | | | 0.181 | | 0.185 | 0.184 |
| 293.15 | 0.178 | | 0.174 | | 0.177 | 0.179 |
| 298.15 | | | 0.168 | | 0.172 | 0.173 |
| 303.15 | | 0.174 | 0.162 | | 0.168 | 0.169 |
| 308.15 | | | 0.157 | | | 0.164 |
| 313.15 | | | 0.153 | | | 0.160 |
| 318.15 | | | 0.148 | | | 0.156 |
| 322.55 | | 0.158 | | | | 0.152 |
| 323.15 | | | 0.145 | | | 0.152 |
| 328.15 | | | 0.142 | | | 0.148 |
| 333.15 | | | 0.140 | | | 0.145 |
| 338.15 | | | 0.138 | | | 0.142 |
| 343.15 | | | 0.136 | | | 0.139 |
| 344.85 | 0.152 | | | | | 0.138 |
| 348.15 | | | 0.133 | | | 0.136 |

The value of Czerski and Czaplinski (6) at 273.15 K appears to be too large and is classed as doubtful. The data of Mishnina, Audeeva, Bozkovakaya (8) are obviously smoothed data. It appears their values are smoothed from both their own data and that of Morrison and Billett. However, their data along with that of Euken and Hertzberg (9), Morrison and Billett (7), and Ben-Naim and Yaacobi (3) are all classed as tentative.

The tentative data were fitted by a linear regression in which the $k_{sc\alpha}$ values of Mishnina *et al.* were weighted one and all the other data were weighted two. The resulting equation, from which the smoothed values above were calculated, is

$$\log k_{sc\alpha} = -1.5034 + 221.37/(T/K)$$

The smoothed values are classed as tentative, but it is felt that they are more reliable than $k_{sc\alpha}$ values reported by any one of the laboratories.

99(2) Ethane + Sodium bromide [7647-15-6] + Water
99(3) Ethane + Sodium iodide [7681-82-5] + Water

Ben-Naim and Yaacobi (3) have reported the solubility of ethane in the two aqueous sodium halide solutions at a concentration of one mol $dm^{-3}$. The values of $k_{sc\alpha}(k_{scL})$ are classed as tentative.

| $T/K$ | $k_{sc\alpha}/dm^3\ mol^{-1}$ | |
|---|---|---|
| | NaBr | NaI |
| 283.15 | 0.182 | 0.166 |
| 288.15 | 0.175 | 0.154 |
| 293.15 | 0.167 | 0.145 |
| 298.15 | 0.160 | 0.138 |
| 303.15 | 0.163 | 0.135 |

99(4) Ethane + Sodium sulfate [7757-82-6]
+ sulfuric acid [7664-93-6] + Water

The single measurement of Kobe and Kenton (10) in an aqueous solution which is 1.76 mol $kg^{-1}$ $Na_2SO_4$ and 0.90 mol $kg^{-1}$ $H_2SO_4$ is classed as tentative. No $k_{sc\alpha}$ value was calculated.

99(5) Ethane + Sodium dodecyl sulfate [151-21-3] + Water
99(6) Ethane + Sodium dodecyl sulfate [151-21-3]
+ sodium chloride [7647-14-5] + Water

Matheson and King (11) and Hoskins and King (12) have reported ethane solubilities in these solutions. The sodium dodecyl sulfate micellular solutions salt-in ethane strongly. The systems appear to be too complicated to be described by a simple Sechenov salt effect parameter and none has been calculated. The data are classed as tentative.

In the sodium chloride solution the authors assume the sodium chloride contributes to salting-out, and that the effect can be treated as if sodium chloride alone were present.

100(1) Ethane + Potassium chloride [7447-40-7] + Water

Ben-Naim and Yaacobi (3) measured the solubility of ethane in water and 1.0 mol $dm^{-3}$ KCl solution. The data are classed as tentative. Values of $k_{sc\alpha}(k_{scL})$ from their data are

| $T/K$ | 283.15 | 288.15 | 293.15 | 298.15 | 303.15 |
|---|---|---|---|---|---|
| $k_{sc\alpha}/dm^3\ mol^{-1}$ | 0.184 | 0.174 | 0.165 | 0.159 | 0.154 |

100(2) Ethane + Potassium iodide [7681-17-8] + Water

Morrison and Billett (7) measured the solubility of ethane in water and 1.0 mol $kg^{-1}$ KI solutions. The data are classed as tentative. Values of $k_{smm}$, $k_{smx}$, and $k_{sc\alpha}$ calculated by the evaluator are found in the table shown on the next page.

COMPONENTS:

(1) Ethane; $C_2H_6$; [74-84-0]

(2) Electrolyte

(3) Water; $H_2O$; [7732-18-5]

EVALUATOR:
H. Lawrence Clever
Department of Chemistry
Emory University
Atlanta, GA 30322 USA

CRITICAL EVALUATION:

| $T$/K | $k_{smm}$/kg mol$^{-1}$ | $k_{smx}$/kg mol$^{-1}$ | $k_{sc\alpha}$/dm$^3$ mol$^{-1}$ |
|---|---|---|---|
| 285.75 | 0.125 | 0.140 | 0.151 |
| 303.15 | 0.101 | 0.116 | 0.127 |
| 322.55 | 0.080 | 0.095 | 0.107 |
| 344.85 | 0.065 | 0.080 | 0.094 |

102(1) Ethane + Cesium chloride [7647-17-8] + Water

Ben-Naim and Yaacobi (3) measured the solubility of ethane in water and in 1.0 mol dm$^{-3}$ CsCl solution. The data are classed as tentative. Values of $k_{sc\alpha}$ ($k_{scL}$) are:

| $T$/K | 287.15 | 288.15 | 293.15 | 298.15 | 303.15 |
|---|---|---|---|---|---|
| $k_{sc\alpha}$/dm$^3$ mol$^{-1}$ | 0.164 | 0.159 | 0.151 | 0.141 | 0.128 |

References:

1. Long, F. A.; McDevit, W. F. *Chem. Rev.* 1952, *51*, 119.

2. Rudakov, E. S.; Lutsyk, A. I. *Zh. Fiz. Khim.* 1979, *53*, 1298 - 1300.

3. Ben-Naim, A.; Yaacobi, M. *J. Phys. Chem.* 1974, *78*, 170 - 5.

4. Wetlaufer, D. B.; Malik, S. K.; Stoller, L.; Coffin, R. L. *J. Am. Chem. Soc.* 1964, *86*, 508 - 14.

5. Wen, W.-Y.; Hung, J. H. *J. Phys. Chem.* 1970, *74*, 170 - 80.

6. Czerski, L.; Czaplinski, A. *Ann. Soc. Chim. Polonorum (Poland)* 1962, *36*, 1827 - 34.

7. Morrison, T. J.; Billett, F. *J. Chem. Soc.* 1952, 3819 - 22.

8. Mishnina, T. A.; Avdeeva, O. I.; Bozhovakaya, T. K. *Materialy Vses. Nauchn. Issled. Geol. Inst.* 1961, *46*, 93 - 110.

9. Eucken, A.; Hertzberg, G. *Z. Phys. Chem. (Leipzig)* 1950, *195*, 1 - 23.

10. Kobe, K. A.; Kenton, F. H. *Ind. Eng. Chem., Anal. Ed.* 1938, *10*, 76 - 7.

11. Matheson, I. B. C.; King, A. D. *J. Coll. Interface Sci.* 1978, *66*, 464 - 9.

12. Hoskins, J. C.; King, A. D. *J. Coll. Interface Sci.* 1981, *82*, 264 - 7.

NOTE: Ethane + Potassium salt of N,N-dimethyl-glycine [17647-86-8] + water

Leuhddemann *et al.* (13) report the solubility of ethane in water and in the salt solution of density $\rho$/g cm$^{-3}$ = 1.17 at 293 to 303K. The salt concentration is not given. The solubility of ethane in water compares well with the recommended value at 298 K. The data are classed as tentative.

13. Leuhddemann, R.; Noddes, G.; Schwarz, H.-G., *Oil Gas J.* 1959, *57* (No. 32), 100, 102, 104.

COMPONENTS:

(1) Ethane; $C_2H_6$; [74-84-0]

(2) Sulfuric acid; $H_2SO_4$; [7664-93-9]

(3) Water; $H_2O$; [7732-18-5]

ORIGINAL MEASUREMENTS:

Rudakov, E.S.; Lutsyk, A.I.

*Zh. Fiz. Khim.*, 1979, *53*, 1298-1300.

*Russ. J. Phys. Chem.* 1979, *53*, 731-733.

VARIABLES:

$T$/K: 298.15, 363.15
$H_2SO_4$/wt.%: 80.0 - 97.7

PREPARED BY:

W. Hayduk

EXPERIMENTAL VALUES:

| $t$/°C | $T$/K | Solvent wt.% $H_2SO_4$ [1] | Partition coefficient[1] $k/cm^3 cm_1^{-3}$ | Ostwald coefficient[2] $L/cm^3 cm^{-3}$ | Bunsen coefficient[2] $\alpha/cm^3 (STP) cm^{-3} atm^{-1}$ |
|---|---|---|---|---|---|
| 25.0 | 298.15 | 80.0 | 87 | 0.0115 | 0.0105 |
| 90.0 | 363.15 | 93.0 | 30 | 0.0333 | 0.0251 |
| 90.0 | 363.15 | 97.7 | 15 | 0.0667 | 0.0501 |

[1] From original data.

[2] Ostwald coefficient and Bunsen coefficient calculated by compiler on basis that partition coefficient is equivalent to the inverse of the Ostwald coefficient and assuming that the ideal gas law applies.

AUXILIARY INFORMATION

METHOD/APPARATUS/PROCEDURE:

Gas chromatographic method used to evaluate partition coefficients. Reactor containing gas and acid solution mechanically shaken. After phase separation a measured volume of gas introduced into carrier gas for analysis. An equal volume of solution placed into a gas stripping cell for complete stripping of the ethane by the carrier gas. The ratio of areas under the ethane peaks used to determine the solubility. Actual equilibrium pressure not specified.

SOURCE AND PURITY OF MATERIALS:

Sources and purities not specified.

ESTIMATED ERROR:

$\delta k/k = 0.10$
(authors)

REFERENCES:

**COMPONENTS:**

(1) Ethane; $C_2H_6$; [74-84-0]

(2) Ammonium chloride; $NH_4Cl$; [12125-02-9]

(3) Water; $H_2O$; [7732-18-5]

**ORIGINAL MEASUREMENTS:**

Ben-Naim, A.; Yaacobi, M.

*J. Phys. Chem.* 1974, *78*, 170-5.

**VARIABLES:**

$P$/KPa: 101.325 (1 atm)
$T$/K: 283.15-303.15
$c_2$/mol dm$^{-3}$: 1.0

**PREPARED BY:**

C.L. Young

**EXPERIMENTAL VALUES:**

| $T$/K | Conc. of ammonium chloride/mol dm$^{-3}$ | Ostwald coefficient,* $L$ |
|---|---|---|
| 283.15 | 1.0 | 0.05236 |
| 288.15 | | 0.04596 |
| 293.15 | | 0.04036 |
| 298.15 | | 0.03548 |
| 303.15 | | 0.03120 |

* Smoothed values obtained from

$$kT \ln L = 677.2 - 4.078\ (T/K) + 0.04356\ (T/K)^2\ \text{cal mol}^{-1}$$

where k is in units of cal mol$^{-1}$ K$^{-1}$

**AUXILIARY INFORMATION**

**METHOD/APPARATUS/PROCEDURE:**

The apparatus was similar to that described by Ben-Naim and Baer (1) and Wen and Hung (2). It consists of three main parts, a dissolution cell of 300 to 600 cm$^3$ capacity, a gas volume measuring column, and a manometer. The solvent is degassed in the dissolution cell, the gas is introduced and dissolved, while the liquid is kept stirred by a magnetic stirrer immersed in the water bath. Dissolution of the gas results in the change in the height of a column of mercury which is measured by a cathetometer.

**SOURCE AND PURITY OF MATERIALS:**

1. Matheson sample, purity 99.9 mol per cent.
2. AR grade.
3. Deionised, doubly distilled.

**ESTIMATED ERROR:**

$\delta T/K = \pm 0.01$; $\delta L/L = \pm 0.005$.
(estimated by compiler).

**REFERENCES:**

1. Ben-Naim, A.; Baer, S. *Trans. Faraday Soc.* 1963, *59*, 2735.
2. Wen, W.-Y.; Hung, J.H. *J. Phys. Chem.* 1970, *74*, 170.

**COMPONENTS:**

(1) Ethane; $C_2H_4$; [74-84-0]

(2) Ammonium bromide; $NH_4Br$; [12124-97-9]

(3) Water; $H_2O$; [7732-18-5]

**ORIGINAL MEASUREMENTS:**

Wen, W.-Y.; Hung, J. H.

*J. Phys. Chem.* 1970, *74*, 170 - 180.

**VARIABLES:**

$T$/K: 278.15 - 308.15
$P$/kPa: 101.325 (1 atm)
$m_2$/mol $kg^{-1}$: 0 - 0.676

**PREPARED BY:**

H. L. Clever

**EXPERIMENTAL VALUES:**

| $T$/K | Salt Molality $m_2$/mol $kg^{-1}$ | Ethane Solubility $S_1/cm^3$ (STP) $kg^{-1}$ | Setschenow Constant[1] k/kg $mol^{-1}$ |
|---|---|---|---|
| 278.15 | 0 | 80.19 ± 0.23 | +0.082 |
| | 0.205 | 77.24 | |
| | 0.644 | 70.96 | |
| 288.15 | 0 | 55.55 ± 0.15 | +0.073 |
| | 0.108 | 54.51 | |
| | 0.214 | 53.39 | |
| | 0.409 | 51.78 | |
| | 0.663 | 49.49 | |
| 298.15 | 0 | 41.20 ± 0.12 | +0.065 |
| | 0.104 | 40.51 | |
| | 0.218 | 39.56 | |
| | 0.486 | 38.81 | |
| | 0.676 | 37.53 | |
| 308.15 | 0 | 32.27 ± 0.10 | +0.056 |
| | 0.109 | 31.60 | |
| | 0.214 | 31.20 | |
| | 0.415 | 29.99 | |
| | 0.672 | 29.47 | |

[1] Setschenow constant, k/kg $mol^{-1}$ = $(1/(m_2/\text{mol kg}^{-1})) \log (S_1^\circ/S_1)$

The authors specify the value of the constant for $m_2$/mol $kg^{-1}$ = 0.1.

**AUXILIARY INFORMATION**

**METHOD/APPARATUS/PROCEDURE:**

The apparatus was similar to that described by Ben-Naim and Baer (1). Teflon needle valves were used in place of stopcocks.

The apparatus consists of three main parts, a dissolution cell of 300 to 600 $cm^3$ capacity, a gas volume measuring column, and a manometer.

The solvent is degassed in the dissolution cell, the gas is introduced and dissolved while the liquid is kept stirred by a magnetic stirrer immersed in the water bath. Dissolution of the gas results in the change in the height of a column of mercury which is measured by a cathetometer.

**SOURCE AND PURITY OF MATERIALS:**

(1) Ethane. Matheson Co. Stated to be better than 99.9 per cent pure.

(2) Ammonium bromide. Baker Chemical Co. Analyzed reagent grade. Used as received.

(3) Water. Distilled from an all Pyrex apparatus. Specific conductivity 1.5 x $10^{-6}$ $(\text{ohm cm})^{-1}$.

**ESTIMATED ERROR:**

$\delta T/K = \pm 0.005$
$\delta S_1/S_1 = \pm 0.003$

**REFERENCES:**

1. Ben-Naim, A.; Baer, S. *Trans. Faraday Soc.* 1963, *59*, 2735.

**COMPONENTS:**

(1) Ethane; $C_2H_6$; [74-84-0]

(2) N,N,N-Trimethylmethanaminium bromide or tetramethylammonium bromide; $C_4H_{12}NBr$; [64-20-0]

(3) Water; $H_2O$; [7732-18-5]

**ORIGINAL MEASUREMENTS:**

Wen, W.-Y.; Hung, J. H.

*J. Phys. Chem.* 1970, *74*, 170 - 180.

**VARIABLES:**

$T$/K: 278.15 - 308.15

$P$/kPa: 101.325 (1 atm)

$m_2$/mol kg$^{-1}$: 0 - 0.325

**PREPARED BY:**

H. L. Clever

**EXPERIMENTAL VALUES:**

| $T$/K | Salt Molality $m_2$/mol kg$^{-1}$ | Ethane Solubility $S_1$/cm$^3$ (STP) kg$^{-1}$ | Setschenow Constant[1] k/kg mol$^{-1}$ |
|---|---|---|---|
| 278.15 | 0 | 80.19 ± 0.23 | -0.016 |
| | 0.169 | 80.65 | |
| 288.15 | 0 | 55.55 ± 0.15 | -0.028 |
| | 0.170 | 55.84 | |
| | 0.310 | 56.39 | |
| 298.15 | 0 | 41.20 ± 0.12 | -0.040 |
| | 0.165 | 41.73 | |
| | 0.325 | 42.46 | |
| 308.15 | 0 | 32.27 ± 0.10 | -0.052 |
| | 0.184 | 33.02 | |

[1] Setschenow constant, k/kg mol$^{-1}$ = $(1/(m_2/\text{mol kg}^{-1}))\log(S_1^\circ/S_1)$

The authors specify the value of the constant for $m_2$/mol kg$^{-1}$ = 0.1.

**AUXILIARY INFORMATION**

**METHOD/APPARATUS/PROCEDURE:**

The apparatus was similar to that described by Ben-Naim and Baer (1). Teflon needle valves were used in place of stopcocks.

The apparatus consists of three main parts, a dissolution cell of 300 to 600 cm$^3$ capacity, a gas volume measuring column, and a manometer.

The solvent is degassed in the dissolution cell, the gas is introduced and dissolved while the liquid is kept stirred by a magnetic stirrer immersed in the water bath. Dissolution of the gas results in the change in the height of a column of mercury which is measured by a cathetometer.

**SOURCE AND PURITY OF MATERIALS:**

(1) Ethane. Matheson Co. Stated to be better than 99.9 per cent pure.

(2) Tetramethylammonium bromide. Eastman Kodak Co. Recrystallized and analyzed. Better than 99.9 per cent pure.

(3) Water. Distilled from an all Pyrex apparatus. Specific conductivity 1.5 x 10$^{-6}$ (ohm cm)$^{-1}$.

**ESTIMATED ERROR:**

$\delta T$/K = ±0.005

$\delta S_1/S_1$ = ±0.003

**REFERENCES:**

1. Ben-Naim, A.; Baer, S. *Trans. Faraday Soc.* 1963, *59*, 2735.

**COMPONENTS:**

(1) Ethane; $C_2H_6$; [74-84-0]

(2) N,N,N-Triethylethanaminium bromide or tetraethylammonium bromide; $C_8H_{20}NBr$; [71-91-0]

(3) Water; $H_2O$; [7732-18-5]

**ORIGINAL MEASUREMENTS:**

Wen, W.-Y.; Hung, J. H.

*J. Phys. Chem.* 1970, *74*, 170 - 180.

**VARIABLES:**

$T$/K: 278.15 - 308.15

$P$/kPa: 101.325 (1 atm)

$m_2$/mol kg$^{-1}$: 0 - 0.436

**PREPARED BY:**

H. L. Clever

**EXPERIMENTAL VALUES:**

| $T$/K | Salt Molality $m_2$/mol kg$^{-1}$ | Ethane Solubility $S_1$/cm$^3$ (STP) kg$^{-1}$ | Setschenow Constant[1] k/kg mol$^{-1}$ |
|---|---|---|---|
| 278.15 | 0 | 80.19 ± 0.23 | -0.082 |
| | 0.116 | 81.90 | |
| | 0.436 | 85.56 | |
| 288.15 | 0 | 55.55 ± 0.15 | -0.095 |
| | 0.152 | 56.96 | |
| | 0.436 | 60.05 | |
| 298.15 | 0 | 41.20 ± 0.12 | -0.117 |
| | 0.161 | 42.98 | |
| | 0.428 | 45.86 | |
| 308.15 | 0 | 32.27 ± 0.10 | -0.147 |
| | 0.098 | 33.41 | |
| | 0.423 | 36.62 | |

[1] Setschenow constant, k/kg mol$^{-1}$ = $(1/(m_2/\text{mol kg}^{-1}))\log(S_1^\circ/S_1)$

The authors specify the value of the constant for $m_2$/mol kg$^{-1}$ = 0.1.

**AUXILIARY INFORMATION**

**METHOD/APPARATUS/PROCEDURE:**

The apparatus was similar to that described by Ben-Naim and Baer (1). Teflon needle valves were used in place of stopcocks.

The apparatus consists of three main parts, a dissolution cell of 300 to 600 cm$^3$ capacity, a gas volume measuring column, and a manometer.

The solvent is degassed in the dissolution cell, the gas is introduced and dissolved while the liquid is kept stirred by a magnetic stirrer immersed in the water bath. Dissolution of the gas results in the change in the height of a column of mercury which is measured by a cathetometer.

**SOURCE AND PURITY OF MATERIALS:**

(1) Ethane. Matheson Co. Stated to be better than 99.9 per cent pure.

(2) Tetraethylammonium bromide. Eastman Kodak Co. Recrystallized and analyzed. Better than 99.9 per cent pure.

(3) Water. Distilled from an all Pyrex apparatus. Specific conductivity 1.5 x $10^{-6}$ (ohm cm)$^{-1}$.

**ESTIMATED ERROR:**

$\delta T$/K = ±0.005

$\delta S_1/S_1$ = ±0.003

**REFERENCES:**

1. Ben-Naim, A.; Baer, S. *Trans. Faraday Soc.* 1963, *59*, 2735.

**COMPONENTS:**

(1) Ethane; $C_2H_6$; [74-84-0]

(2) N,N,N-Tripropylpropanaminium bromide or tetrapropylammonium bromide; $C_{12}H_{28}NBr$; [1941-30-6]

(3) Water; $H_2O$; [7732-18-5]

**ORIGINAL MEASUREMENTS:**

Wen, W.-Y.; Hung, J. H.

*J. Phys. Chem.* 1970, *74*, 170 - 180.

**VARIABLES:**

$T$/K: 278.15 - 308.15

$P$/kPa: 101.325 (1 atm)

$m_2$/mol kg$^{-1}$: 0 - 0.805

**PREPARED BY:**

H. L. Clever

**EXPERIMENTAL VALUES:**

| $T$/K | Salt Molality $m_2$/mol kg$^{-1}$ | Ethane Solubility $S_1$/cm$^3$ (STP) kg$^{-1}$ | Setschenow Constant[1] k/kg mol$^{-1}$ |
|---|---|---|---|
| 278.15 | 0 | 80.19 ± 0.23 | -0.075 |
| | 0.102 | 81.59 | |
| | 0.208 | 82.73 | |
| | 0.267 | 82.96 | |
| | 0.461 | 83.92 | |
| | 0.470 | 83.70 | |
| | 0.805 | 82.37 | |
| 288.15 | 0 | 55.55 ± 0.15 | -0.105 |
| | 0.280 | 58.85 | |
| | 0.410 | 59.97 | |
| | 0.451 | 60.74 | |
| | 0.771 | 61.82 | |
| 298.15 | 0 | 41.20 ± 0.12 | -0.141 |
| | 0.270 | 44.70 | |
| | 0.270 | 44.79 | |
| | 0.410 | 46.13 | |
| | 0.461 | 46.80 | |
| | 0.749 | ( )[2] | |
| 308.15 | 0 | 32.27 ± 0.10 | -0.190 |
| | 0.255 | 35.60 | |
| | 0.462 | 37.66 | |
| | 0.736 | 40.98 | |

**AUXILIARY INFORMATION**

**METHOD/APPARATUS/PROCEDURE:**

The apparatus was similar to that described by Ben-Naim and Baer (1). Teflon needle valves were used in place of stopcocks.

The apparatus consists of three main parts, a dissolution cell of 300 to 600 cm$^3$ capacity, a gas volume measuring column, and a manometer.

The solvent is degassed in the dissolution cell, the gas is introduced and dissolved while the liquid is kept stirred by a magnetic stirrer immersed in the water bath. Dissolution of the gas results in the change in the height of a column of mercury which is measured by a cathetometer.

**SOURCE AND PURITY OF MATERIALS:**

(1) Ethane. Matheson Co. Stated to be better than 99.9 per cent pure.

(2) Tetrapropylammonium bromide. Eastman Kodak Co. Recrystallized and analyzed. Better than 99.9 per cent pure.

(3) Water. Distilled from an all Pyrex apparatus. Specific conductivity 1.5 x 10$^{-6}$ (ohm cm)$^{-1}$.

**ESTIMATED ERROR:**

$\delta T$/K = ±0.005

$\delta S_1/S_1$ = ±0.003

**REFERENCES:**

1. Ben-Naim, A.; Baer, S. *Trans. Faraday Soc.* 1963, *59*, 2735.

[1] Setschenow constant, k/kg mol$^{-1}$ = $(1/(m_2/\text{mol kg}^{-1})) \log (S_1^\circ/S_1)$

The authors specify the value of the constant for $m_2$/mol kg$^{-1}$ = 0.1.

[2] Solubility value missing in original paper.

**COMPONENTS:**

(1) Ethane; $C_2H_6$; [74-84-0]

(2) N,N,N-Tributylbutanaminium bromide or tetrabutylammonium bromide; $C_{16}H_{36}NBr$; [1643-19-2]

(3) Water; $H_2O$; [7732-18-5]

**ORIGINAL MEASUREMENTS:**

Wen, W.-Y.; Hung, J. H.

*J. Phys. Chem.* 1970, *74*, 170 - 180.

**VARIABLES:**

$T$/K: 278.15 - 308.15

$P$/kPa: 101.325 (1 atm)

$m_2$/mol kg$^{-1}$: 0 - 0.304

**PREPARED BY:**

H. L. Clever

**EXPERIMENTAL VALUES:**

| $T$/K | Salt Molality $m_2$/mol kg$^{-1}$ | Ethane Solubility $S_1$/cm$^3$ (STP) kg$^{-1}$ | Setschenow Constant[1] k/kg mol$^{-1}$ |
|---|---|---|---|
| 278.15 | 0 | 80.19 ± 0.23 | -0.063 |
| | 0.099 | 81.15 | |
| | 0.193 | 82.03 | |
| 288.15 | 0 | 55.55 ± 0.15 | -0.101 |
| | 0.193 | 58.18 | |
| | 0.304 | 59.30 | |
| 298.15 | 0 | 41.20 ± 0.12 | -0.155 |
| | 0.165 | 43.75 | |
| | 0.290 | 45.45 | |
| 308.15 | 0 | 32.27 ± 0.10 | -0.225 |
| | 0.103 | 34.05 | |
| | 0.205 | 35.90 | |

[1] Setschenow constant, k/kg mol$^{-1}$ = $(1/(m_2/\text{mol kg}^{-1}))\log(S_1^\circ/S_1)$

The authors specify the value of the constant for $m_2$/mol kg$^{-1}$ = 0.1.

**AUXILIARY INFORMATION**

**METHOD/APPARATUS/PROCEDURE:**

The apparatus was similar to that described by Ben-Naim and Baer (1). Teflon needle valves were used in place of stopcocks.

The apparatus consists of three main parts, a dissolution cell of 300 to 600 cm$^3$ capacity, a gas volume measuring column, and a manometer.

The solvent is degassed in the dissolution cell, the gas is introduced and dissolved while the liquid is kept stirred by a magnetic stirrer immersed in the water bath. Dissolution of the gas results in the change in the height of a column of mercury which is measured by a cathetometer.

**SOURCE AND PURITY OF MATERIALS:**

(1) Ethane. Matheson Co. Stated to be better than 99.9 per cent pure.

(2) Tetrabutylammonium bromide. Eastman Kodak Co. Recrystallized and analyzed. Better than 99.9 per cent pure.

(3) Water. Distilled from an all Pyrex apparatus. Specific conductivity 1.5 x $10^{-6}$ (ohm cm)$^{-1}$.

**ESTIMATED ERROR:**

$\delta T$/K = ±0.005

$\delta S_1/S_1$ = ±0.003

**REFERENCES:**

1. Ben-Naim, A.; Baer, S. *Trans. Faraday Soc.* 1963, *59*, 2735.

**COMPONENTS:**

(1) Ethane; $C_2H_6$; [74-84-0]

(2) Guanidine monohydrochloride (Guanidinium chloride); $CH_6ClN_3$;

(3) Water; $H_2O$; [7732-18-5]

**ORIGINAL MEASUREMENTS:**

Wetlaufer, D. B.; Malik, S. K.; Stoller, L.; Coffin, R. L.

*J. Am. Chem. Soc.* 1964, *86*, 508-514.

**VARIABLES:**

$P$/KPa: 101.325

$T$/K: 278.2-318.2

$c_2$/mol dm$^{-3}$: 4.87

**PREPARED BY:**

C. L. Young

**EXPERIMENTAL VALUES:**

| $T$/K | Conc. of guanidinium chloride in soln. $c_2$/mol dm$^{-3}$ | $10^3$ Conc. of ethane† in soln. $c_1$/mol dm$^{-3}$ | Mole fraction* of ethane $x_{C_2H_6}$ |
|---|---|---|---|
| 278.2 | 4.87 | 2.28 | 0.0000556 |
| 298.2 | 4.87 | 1.54 | 0.0000375 |
| 318.2 | 4.87 | 1.22 | 0.0000297 |

† at a partial pressure of 101.3 kPa.

* calculated by compiler.

**AUXILIARY INFORMATION**

**METHOD/APPARATUS/PROCEDURE:**

Modified Van Slyke-Neill apparatus fitted with a magnetic stirrer. Solution was saturated with gas and then sample transferred to the Van Slyke extraction chamber.

**SOURCE AND PURITY OF MATERIALS:**

1. Matheson c.p. grade, purity 99 mole per cent or better.
2. Distilled.
3. Prepared from the action of reagent grade hydrochloric acid on twice or three times recrystallized guanidinium carbonate.

**ESTIMATED ERROR:**

$\delta T$/K = ±0.05; $\delta x_{C_2H_6}$ = ±2%.

**REFERENCES:**

**COMPONENTS:**

(1) Ethane; $C_2H_6$; [74-84-0]

(2) 2-Hydroxy-N,N,N-tris(2-hydroxyethyl)-ethanaminium bromide or tetraethanolammonium bromide; $C_8H_{20}NO_4Br$; [4328-04-5]

(3) Water; $H_2O$; [7732-18-5]

**ORIGINAL MEASUREMENTS:**

Wen, W.-Y.; Hung, J. H.

*J. Phys. Chem.* 1970, *74*, 170 - 180.

**VARIABLES:**

$T$/K: 278.15 - 308.15

$P$/kPa: 101.325 (1 atm)

$m_2$/mol kg$^{-1}$: 0 - 0.577

**PREPARED BY:**

H. L. Clever

**EXPERIMENTAL VALUES:**

| $T$/K | Salt Molality $m_2$/mol kg$^{-1}$ | Ethane Solubility $S_1$/cm$^3$ (STP) kg$^{-1}$ | Setschenow Constant[1] k/kg mol$^{-1}$ |
|---|---|---|---|
| 278.15 | 0 | 80.19 ± 0.23 | +0.014 |
| | 0.092 | 79.93 | |
| | 0.184 | 79.67 | |
| | 0.355 | 79.27 | |
| | 0.526 | 78.86 | |
| 288.15 | 0 | 55.55 ± 0.15 | +0.002 |
| | 0.103 | 55.43 | |
| | 0.202 | 55.53 | |
| | 0.401 | 55.62 | |
| | 0.577 | 55.73 | |
| 298.15 | 0 | 41.20 ± 0.12 | -0.013 |
| | 0.099 | 41.44 | |
| | 0.198 | 41.46 | |
| | 0.386 | 41.76 | |
| | 0.565 | 42.16 | |
| 308.15 | 0 | 32.27 ± 0.10 | -0.038 |
| | 0.096 | 32.41 | |
| | 0.154 | 32.74 | |
| | 0.193 | 32.96 | |
| | 0.380 | 33.56 | |
| | 0.558 | 33.81 | |

**AUXILIARY INFORMATION**

**METHOD/APPARATUS/PROCEDURE:**

The apparatus was similar to that described by Ben-Naim and Baer (1). Teflon needle valves were used in place of stopcocks.

The apparatus consists of three main parts, a dissolution cell of 300 to 600 cm$^3$ capacity, a gas volume measuring column, and a manometer.

The solvent is degassed in the dissolution cell, the gas is introduced and dissolved while the liquid is kept stirred by a magnetic stirrer immersed in the water bath. Dissolution of the gas results in the change in the height of a column of mercury which is measured by a cathetometer.

**SOURCE AND PURITY OF MATERIALS:**

(1) Ethane. Matheson Co. Stated to be better than 99.9 per cent pure.

(2) Tetraethanolammonium bromide. Prepared and analyzed. Better than 99.9 per cent pure. m.p., $t$/°C 102.

(3) Water. Distilled from an all Pyrex apparatus. Specific conductivity 1.5 x 10$^{-6}$ (ohm cm)$^{-1}$.

**ESTIMATED ERROR:**

$\delta T$/K = ±0.005

$\delta S_1/S_1$ = ±0.003

**REFERENCES:**

1. Ben-Naim, A.; Baer, S. *Trans. Faraday Soc.* 1963, *59*, 2735.

[1] Setschenow constant, k/kg mol$^{-1}$ = (1/($m_2$/mol kg$^{-1}$)) log ($S_1^\circ/S_1$)

The authors specify the value of the constant for $m_2$/mol kg$^{-1}$ = 0.1.

**COMPONENTS:**

(1) Ethane; $C_2H_6$; [74-84-0]

(2) Calcium chloride; $CaCl_2$; [10043-52-4]

(3) Water; $H_2O$; [7732-18-5]

**ORIGINAL MEASUREMENTS:**

Czerski, L.; Czaplinski, A.

*Ann. Soc. Chim. Polonorum (Poland)* 1962, *36*, 1827-1834.

**VARIABLES:**

$T$/K: 273.15
$P$/kPa: 101.3 - 1600
$c_2$/mol dm$^{-3}$: 0.5 - 1.5

**PREPARED BY:**

W. Hayduk

**EXPERIMENTAL VALUES:**

| Concentration of salt $x_2$/mol dm$^{-3}$ | Pressure[1], $P$/atm 1.0 | 3.0 | 5.0 | 6.4 | 7.0 | 8.2 | 9.9 | 10.3 | 14.6 | 15.8 |
|---|---|---|---|---|---|---|---|---|---|---|
| | Solubility[1], $S$/cm$^3$(STP)dm$^{-3}$ solution | | | | | | | | | |
| 0.5 | 53 | 158 | 263 | 336 | 367 | - | - | - | - | - |
| 1.0 | 33 | 100 | 168 | 215 | 235 | 274 | 332 | - | - | - |
| 1.5 | 22 | 66 | 110 | 140 | 152 | 178 | 216 | 225 | 318 | 344 |

| Concentration of salt $x_2$/mol dm$^{-3}$ | Solubility[2], $S_1$/cm$^3$(STP)dm$^{-3}$ | Salt parameter[3], $(1/c_2)\log(S_1^\circ/S_1)$ |
|---|---|---|
| 0.5 | 52.6 | 0.3467 |
| 1.0 | 33.6 | 0.3680 |
| 1.5 | 21.8 | 0.3705 |

[1] Original data, at 0.0°C.
[2] Calculated by authors assuming Henry's law applies and obtaining average value of Henry's constant at a gas partial pressure of 101.3 kPa.
[3] Salt effect parameter calculated by compiler using salt concentration expressed as $c_2$/mol dm$^{-3}$ solution and solubility of ethane in water as determined by authors: $S_1^\circ$ = 78.4 cm$^3$(STP)dm$^{-3}$.

AUXILIARY INFORMATION

**METHOD/APPARATUS/PROCEDURE:**

Apparatus consists of a contact chamber agitated by a rocking device. Gas incrementally added from a second smaller chamber of known volume while observing pressure change with each gas addition. Equilibrium established in 2-3 h.

**SOURCE AND PURITY OF MATERIALS:**

Source, purities not available.

**ESTIMATED ERROR:**

$\delta T$/K = 0.05
$\delta c_2/c_2$ = 0.02
$\delta P/P$ = 0.02
(estimated by compiler)

**REFERENCES:**

**COMPONENTS:**

(1) Ethane; $C_2H_6$; [74-84-0]

(2) Lithium chloride; LiCl; [7447-41-8]

(3) Water; $H_2O$; [7732-18-5]

**ORIGINAL MEASUREMENTS:**

Morrison, T. J.; Billett, F.

*J. Chem. Soc.* 1952, 3819 - 3822.

**VARIABLES:**

$T$/K: 285.75 - 344.85
$p$/kPa: 101.325 (1 atm)

**PREPARED BY:**

H. L. Clever

**EXPERIMENTAL VALUES:**

| Temperature | | | Salt Effect Parameters | |
|---|---|---|---|---|
| t/°C | $T$/K | $1/(T/K)$ | $(1/m_2)\log(S°/S)$[1] | $(1/m_2)\log(x°/x)$ |
| 12.6 | 285.75 | 0.0035 | 0.155 | 0.170 |
| 30.0 | 303.15 | 0.0033 | 0.124 | 0.139 |
| 49.4 | 322.55 | 0.0031 | 0.110 | 0.125 |
| 71.7 | 344.85 | 0.0029 | 0.107 | 0.122 |

[1] The authors used (1/c)log(S°/S) with c defined as g eq salt per kg of water. For the 1-1 electrolyte the compiler changed the c to an m for $m_2$/mol $kg^{-1}$. The ethane solubility S is $cm^3$ (STP) $kg^{-1}$.

The salt effect parameters were calculated from two measurements. The solubility of ethane in water, S°, and in the one molal salt solution, S. Only the solubility of the ethane in water, and the value of the salt effect parameter are given in the paper. The solubility values in the salt solution are not given.

The compiler calculated the values of the salt effect parameter using the mole fraction gas solubility ratio.

AUXILIARY INFORMATION

**METHOD/APPARATUS/PROCEDURE:**

The degassed solvent flows in a thin film down an absorption helix containing the ethane gas plus solvent vapor at a total pressure of one atmosphere. The volume of gas absorbed is measured in an attached buret system (1).

**SOURCE AND PURITY OF MATERIALS:**

(1) Ethane. Prepared from Grignard reagent.

(2) Lithium chloride. "AnalaR" material.

(3) Water. No information given.

**ESTIMATED ERROR:**

$\delta k/kg^{-1}$ mol = 0.010

**REFERENCES:**

1. Morrison, T. J.; Billett, F. *J. Chem. Soc.* 1948, 2033.

COMPONENTS:

(1) Ethane; $C_2H_6$; [74-84-0]

(2) Lithium chloride; LiCl; [7447-41-8]

(3) Water; $H_2O$; [7732-18-5]

ORIGINAL MEASUREMENTS:

Ben-Naim, A.; Yaacobi, M.

*J. Phys. Chem.* 1974, *78*, 170-5.

VARIABLES:

$T$/K : 283.15-303.15

$c_2$/mol dm$^{-3}$ : 1.0

PREPARED BY:

C.L. Young

EXPERIMENTAL VALUES:

| $T$/K | Conc. of lithium chloride/mol dm$^{-3}$ | Ostwald coefficient,* $L$ |
|---|---|---|
| 283.15 | 1.0 | 0.04920 |
| 288.15 | | 0.04278 |
| 293.15 | | 0.03768 |
| 298.15 | | 0.03361 |
| 303.15 | | 0.03033 |

* Smoothed values obtained from

$kT \ln L = 12.418 - 77.186\ (T/K) + 0.09657\ (T/K)^2$ cal mol$^{-1}$
where k is in units of cal mol$^{-1}$ K$^{-1}$

AUXILIARY INFORMATION

METHOD/APPARATUS/PROCEDURE:

The apparatus was similar to that described by Ben-Naim and Baer (1) and Wen and Hung (2). It consists of three main parts, a dissolution cell of 300 to 600 cm$^3$ capacity, a gas volume measuring column, and a manometer. The solvent is degassed in the dissolution cell, the gas is introduced and dissolved while the liquid is kept stirred by a magnetic stirrer immersed in the water bath. Dissolution of the gas results in the change in the height of a column of mercury which is measured by a cathetometer.

SOURCE AND PURITY OF MATERIALS:

1. Matheson sample, purity 99.9 mol per cent.
2. AR grade
3. Deionised, doubly distilled.

ESTIMATED ERROR:

$\delta T$/K = ±0.01; $\delta L/L$ = ±0.005 (estimated by compiler).

REFERENCES:

1. Ben-Naim, A.; Baer, S. *Trans. Faraday Soc.* 1963, *59*, 2735.

2. Wen, W.-Y.; Hung, J.H. *J. Phys. Chem.* 1970, *74*, 170.

**COMPONENTS:**

(1) Ethane; $C_2H_6$; [74-84-0]

(2) Sodium chloride; NaCl; [7647-14-5]

(3) Water; $H_2O$; [7732-18-5]

**ORIGINAL MEASUREMENTS:**

Eucken, A.; Hertzberg, G.

*Z. Physik. Chem.* 1950, *195*, 1 - 23.

**VARIABLES:**

$T$/K = 273.15, 293.15
$P$/kPa = 101.325 (1 atm)
$m_2$/mol kg$^{-1}$ = 0 - 2.95

**PREPARED BY:**

P. L. Long
H. L. Clever

**EXPERIMENTAL VALUES:**

| $T$/K | Sodium Chloride $m_2$/mol kg$^{-1}$ | Ostwald Coefficient $L$/cm$^3$cm$^{-3}$ | Setschenow Constant[1] $k_{smL} = (1/m)\log(L°/L)$ |
|---|---|---|---|
| 273.15 | 0 | 0.0987 | |
| | 1.10 | 0.0597 | 0.198 |
| | 2.13 | 0.0384 | 0.193 |
| | 2.95 | 0.0250 | 0.202 |
| 293.15 | 0 | 0.0509 | |
| | 0.57 | 0.0405 | 0.174 |
| | 1.10 | 0.0328 | 0.173 |
| | 1.79 | 0.0245 | 0.177 |

[1] salt effect parameter, $k_{smL}$/kg mol$^{-1}$.

**AUXILIARY INFORMATION**

**METHOD/APPARATUS/PROCEDURE:**

Gas absorption. The apparatus consists of a gas buret and an absorption flask connected by a capillary tube. The whole apparatus is shaken. The capillary tube is a 2 m long glass helix. An amount of gas is measured at STP and placed in the gas buret. After shaking, the difference from the original amount of gas placed in the gas buret is determined.

**SOURCE AND PURITY OF MATERIALS:**

Components. No information given.

**ESTIMATED ERROR:**

$\delta L/L = 0.01$ (authors)

**REFERENCES:**

**COMPONENTS:**

(1) Ethane; $C_2H_6$; [74-84-0]

(2) Sodium chloride; NaCl; [7647-14-5]

(3) Water; $H_2O$; [7732-18-5]

**ORIGINAL MEASUREMENTS:**

Morrison, T. J.; Billett, F.

*J. Chem. Soc.* 1952, 3819 - 3822.

**VARIABLES:**

$T$/K: 285.75 - 344.85

$p$/kPa: 101.325 (1 atm)

**PREPARED BY:**

H. L. Clever

**EXPERIMENTAL VALUES:**

| Temperature | | | Salt Effect Parameters | |
|---|---|---|---|---|
| t/°C | $T$/K | 1/($T$/K) | $(1/m_2)\log(S°/S)$[1] | $(1/m_2)\log(x°/x)$ |
| 12.6 | 285.75 | 0.0035 | 0.184 | 0.199 |
| 30.0 | 303.15 | 0.0033 | 0.162 | 0.177 |
| 49.4 | 322.55 | 0.0031 | 0.145 | 0.160 |
| 71.7 | 344.85 | 0.0029 | 0.135 | 0.150 |

[1] The authors used (1/c)log(S°/S) with c defined as g eq salt per kg of water. For the 1-1 electrolyte the compiler changed the c to an m for $m_2$/mol $kg^{-1}$. The ethane solubility S is $cm^3$ (STP) $kg^{-1}$.

The salt effect parameters were calculated from two measurements. The solubility of ethane in water, S°, and in the one molal salt solution, S. Only the solubility of the ethane in water, and the value of the salt effect parameter are given in the paper. The solubility values in the salt solution are not given.

The compiler calculated the values of the salt effect parameter using the mole fraction gas solubility ratio.

**AUXILIARY INFORMATION**

**METHOD/APPARATUS/PROCEDURE:**

The degassed solvent flows in a thin film down an absorption helix containing the ethane gas plus solvent vapor at a total pressure of one atmosphere. The volume of gas absorbed is measured in an attached buret system (1).

**SOURCE AND PURITY OF MATERIALS:**

(1) Ethane. Prepared from Grignard reagent.

(2) Sodium chloride. "AnalaR" material.

(3) Water. No information given.

**ESTIMATED ERROR:**

$\delta k/kg^{-1}$ mol = 0.010

**REFERENCES:**

1. Morrison, T. J.; Billett, F. *J. Chem. Soc.* 1948, 2033.

| COMPONENTS: | ORIGINAL MEASUREMENTS: |
|---|---|
| (1) Ethane; $C_2H_6$; [74-84-0]<br>(2) Sodium chloride; NaCl; [7647-14-5]<br>(3) Water; $H_2O$; [7732-18-5] | Mishnina, T. A.; Avdeeva, O. I.; Bozhovakaya, T. K.<br>*Materialy Vses. Nauchn. Issled. Geol. Inst.* 1961, *46*, 93 - 110. |
| VARIABLES: $T$/K = 283.15 - 348.15<br>$p$/kPa = 101.325 (1 atm)<br>$c_2$/mol $dm^{-3}$ = 0 - 5.4 | PREPARED BY:<br>H. L. Clever |

| NaCl $c_2$/mol $dm^{-3}$ | Bunsen Coefficient, $10^3\alpha/cm^3(STP)cm^{-3}atm^{-1}$ | | | | | | | | | | | | Salt Effect Parameter[1] |
|---|---|---|---|---|---|---|---|---|---|---|---|---|---|
| $T$/K | 0.0 | 0.5 | 1.0 | 1.5 | 2.0 | 2.5 | 3.0 | 3.5 | 4.0 | 4.5 | 5.0 | 5.4 | $k_{sc\alpha}$ |
| 283.15 | 57.6 | 46.2 | 37.4 | 30.0 | 24.2 | 19.5 | 15.7 | 12.6 | 10.2 | 8.3 | 6.7 | 5.6 | 0.188 |
| 288.15 | 49.2 | 40.0 | 32.4 | 26.3 | 21.3 | 17.5 | 14.1 | 11.4 | 9.3 | 7.6 | 6.2 | 5.2 | 0.181 |
| 293.15 | 43.2 | 35.2 | 28.9 | 23.8 | 19.4 | 15.9 | 13.0 | 10.6 | 8.7 | 7.2 | 5.8 | 5.0 | 0.174 |
| 298.15 | 38.1 | 31.2 | 25.9 | 21.3 | 17.6 | 14.4 | 11.9 | 9.8 | 8.1 | 6.7 | 5.6 | 4.7 | 0.168 |
| 303.15 | 34.0 | 28.0 | 23.4 | 19.4 | 16.1 | 13.4 | 11.1 | 9.2 | 7.6 | 6.3 | 5.4 | 4.5 | 0.162 |
| 308.15 | 30.2 | 25.2 | 21.0 | 17.6 | 14.6 | 12.7 | 10.5 | 8.8 | 7.3 | 6.1 | 5.2 | 4.4 | 0.157 |
| 313.15 | 28.6 | 24.1 | 20.2 | 17.1 | 14.2 | 12.0 | 10.0 | 8.4 | 7.0 | 5.9 | 5.0 | 4.3 | 0.153 |
| 318.15 | 26.6 | 22.4 | 18.9 | 16.0 | 13.4 | 11.4 | 9.6 | 8.1 | 6.8 | 5.8 | 4.8 | 4.2 | 0.148 |
| 323.15 | 24.9 | 21.0 | 17.8 | 15.1 | 12.8 | 10.8 | 9.2 | 7.8 | 6.6 | 5.6 | 4.7 | 4.1 | 0.145 |
| 328.15 | 23.6 | 20.0 | 17.0 | 14.4 | 12.3 | 10.4 | 8.8 | 7.5 | 6.4 | 5.4 | 4.6 | 4.0 | 0.142 |
| 333.15 | 22.4 | 19.0 | 16.2 | 13.8 | 11.8 | 10.0 | 8.5 | 7.3 | 6.2 | 5.3 | 4.4 | 3.9 | 0.140 |
| 338.15 | 21.4 | 18.3 | 15.6 | 13.3 | 11.3 | 9.7 | 8.3 | 7.0 | 6.1 | 5.2 | 4.3 | 3.8 | 0.138 |
| 343.15 | 20.6 | 17.6 | 15.0 | 12.8 | 11.0 | 9.4 | 8.0 | 6.8 | 5.9 | 5.0 | 4.2 | 3.8 | 0.136 |
| 348.15 | 20.2 | 17.3 | 14.9 | 12.7 | 10.9 | 9.4 | 8.0 | 6.8 | 5.9 | 5.0 | 4.2 | 3.8 | 0.133 |

The table of smoothed Bunsen coefficients of ethane dissolved in aqueous sodium chloride solutions was prepared by the authors. The complete source of data for the table is not clear. The data of Morrison and Johnstone (*J. Chem. Soc.* 1954, 3441) are mentioned. A 1958 report of A. A. Cherepinnikov, mentioned in the paper, was not available to us.

[1] salt effect parameter, $k_{sc\alpha}/dm^3 mol^{-1}$.

**COMPONENTS:**

(1) Ethane; $C_2H_6$; [74-84-0]

(2) Sodium Chloride; NaCl; [7647-14-5]

(3) Water; $H_2O$; [7732-18-5]

**ORIGINAL MEASUREMENTS:**

Czerski, L.; Czaplinski, A.

*Ann. Soc. Chim. Polonorum (Poland)* 1962, *36*, 1827-1834.

**VARIABLES:**

$T$/K: 273.15
$P$/kPa: 101.3 - 1600
$c_2$/mol dm$^{-3}$: 0.5 - 2.0

**PREPARED BY:**

W. Hayduk

**EXPERIMENTAL VALUES:**

| Concentration of salt $c_2$/mol dm$^{-3}$ | Pressure[1], $P$/atm 1.0 | 3.0 | 5.0 | 6.4 | 7.0 | 8.2 | 9.9 | 10.3 | 14.6 | 15.8 |
|---|---|---|---|---|---|---|---|---|---|---|
| | Solubility[1], $S$/cm$^3$(STP)dm$^{-3}$ solution | | | | | | | | | |
| 0.5 | 55 | 167 | 277 | 355 | - | - | - | - | - | - |
| 1.0 | 40 | 124 | 208 | 267 | 293 | 343 | - | - | - | - |
| 1.5 | 32 | 97 | 163 | 208 | 228 | 267 | 292 | 337 | - | - |
| 2.0 | 20 | 63 | 104 | 133 | 147 | 172 | 187 | 216 | 307 | 333 |

| $c_2$/mol dm$^{-3}$ | Solubility[2], $S_1$/cm$^3$(STP)dm$^{-3}$ | Salt parameter[3], $(1/c_2)\log(S_1^\circ/S_1)$ |
|---|---|---|
| 0.5 | 55.4 | 0.3016 |
| 1.0 | 41.6 | 0.2752 |
| 1.5 | 32.6 | 0.2541 |
| 2.0 | 21.1 | 0.2850 |

[1] Original data, at 0.0°C.
[2] Calculated by authors assuming Henry's law applies and obtaining average value of Henry's constant at a gas partial pressure of 101.3 kPa.
[3] Salt effect parameter calculated by compiler using salt concentration expressed as $c_2$/mol dm$^{-3}$ solution and solubility of ethane in water as determined by authors: $S_1^\circ$ = 78.4 cm$^3$(STP)dm$^{-3}$.

**AUXILIARY INFORMATION**

**METHOD/APPARATUS/PROCEDURE:**

Apparatus consists of a contact chamber agitated by a rocking device. Gas incrementally added from a second smaller chamber of known volume while observing pressure change with each gas addition. Equilibrium established in 2-3 h.

**SOURCE AND PURITY OF MATERIALS:**

Source, purities not available.

**ESTIMATED ERROR:**

$\delta T$/K = 0.05
$\delta c_2/c_2$ = 0.02
$\delta P/P$ = 0.02
(estimated by compiler)

**REFERENCES:**

COMPONENTS:

(1) Ethane; $C_2H_6$; [74-84-0]

(2) Sodium chloride; NaCl; [7647-14-5]

(3) Water; $H_2O$; [7732-18-5]

ORIGINAL MEASUREMENTS:

Ben-Naim, A.; Yaacobi, M.

*J. Phys. Chem.* 1974, *78*, 170-5

VARIABLES:

$T$/K: 283.15-303.15

$c_2$/mol dm$^{-3}$: 2.0

PREPARED BY:

C.L. Young

EXPERIMENTAL VALUES:

| $T$/K | Conc. of sodium chloride/mol dm$^{-3}$ | Ostwald coefficient,* $L$ |
|---|---|---|
| 283.15 | 0.25 | 0.06151 |
| 288.15 | | 0.05341 |
| 293.15 | | 0.04678 |
| 298.15 | | 0.04131 |
| 303.15 | | 0.03676 |
| 283.15 | 0.50 | 0.05579 |
| 288.15 | | 0.04772 |
| 293.15 | | 0.04156 |
| 298.15 | | 0.03680 |
| 303.15 | | 0.03311 |
| 283.15 | 1.0 | 0.04367 |
| 288.15 | | 0.03795 |
| 293.15 | | 0.03360 |
| 298.15 | | 0.03029 |
| 303.15 | | 0.02777 |
| 283.15 | 2.0 | 0.02862 |
| 288.15 | | 0.02536 |
| 293.15 | | 0.02282 |
| 298.15 | | 0.02083 |
| 303.15 | | 0.01929 |

* Smoothed values obtained from

$kT \ln L = 8,164.1 - 46.822\ (T/K) + 0.04395\ (T/K)^2$ cal mol$^{-1}$;

$kT \ln L = 16,759.7 - 105.52\ (T/K) + 0.14339\ (T/K)^2$ cal mol$^{-1}$;

$kT \ln L = 17,910.6 - 115.82\ (T/K) + 0.16366\ (T/K)^2$ cal mol$^{-1}$;

$kT \ln L = 14,678.7 - 96.230\ (T/K) + 0.13182\ (T/K)^2$ cal mol$^{-1}$; where k is in units of cal mol$^{-1}$ K$^{-1}$) for concentration of 0.25, 0.50, 1.0, 2.0 mol l$^{-1}$ respectively.

AUXILIARY INFORMATION

METHOD/APPARATUS/PROCEDURE:

The apparatus was similar to that described by Ben-Naim and Baer (1) and Wen and Hung (2). It consists of three main parts, a dissolution cell of 300 to 600 cm$^3$ capacity, a gas volume measuring column, and a manometer. The solvent is degassed in the dissolution cell, the gas is introduced and dissolved while the liquid is kept stirred by a magnetic stirrer immersed in the water bath. Dissolution of the gas results in the change in the height of a column of mercury which is measured by a cathetometer.

SOURCE AND PURITY OF MATERIALS:

1. Matheson sample, purity 99.9 mol per cent.
2. AR grade.
3. Deionised, doubly distilled.

ESTIMATED ERROR:

$\delta T/K = \pm 0.01$; $\delta L/L = \pm 0.005$. (estimated by compiler.)

REFERENCES:


1. Ben-Naim, A.; Baer, S. *Trans. Faraday Soc.* 1963, *59*, 2735.
2. Wen, W.-Y.; Hung, J.H. *J. Phys. Chem.* 1970, *74*, 170.

COMPONENTS:

(1) Ethane; $C_2H_6$; [74-84-0]

(2) Sodium bromide; NaBr; [7647-15-6]

(3) Water; $H_2O$; [7732-18-5]

ORIGINAL MEASUREMENTS:

Ben-Naim, A.; Yaacobi, M.

*J. Phys. Chem.* 1974, *78*, 170-5

VARIABLES:

$T$/K : 283.15-303.15

$c_2$/mol $dm^{-3}$: 1.0

PREPARED BY:

C.L. Young

EXPERIMENTAL VALUES:

| $T$/K | Conc. of sodium bromide/mol $dm^{-3}$ | Ostwald coefficient,* $L$ |
|---|---|---|
| 283.15 | 1.0 | 0.04540 |
| 288.15 | | 0.03955 |
| 293.15 | | 0.03497 |
| 298.15 | | 0.03135 |
| 303.15 | | 0.02848 |

* Smoothed values obtained from

$kT \ln L = 13{,}998 - 88.641\ (T/K) + 0.11675\ (T/K)^2$ cal $mol^{-1}$
where k is in units of cal $mol^{-1}$ $K^{-1}$

AUXILIARY INFORMATION

METHOD/APPARATUS/PROCEDURE:

The apparatus was similar to that described by Ben-Naim and Baer (1) and Wen and Hung (2). It consists of three main parts, a dissolution cell of 300 to 600 $cm^3$ capacity, a gas volume measuring column, and a manometer. The solvent is degassed in the dissolution cell, the gas is introduced and dissolved while the liquid is kept stirred by a magnetic stirrer immersed in the water bath. Dissolution of the gas results in the change in the height of a column of mercury which is measured by a cathetometer.

SOURCE AND PURITY OF MATERIALS:

1. Matheson sample, purity 99.9 mol per cent.
2. AR grade.
3. Deionised, doubly distilled.

ESTIMATED ERROR:

$\delta T/K = \pm 0.01$; $\delta L/L = \pm 0.005$; (estimated by compiler).

REFERENCES:

1. Ben-Naim, A.; Baer, S. *Trans. Faraday Soc.* 1963, *59*, 2735.

2. Wen, W.-Y.; Hung, J.H. *J. Phys. Chem.* 1974 , *74*, 170.

**COMPONENTS:**

(1) Ethane; $C_2H_6$; [74-84-0]

(2) Sodium iodide; NaI; [7681-82-5]

(3) Water; $H_2O$; [7732-18-5]

**ORIGINAL MEASUREMENTS:**

Ben-Naim, A.; Yaacobi, M.

*J. Phys. Chem.* 1974, *78*, 170-5

**VARIABLES:**

$T$/K : 283.15-303.15

$c_2$/mol dm$^{-3}$

**PREPARED BY:**

C.L. Young

**EXPERIMENTAL VALUES:**

| $T$/K | Conc. of sodium iodide/mol dm$^{-3}$ | Ostwald coefficient,* $L$ |
|---|---|---|
| 283.15 | 1.0 | 0.04713 |
| 288.15 | | 0.04147 |
| 293.15 | | 0.03683 |
| 298.15 | | 0.03300 |
| 303.15 | | 0.02981 |

* Smoothed values obtained from

$$kT \ln L = 8{,}825.0 - 53.458\ (T/K) + 0.05728\ (T/K)^2 \text{ cal mol}^{-1}$$

where k is in units of cal mol$^{-1}$ K$^{-1}$

**AUXILIARY INFORMATION**

**METHOD/APPARATUS/PROCEDURE:**

The apparatus was similar to that described by Ben-Naim and Baer (1) and Wen and Hung (2). It consists of three main parts, a dissolution cell of 300 to 600 cm$^3$ capacity, a gas volume measuring column, and a manometer. The solvent is degassed in the dissolution cell, the gas is introduced and dissolved while the liquid is kept stirred by a magnetic stirrer immersed in the water bath. Dissolution of the gas results in the change in the height of a column of mercury which is measured by a cathetomer.

**SOURCE AND PURITY OF MATERIALS:**

1. Matheson sample, purity 99.9 mol per cent.

2. AR grade

3. Deionised, doubly distilled.

**ESTIMATED ERROR:**

$\delta T$/K = ±0.01; $\delta L/L$ = ±0.005; (estimated by compiler).

**REFERENCES:**


1. Ben-Naim, A.; Baer, S. *Trans. Faraday Soc.* 1963, *59*, 2735.

2. Wen, W.-Y.; Hung, J.H. *J. Phys. Chem.* 1970, *74*, 170.

**COMPONENTS:**

(1) Ethane; $C_2H_6$; [74-84-0]

(2) Sulfuric acid; $H_2SO_4$; [7664-93-9]

(3) Sodium sulfate; $Na_2SO_4$; [7757-82-6]

(4) Water; $H_2O$; [7732-18-5]

**ORIGINAL MEASUREMENTS:**

Kobe, K. A.; Kenton, F. H.

*Ind. Eng. Chem., Anal. Ed.* 1938, *10*, 76 - 77.

**VARIABLES:**

$T$/K: 298.15

$p_1$/kPa: 101.325 (1 atm)

**PREPARED BY:**

P. L. Long
H. L. Clever

**EXPERIMENTAL VALUES:**

| Temperature $t/^{\circ}C$ | Temperature $T/K$ | Solvent Volume $V/cm^3$ | Ethane Volume Absorbed $v_1/cm^3$ | Bunsen Coefficient $\alpha/cm^3(STP)cm^{-3}atm^{-1}$ | Ostwald Coefficient $L/cm^3cm^{-3}$ |
|---|---|---|---|---|---|
| 25 | 298.15 | 49.54 | 0.54 | | |
| | | 49.54 | 0.53 | 0.0099 | 0.0108 |

The solvent is a mixture of 800 g $H_2O$
200 g $Na_2SO_4$ (anhydrous)
40 ml $H_2SO_4$ (Conc., 36 normal)

Thus the molality of the solution is

$$m_2/\text{mol kg}^{-1} = 0.90 \quad (H_2SO_4)$$

$$m_3/\text{mol kg}^{-1} = 1.76 \quad (Na_2SO_4)$$

**AUXILIARY INFORMATION**

**METHOD/APPARATUS/PROCEDURE:**

The apparatus is described in detail in an earlier paper (1). The apparatus consists of a gas buret, a pressure compensator, and a 200 $cm^3$ absorption bulb and mercury leveling bulb. The absorption bulb is attached to a shaking mechanism.

The solvent and the gas are placed in the absorption bulb. The bulb is shaken until equilibrium is reached. The remaining gas is returned to the buret. The difference in the final and initial volumes is taken as the volume of gas absorbed.

**SOURCE AND PURITY OF MATERIALS:**

(1) Ethane. Source not given. Purity stated to be $99^+$ per cent.

(2, 3) Sulfuric acid and sodium sulfate. Sources not given. Analytical grade.

(4) Water. Distilled.

**ESTIMATED ERROR:**

$\delta\alpha/cm^3 = \pm 0.001$ (authors)

**REFERENCES:**

1. Kobe, K. A.; Williams, J. S. *Ind. Eng. Chem., Anal. Ed.* 1935, *7*, 37.

**COMPONENTS:**

(1) Ethane; $C_2H_6$; [74-84-0]

(2) Sulfuric acid monododecyl ester sodium salt (sodium dodecyl sulfate or SDS)' $C_{12}H_{26}O_4S.Na$; [151-21-3]

(3) Water; $H_2O$; [7732-18-5]

**ORIGINAL MEASUREMENTS:**

Matheson, I.B.C; King, A.D.

*J. Coll. Interface Sci.* 1978, *66*, 464 - 469.

**VARIABLES:**

$T$/K: 298.15

$p$/kPa: 124.1-689.5

SDS/mol $kg^{-1}$ $H_2O$: 0-0.300

**PREPARED BY:**

H.L. Clever

**EXPERIMENTAL VALUES:**

| $T$/K | Sulfuric acid monodecyl ester sodium salt $m_2$/mol $kg^{-1}$ | Pressure pounds per square inch,gauge $p$/psig | Volume gas evolved $V_1/cm^3$ | Ambient Pressure $p$/mmHg | Ambient Temperature $t$/°C | Henry's constant $10^3$K/mol $kg^{-1}atm^{-1}$ |
|---|---|---|---|---|---|---|
| 298.15 | 0 | 34.2 | 10.6 | 751.0 | 23.1 | |
| | | 53.5 | 16.1 | 747.8 | 23.9 | |
| | | 57.9 | 16.8 | 750.0 | 24.0 | |
| | | 100.0 | 31.6 | 752.8 | 23.2 | 1.76±0.06 |
| | 0.150 | 28.1 | 16.6 | 748.9 | 23.2 | |
| | | 39.3 | 23.7 | 750.1 | 23.5 | |
| | | 50.2 | 29.9 | 752.6 | 22.9 | 3.46±0.03 |
| | 0.300 | 18.0 | 15.3 | 747.6 | 22.5 | |
| | | 23.6 | 21.2 | 749.6 | 22.0 | |
| | | 37.1 | 30.9 | 750.9 | 21.5 | 4.99±0.17 |

**AUXILIARY INFORMATION**

**METHOD/APPARATUS/PROCEDURE:**

The solvent solution consisting of 100 g of aqueous colloidal electrolyte is contained in a glass-lined brass equilibrium cell resting on a magnetic stirrer. The solution is degassed by evacuation and stirring. Gas is introduced at pressures above atmospheric and equilibration is continued for at least five hours. Subsequently as the pressure is released to a lower pressure, the gas evolved from the supersaturated solution is collected at atmospheric pressure and ambient temperature in a Warburg manometer, and its volume measured. Corrections are made for gas lost during the venting procedure, the differences in temperature and pressure, and the water vapor pressure in the calculation of Henry's constant.

**SOURCE AND PURITY OF MATERIALS:**

1. Source not given. Chemically pure or the equivalent of 99.0 mole percent purity.
2. Sulfuric acid monodecyl ester sodium salt. Aldrich Chemical Co., Inc. Recrystallized from ethanol and dried *in vacuo*.
3. Laboratory distilled.

**ESTIMATED ERROR:**

$\delta K/K = 0.02$

**REFERENCES:**

**COMPONENTS:**

(1) Ethane; $C_2H_6$; [74-84-0]

(2) Sufluric acid monododecyl ester sodium salt or sodium dodecyl sulfate or SDS; $C_{12}H_{26}O_4S.Na$; [151-21-3]

(3) Sodium chloride; NaCl; [7647-18-5]

(4) Water; $H_2O$; [7732-18-5]

**ORIGINAL MEASUREMENTS:**

Hoskins, J.C.; King, A.D.

*J. Coll. Interface Sci.* 1981, *82*, 264 - 267.

**VARIABLES:**

$T$/K: 298.15

$m_2$/mol kg$^{-1}$: 0.1-0.3

$m_3$/mol kg$^{-1}$: 0-0.9

**PREPARED BY:**

H.L. Clever

**EXPERIMENTAL VALUES:**

| | Sodium Dodecyl sulfate | Sodium Chloride | Ethane Solubility 10 $K$/mol kg$^{-1}$ atm$^{-1}$ | | |
|---|---|---|---|---|---|
| $T$/K | $m_2$/mol kg$^{-1}$ | $m_3$/mol kg$^{-1}$ | $10^3K$ | $10^3K^0$ | $10^3(K - K^0)$ |
| 298.15 | 0.1 | 0.00 | 2.98 | 1.80 | 1.18 |
| | | 0.03 | 3.16 | 1.78 | 1.38 |
| | | 0.06 | 3.19 | 1.76 | 1.43 |
| | | 0.10 | 3.16 | 1.73 | 1.43 |
| | | 0.20 | 3.04 | 1.67 | 1.37 |
| | | 0.30 | 3.01 | 1.60 | 1.41 |
| | | 0.40 | 3.00 | 1.54 | 1.46 |
| | | 0.50 | 2.93 | 1.48 | 1.45 |
| | | 0.55 | 2.90 | 1.46 | 1.44 |
| | | 0.60 | 2.85 | 1.43 | 1.42 |
| | | 0.65 | 2.85 | 1.40 | 1.45 |
| | | 0.70 | 2.68 | 1.37 | 1.31 |
| | | 0.75 | 2.79 | 1.35 | 1.44 |
| | | 0.80 | 2.56 | 1.32 | 1.24 |
| | | 0.85 | 2.49 | 1.30 | 1.19 |
| | | 0.90 | - | | |
| | 0.2 | 0.00 | 4.16 | 1.80 | 2.36 |
| | | 0.03 | 4.24 | 1.78 | 2.46 |
| | | 0.06 | 4.38 | 1.76 | 2.62 |
| | | 0.10 | 4.37 | 1.73 | 2.64 |
| | | 0.20 | 4.22 | 1.67 | 2.55 |
| | | 0.30 | 4.16 | 1.60 | 2.56 |

continued...

**AUXILIARY INFORMATION**

**METHOD/APPARATUS/PROCEDURE:**

The solvent solution is contained in a glass-lined brass equilibrium cell resting on a magnetic stirrer. The solution is degassed by evacuation and stirring. Gas is introduced at pressures above atmospheric and equilibration is continued for at least five hours. Subsequently as the pressure is released to a lower pressure, the gas evolved from the supersaturated solution is collected at atmospheric pressure and ambient temperature in a Warburg manometer and its volume measured. Corrections are made for gas lost during the venting procedure, the differences in temperature and pressure, and the water vapor pressure in the calculation of Henry's constant.

Details in reference (1).

**SOURCE AND PURITY OF MATERIALS:**

1. Matheson Co., Minimum purity 99.0 per cent.
2. Aldrich Chem. Co., Recrystallized twice from 2-propanol, dried *in vacuo*. Analysis of the purified product showed 74% $C_{12}$ sulfate, 22% $C_{14}$ sulfate, and 4% $C_{16}$ sulfate.
3. Baker "Analyzed" reagent grade.
4. Distilled.

**ESTIMATED ERROR:**

$\delta K/K = \pm 0.02$

**REFERENCES:**


1. Matheson, I.B.C.; King, A.D.

   *J. Coll. Interface Sci.* 1978, *66*, 464.

| COMPONENTS: | ORIGINAL MEASUREMENTS: |
|---|---|
| (1) Ethane; $C_2H_6$; [74-84-0]<br>(2) Sulfuric acid monododecyl ester sodium salt or sodium dodecyl sulfate or SDS; $C_{12}H_{26}O_4S.Na$; [151-21-3]<br>(3) Sodium chloride; NaCl; [7647-18-5]<br>(4) Water; $H_2O$; [7732-18-5] | Hoskins, J.C.; King, A.D.<br>*J. Coll. Interface Sci.* 1981, *82*, 264 - 267. |
| VARIABLES: | PREPARED BY: |
| $T$/K: 298.15<br>$m_2$/mol kg$^{-1}$: 0.1-0.3<br>$m_3$/mol kg$^{-1}$: 0-0.9 | H.L. Clever |

EXPERIMENTAL VALUES:

continued

| | Sodium Dodecyl sulfate | Sodium Chloride | Ethane Solubility 10 $K$/mol kg$^{-1}$ atm$^{-1}$ | | |
|---|---|---|---|---|---|
| $T$/K | $m_2$/mol kg$^{-1}$ | $m_3$/mol kg$^{-1}$ | $10^3K$ | $10^3K^0$ | $10^3(K - K^0)$ |
| 298.15 | 0.2 | 0.40 | 4.09 | 1.54 | 2.55 |
| | | 0.50 | 4.07 | 1.48 | 2.59 |
| | | 0.60 | 4.07 | 1.43 | 2.64 |
| | | 0.65 | 3.91 | 1.40 | 2.51 |
| | | 0.70 | 3.87 | 1.37 | 2.50 |
| | | 0.75 | - | | |
| | 0.3 | 0.00 | 5.35 | 1.80 | 3.55 |
| | | 0.03 | 5.31 | 1.78 | 3.53 |
| | | 0.06 | 5.47 | 1.76 | 3.71 |
| | | 0.10 | 5.48 | 1.73 | 3.75 |
| | | 0.20 | 5.37 | 1.67 | 3.70 |
| | | 0.30 | 5.38 | 1.60 | 3.78 |
| | | 0.40 | 5.27 | 1.54 | 3.73 |
| | | 0.50 | 5.20 | 1.48 | 3.72 |
| | | 0.55 | 5.17 | 1.45 | 3.72 |

The ethane solubility is given as a Henry's constant in the form $K$/mol kg$^{-1}$ atm$^{-1}$ = ($m_1$/mol kg$^{-1}$)/($p_1$/atm). The average error in $10^3K$ is ± 0.05.

Under the ethane solubility heading above the first column is the solubility of ethane in the solution, the second column is the calculated solubility of ethane in a solution of sodium chloride alone, and the third column is the calculated enhancement of solubility due to the sodium dodecyl sulfate.

The authors calculated the second column values using a Sechenov constant of 0.168 derived from data found in Morrison, T.J.; Billett, F. *J. Chem. Soc.* 1952, 3819.

**COMPONENTS:**

1. Ethane; $C_2H_6$; [74-84-0]
2. Potassium chloride; KCl; [7447-40-7]
3. Water; $H_2O$; [7732-18-5]

**ORIGINAL MEASUREMENTS:**

Ben-Naim, A.; Yaacobi, M.

*J. Phys. Chem.* 1974, *78*, 170-5.

**VARIABLES:**

$T$/K: 283.15-303.15

$c_2$/mol $dm^{-3}$: 1.0

**PREPARED BY:**

C.L. Young

**EXPERIMENTAL VALUES:**

| $T$/K | Conc. of potassium chloride/mol $dm^{-3}$ | Ostwald coefficient,* $L$ |
|---|---|---|
| 283.15 | 1.0 | 0.04522 |
| 288.15 | | 0.03963 |
| 293.15 | | 0.03511 |
| 298.15 | | 0.03144 |
| 303.15 | | 0.02843 |

* Smoothed values obtained from

$$kT \ln L = 10{,}431.2 - 64.337\ (T/K) + 0.07538\ (T/K)^2 \text{ cal mol}^{-1}$$

where k is in units of cal $mol^{-1}$ $K^{-1}$

**AUXILIARY INFORMATION**

**METHOD/APPARATUS/PROCEDURE:**

The apparatus was similar to that described by Ben-Naim and Baer (1) and Wen and Hung (2). It consists of three main parts, a dissolution cell of 300 to 600 $cm^3$ capacity, a gas volume measuring column, and a manometer. The solvent is degassed in the dissolution cell, the gas is introduced and dissolved while the liquid is kept stirred by a magnetic stirrer immersed in the water bath. Dissolution of the gas results in the change in the height of a column of mercury which is measured by a cathetometer.

**SOURCE AND PURITY OF MATERIALS:**

1. Matheson sample, purity 99.9 mol per cent.
2. AR grade
3. Deionised, doubly distilled.

**ESTIMATED ERROR:**

$\delta T$/K = ±0.01; $\delta L/L$ = ±0.005.
(estimated by compiler).

**REFERENCES:**


1. Ben-Naim, A.; Baer, S. *Trans. Faraday Soc.* 1963, *59*, 2735.
2. Wen, W.-Y.; Hung, J.H. *J. Phys. Chem.* 1970, *74*, 170.

**COMPONENTS:**

(1) Ethane; $C_2H_6$; [74-84-0]

(2) Potassium iodide; KI; [7681-11-0]

(3) Water; $H_2O$; [7732-18-5]

**ORIGINAL MEASUREMENTS:**

Morrison, T. J.; Billett, F.

*J. Chem. Soc.* 1952, 3819 - 3822.

**VARIABLES:**

$T$/K: 285.75 - 344.85

$p$/kPa: 101.325 (1 atm)

**PREPARED BY:**

H. L. Clever

**EXPERIMENTAL VALUES:**

| Temperature | | | Salt Effect Parameters | |
|---|---|---|---|---|
| t/°C | $T$/K | 1/($T$/K) | $(1/m_2)\log(S°/S)$[1] | $(1/m_2)\log(x°/x)$ |
| 12.6 | 285.75 | 0.0035 | 0.125 | 0.140 |
| 30.0 | 303.15 | 0.0033 | 0.101 | 0.116 |
| 49.4 | 322.55 | 0.0031 | 0.080 | 0.095 |
| 71.7 | 344.85 | 0.0029 | 0.065 | 0.080 |

[1] The authors used (1/c)log(S°/S) with c defined as g eq salt per kg of water. For the 1-1 electrolyte the compiler changed the c to an m for $m_2$/mol $kg^{-1}$. The ethane solubility S is $cm^3$ (STP) $kg^{-1}$.

The salt effect parameters were calculated from two measurements. The solubility of ethane in water, S°, and in the one molal salt solution, S. Only the solubility of the ethane in water, and the value of the salt effect parameter are given in the paper. The solubility values in the salt solution are not given.

The compiler calculated the values of the salt effect parameter using the mole fraction gas solubility ratio.

AUXILIARY INFORMATION

**METHOD/APPARATUS/PROCEDURE:**

The degassed solvent flows in a thin film down an absorption helix containing the ethane gas plus solvent vapor at a total pressure of one atmosphere. The volume of gas absorbed is measured in an attached buret system (1).

**SOURCE AND PURITY OF MATERIALS:**

(1) Ethane. Prepared from Grignard reagent.

(2) Potassium iodide; "AnalaR" material.

(3) Water. No information given.

**ESTIMATED ERROR:**

$\delta k/kg^{-1}$ mol = 0.010

**REFERENCES:**

1. Morrison, T. J.; Billett, F. *J. Chem. Soc.* 1948, 2033.

**COMPONENTS:**

1. Ethane; $C_2H_6$; [74-84-0]
2. Cesium chloride; CsCl; [7647-17-8]
3. Water; $H_2O$; [7732-18-5]

**ORIGINAL MEASUREMENTS:**

Ben-Naim, A.; Yaacobi, M.

*J. Phys. Chem.* 1974, *78*, 170-5.

**VARIABLES:**

$P$/KPa: 101.325 (1 atm)
$T$/K: 283.15-303.15
$c_2$/mol $dm^{-3}$: 1.0

**PREPARED BY:**

C.L. Young

**EXPERIMENTAL VALUES:**

| $T$/K | Conc. of cesium chloride/mol $dm^{-3}$ | Ostwald coefficient,* $L$ |
|---|---|---|
| 283.15 | 1.0 | 0.04736 |
| 288.15 | | 0.04100 |
| 293.15 | | 0.03627 |
| 298.15 | | 0.03277 |
| 303.15 | | 0.03020 |

* Smoothed values obtained from

$$kT \ln L = 20{,}843 - 135.77\,(T/K) + 0.19810\,(T/K)^2 \text{ cal mol}^{-1}$$

where k is in units of cal $mol^{-1}$ $K^{-1}$

AUXILIARY INFORMATION

**METHOD/APPARATUS/PROCEDURE:**

The apparatus was similar to that described by Ben-Naim and Baer (1) and Wen and Hung (2). It consists of three main parts, a dissolution cell of 300 to 600 $cm^3$ capacity, a gas volume measuring column, and a manometer. The solvent is degassed in the dissolution cell, the gas is introduced and dissolved while the liquid is kept stirred by a magnetic stirrer immersed in the water bath. Dissolution of the gas results in the change in the height of a column of mercury which is measured by a cathetometer.

**SOURCE AND PURITY OF MATERIALS:**

1. Matheson sample, purity 99.9 mol per cent.
2. AR grade.
3. Deionised, doubly distilled.

**ESTIMATED ERROR:**

$\delta T$/K = ±0.01; $\delta L/L$ = ±0.005. (estimated by compiler).

**REFERENCES:**

1. Ben-Naim, A.; Baer, S. *Trans. Faraday Soc.* 1963, *59*, 2735.
2. Wen, W.-Y.; Hung, J.H. *J. Phys. Chem.* 1970, *74*, 170.

**COMPONENTS:**

(1) Ethane; $C_2H_6$; [74-84-0]

(2) Potassium salt of N,N-dimethylglycine; $KC_4H_8NO_2$; [17647-86-8]

(3) Water; $H_2O$; [7732-18-5]

**ORIGINAL MEASUREMENTS:**

Leuhddemann, R.; Noddes, G.; Schwarz, H.-G.

*Oil Gas J.* 1959, *57* (No.32), 100, 102, 104.

**VARIABLES:**

$T$/K = 293 - 303
$p_1$/kPa = 101.3 (1 atm)

**PREPARED BY:**

H. L. Clever

**EXPERIMENTAL VALUES:**

| $T$/K | Bunsen Coefficient $\alpha$/$cm^3$(STP)$cm^{-3}atm^{-1}$ |
|---|---|
| Water | |
| 293 - 303 | 0.041 |
| Solution of density $\rho^{293}$/g $cm^{-3}$ = 1.17. | |
| 293 - 303 | 0.018 |

The concentration of the water + potassium salt of N,N-dimethyl glycine is not given, only the density is given. The solution has a pH of 10-12.

In the paper the solution is identified as "Alkazid Dik".

Other names for the potassium salt of the amino acid are:
Potassium N,N-dimethyl amino acetate,
Potassium salt of dimethylglycocoll, and
Potassium salt of dimethylglycine.

**AUXILIARY INFORMATION**

**METHOD/APPARATUS/PROCEDURE:**

No details given.

NOTE: Landolt-Bornstein Tabellen, Volume IV, Part 4.c.1, 1976, pp.404 - 407, credits this paper with data on the ethane + potassium salt of N-methyle-*DL*-alanine + water system. However, the data do not appear in the paper. The solubility reported in Landolt-Bornstein for a solution density 1.17 is about one-half the value for the system above.

**SOURCE AND PURITY OF MATERIALS:**

No information.

**ESTIMATED ERROR:**

$\delta\alpha/\alpha = \pm 0.15$ (Compiler)

**REFERENCES:**

| COMPONENTS: | EVALUATOR: |
| --- | --- |
| (1) Ethane; $C_2H_6$; [74-84-0]<br>(2) Sulfuric acid monododecyl ester sodium salt or sodium dodecyl sulfate or SDS; $C_{12}H_{26}O_4S.Na$; [151-21-3]<br>(3) 1-Pentanol; $C_5H_{12}O$; [71-41-0]<br>(4) Water; $H_2O$; [7732-18-5] | H. Lawrence Clever<br>Department of Chemistry<br>Emory University<br>Atlanta, GA 30322 |

CRITICAL EVALUATION:

King and co-workers (1, 2, 3) have reported the solubility of ethane in several solutions that form micelles including the ethane + sodium dodecyl sulfate + 1-pentanol + water system. Their data are classed as tentative. The authors state a standard deviation of 2 per cent, however, there is reason to believe that their values may be too small by as much as five per cent.

Their apparatus consists of a glass liner inside of a thermostated brass bomb which rests on a variable speed magnetic stirrer. The amount of ethane released from an ethane saturated solution is determined as follows: (*i*) the degassed solution is stirred and allowed to equilibrate with ethane gas at some elevated pressure; (*ii*) the stirrer is turned off and the solution is allowed to become still where upon the pressure is released; and (*iii*) after allowing a short time period for thermal equilibration, the now super saturated solution is again stirred and the volume of gas evolved is measured in a Warburg Manometer system at ambient conditions. Corrections for the gas lost during venting and thermal equilibration and for water vapor pressure are made. One obtains the $\Delta n/\Delta p$ ratio as Henry's constant from the number of moles of ethane evolved isothermally during the change from the equilibrium pressure to the ambient pressure.

Two values of the solubility of ethane in water are given. They are equivalent to $(3.17 \pm 0.11) \times 10^{-5}$ (1) and $(3.24 \pm 0.10) \times 10^{-5}$ (2) mole fraction at 298.15 K and 101.325 kPa (1 atm) partial pressure of ethane. The results are 6.8 and 4.7 per cent, respectively, lower than the values recommended by Battino (see pp 1-4 of this volume). A possible reason for the discrepancy is that more ethane is lost during venting and thermal equilibration than is corrected for by the authors.

References:


1. Matheson, I. B. C.; King, A. D. *J. Coll. Interface Sci.* 1978, *66*, 464-469.

2. Hoskins, J. C.; King, A. D. *J. Coll. Interface Sci.* 1981, *82*, 260-263.

3. Hoskins, J. C.; King, A. D. *J. Coll. Interface Sci.* 1981, *82*, 264-267.

**COMPONENTS:**

(1) Ethane; $C_2H_6$; [74-84-0]
(2) Sulfuric acid monododecyl ester sodium salt or sodium dodecyl sulfate or SDS; $C_{12}H_{26}O_4S.Na$; [151-21-3]
(3) 1-Pentanol; $C_5H_{12}O$; [71-41-0]
(4) Water; $H_2O$; [7732-18-5]

**ORIGINAL MEASUREMENTS:**

Hoskins, J.C.; King, A.D.

*J. Coll. Interface Sci.* 1981, *82*, 260-263.

**VARIABLES:**

$T$/K = 298.15
$p_1$/kPa = 101.325
$m_2$/mol kg$^{-1}$ = 0 - 0.40
$m_3$/mol kg$^{-1}$ = 0 - 2.75

**PREPARED BY:**

H.L. Clever

**EXPERIMENTAL VALUES:**

| Sodium Dodecyl Sulfate $m_2$/mol kg$^{-1}$ | 1-Pentanol $m_3$/mol kg$^{-1}$ | Ethane $10^3K$/mol kg$^{-1}$ atm$^{-1}$ | Sodium Dodecyl Sulfate $m_2$/mol kg$^{-1}$ | 1-Pentanol $m_3$/mol kg$^{-1}$ | Ethane $10^3K$/mol kg$^{-1}$ atm$^{-1}$ |
|---|---|---|---|---|---|
| 0 | 0 | 1.80 | 0.25 | 1.00 | 9.97 |
| | 0.1 | 1.81 | | 1.25 | 12.5 |
| | 0.2 | 1.79 | | 1.50 | 14.5 |
| | | | | 1.75 | 15.3 |
| 0.1 | 0 | 2.88 | | 2.00 | 17.6 |
| | 0.1 | 3.17 | | 2.25 | 20.1 |
| | 0.2 | 3.56 | | 2.50 | - |
| | 0.3 | 3.99 | | 2.75 | 26.1 |
| | 0.4 | 4.33 | | | |
| | 0.5 | 5.03 | 0.40 | 0 | 6.31 |
| | 0.6 | 5.55 | | 0.25 | 7.34 |
| | 0.7 | 6.29 | | 0.50 | 9.29 |
| | | | | 0.75 | 10.5 |
| 0.25 | 0 | 4.69 | | 1.00 | 12.3 |
| | 0.25 | 5.83 | | 1.25 | 13.9 |
| | 0.50 | 7.09 | | 1.50 | 15.7 |
| | 0.75 | 8.77 | | 1.75 | 17.9 |

All measurements were made at a temperature of 25°C or 298.15K. The ethane solubility is given as a Henry's constant in the form $K/\text{mol kg}^{-1}\ \text{atm}^{-1} = (m_1/\text{mol kg}^{-1})/(p_1/\text{atm})$.

**AUXILIARY INFORMATION**

**METHOD/APPARATUS/PROCEDURE:**

Described in Critical Evaluation and in more detail in reference (1).

**SOURCE AND PURITY OF MATERIALS:**

1. Matheson Co., Inc. Purity 99.0 per cent.
2. Aldrich Chem. Co., Inc. Recrystallized twice from 2-propanol, dried *in vacuo*. Analysis of the purified product showed 74% $C_{12}$ sulfate, 22% $C_{14}$ sulfate, and 4% $C_{16}$ sulfate.
3. Eastman Kodak Co. Fractionally distilled.
4. Distilled.

**ESTIMATED ERROR:**

$\delta K/K = \pm\ 0.02$

**REFERENCES:**

1. Matheson, I.B.C.; King, A.D. *J. Coll. Interface Sci.* 1978, *66*, 464.

**COMPONENTS:**

(1) Ethane; $C_2H_6$; [74-84-0]
(2) Sulfuric acid monododecyl ester sodium salt or sodium dodecyl sulfate or SDS; $C_{12}H_{26}O_4S.Na$; [151-21-3]
(3) 1-Pentanol; $C_5H_{12}O$; [71-41-0]
(4) Water; $H_2O$; [7732-18-5]

**ORIGINAL MEASUREMENTS:**

Hoskins, J. C.; King, A. D.

*J. Coll. Interface Sci.* 1981, *82*, 260 - 263.

**VARIABLES:**

$T/K = 298.15$
$p_1/kPa = 101.325$
$m_2/mol\ kg^{-1} = 0 - 0.40$
$m_3/mol\ kg^{-1} = 0 - 2.75$

**PREPARED BY:**

H. L. Clever

**EXPERIMENTAL VALUES:**

| Sodium Dodecyl Sulfate $m_2$/mol $kg^{-1}$ | 1-Pentanol $m_3$/mol $kg^{-1}$ | Ethane $10^3K$/mol $kg^{-1}$ $atm^{-1}$ |
|---|---|---|
| 0 | 0 | 1.80 |
| | 0.1 | 1.81 |
| | 0.2 | 1.79 |
| 0.1 | 0 | 2.88 |
| | 0.1 | 3.17 |
| | 0.2 | 3.56 |
| | 0.3 | 3.99 |
| | 0.4 | 4.33 |
| | 0.5 | 5.03 |
| | 0.6 | 5.55 |
| | 0.7 | 6.29 |
| 0.25 | 0 | 4.69 |
| | 0.25 | 5.83 |
| | 0.50 | 7.09 |
| | 0.75 | 8.77 |
| 0.25 | 1.00 | 9.97 |
| | 1.25 | 12.5 |
| | 1.50 | 14.5 |
| | 1.75 | 15.3 |
| | 2.00 | 17.6 |
| | 2.25 | 20.1 |
| | 2.50 | - |
| | 2.75 | 26.1 |
| 0.40 | 0 | 6.31 |
| | 0.25 | 7.34 |
| | 0.50 | 9.29 |
| | 0.75 | 10.5 |
| | 1.00 | 12.3 |
| | 1.25 | 13.9 |
| | 1.50 | 15.7 |
| | 1.75 | 17.9 |

All measurements were made at a temperature of 25 °C or 298.15 K.

The ethane solubility is given as a Henry's constant in the form

$K/mol\ kg^{-1}\ atm^{-1} = (m_1/mol\ kg^{-1})/(p_1/\ atm)$.

**AUXILIARY INFORMATION**

**METHOD/APPARATUS/PROCEDURE:**

The apparatus and procedure were described in detail earlier (1). The amount of ethane released from a solution which has previously been saturated with ethane at a known pressure is determined as follows: (*i*) the solution to be studied is stirred and allowed to equilibrate with ethane at some elevated pressure in a thermostated brass bomb; (*ii*) the stirrer is turned off and the solution is allowed to become still where upon the pressure is released; (*iii*) after allowing a short period of time for thermal equilibration, the now supersaturated solution is again stirred and the volume of gas evolved is measured manometrically under ambient conditions. One obtains the number of ethane moles evolved as the pressure changes isothermally from the equilibrium pressure to ambient pressure. The $\Delta n/\Delta p$ ratio is the Henry's constant. Experiments at several pressures establish Henry's law is obeyed. Corrections for the gas lost during venting and thermal equilibration, and for water vapor pressure are made.

**SOURCE AND PURITY OF MATERIALS:**

(1) Ethane. Matheson Co., Inc. Purity 99.0 per cent.
(2) Sodium dodecyl sulfate. Aldrich Chem. Co., Inc. Recrystallized twice from 2-propanol, dried *in vacuo*. Analysis of the purified product showed 74 % $C_{12}$ sulfate, 22 % $C_{14}$ sulfate, and 4 % $C_{16}$ sulfate.
(3) 1-Propanol. Eastman Kodak Co. Fractionally distilled.
(4) Water. Distilled.

**ESTIMATED ERROR:**

$\delta K/K = \pm\ 0.02$

**REFERENCES:**

1. Matheson, I. B. C.; King, A. D. *J. Coll. Interface Sci.* 1978, *66*, 464.

| COMPONENTS: | EVALUATOR: |
|---|---|
| (1) Ethane; $C_2H_6$; [74-84-0]<br>(2) Aqueous organic solvent solutions | Walter Hayduk<br>Department of Chemical Engineering<br>University of Ottawa<br>Ottawa, Canada K1N 9B4 |

CRITICAL EVALUATION:

Ethane solubilities are available in water containing low concentrations of a second miscible organic component as measured by Ben-Naim and Yaacobi (1). The organic components include 1-propanol, 1,4-dioxane, dimethylsulfoxide and sucrose. The presence of the first three organic components enhances the solubility, while the sucrose inhibits the solubility, when comparing the solubilities with that in water. The data of Yaacobi and Ben-Naim (2,3) for the solubility of ethane in water were favorably compared with the recommended values for that solvent (see Critical Evaluation for water). The same authors (2) reported solubilities in aqueous ethanol for the whole concentration range and for several temperatures near the ambient. This particular solvent solution is of interest because of the large change in solubility with increasing ethanol concentration. The solubility increases more than a hundred-fold when increasing the concentration from 0 to 100% ethanol.

These data are classified as tentative.

Data are available for ethane solubilities in aqueous urea solutions. As well as the data of Yaacobi and Ben-Naim (2,3), those of Wetlaufer et al. (4) for the solubility of ethane in water compare favorably with the recommended values, whereas the data of Wen and Hung (5) are systematically low (see Critical Evaluation for water). When the data of Wen and Hung (5) for ethane solubility in dilute aqueous urea solutions are compared with those of Ben-Naim and Yaacobi (1) they are found to be systematically lower, by as much as 8% for a comparable concentration. The data of Wen and Hung (5) are therefore classified as doubtful. A smaller difference exists between the data of Ben-Naim and Yaacobi (1) and that of Wetlaufer et al. (4) for a concentrated aqueous urea solvent (7 mol $dm^{-3}$) with the latter data indicating a lower ethane solubility. Because it is not possible to say which of the two is more accurate it is suggested that the average of the two is probably the most accurate.

These data are classified as tentative.

REFERENCE


1. Ben-Naim, A ; Yaacobi, M. *J. Phys. Chem.* 1974, *78*, 170-175.
2. Yaacobi, M.; Ben-Naim, A. *J. Soln. Chem.* 1973, *2*, 425-443.
3. Yaacobi, M.; Ben-Naim, A. *J. Phys. Chem.* 1974, *78*, 175-178.
4. Wetlaufer, D.B.; Malik, S.K.; Stoller, L.; Coffin, R.L. *J. Am. Chem. Soc.* 1964, *86*, 508-514.
5. Wen, W-Y.; Hung, J.H. *J. Phys. Chem.* 1970, *74*, 170-180.

**COMPONENTS:**

(1) Ethane; $C_2H_2$; [74-84-0]

(2) 1-Propanol; $C_3H_8O$; [71-23-8]

(3) Water; $H_2O$; [7732-18-5]

**ORIGINAL MEASUREMENTS:**

Ben-Naim, A.; Yaacobi, M.

*J. Phys. Chem.* 1974, *78*, 170-5.

**VARIABLES:**

$P$/KPa: 101.325 (1 atm)
$T$/K: 283.15-303.15
$x_2$/mol fraction: 0.03

**PREPARED BY:**

C.L. Young

**EXPERIMENTAL VALUES:**

| $T$/K | Mole fraction of propanol, $x_{C_3H_8O}$ ** | Ostwald coefficient, $L$ * |
|---|---|---|
| 283.15 | 0.03 | 0.07437 |
| 288.15 | | 0.06546 |
| 293.15 | | 0.05854 |
| 298.15 | | 0.05311 |
| 303.15 | | 0.04886 |

* Smoothed values obtained from

$kT \ln L = 14{,}662.0 - 93.490\,(T/K) + 0.12906\,(T/K)^2$ cal mol$^{-1}$
where k is in units of cal mol$^{-1}$K$^{-1}$

** Mole fraction before saturation with ethane which is virtually the same as the mole fraction after saturation.

**AUXILIARY INFORMATION**

**METHOD/APPARATUS/PROCEDURE:**

The apparatus was similar to that described by Ben-Naim and Baer (1) and Wen and Hung (2). It consists of three main parts, a dissolution cell of 300 to 600 cm$^3$ capacity, a gas volume measuring column, and a manometer. The solvent is degassed in the dissolution cell, the gas is introduced and dissolved while the liquid is kept stirred by a magnetic stirrer immersed in the water bath. Dissolution of the gas results in the change in the height of a column of mercury which is measured by a cathetometer.

**SOURCE AND PURITY OF MATERIALS:**

1. Matheson sample, purity 99.9 mol per cent.
2. CP grade.
3. Deionised, doubly distilled.

**ESTIMATED ERROR:**

$\delta T$/K = ±0.01; $\delta L/L$ = ±0.005. (estimated by compiler).

**REFERENCES:**

1. Ben-Naim, A.; Baer, S. *Trans. Faraday Soc.* 1963, *59*, 2735.
2. Wen, W.-Y.; Hung, J.H. *J. Phys. Chem.* 1970, *74*, 170.

**COMPONENTS:**

(1) Ethane; $C_2H_6$; [74-84-0]

(2) 1,4-Dioxane; $C_4H_8O_2$; [123-91-1]

(3) Water; $H_2O$; [7732-18-5]

**ORIGINAL MEASUREMENTS:**

Ben-Naim, A.; Yaacobi, M.

*J. Phys. Chem.* 1974, *78*, 170-5

**VARIABLES:**

$T$/K: 283.15-303.15
$P$/KPa: 101.325 (1 atm)
$x_2$/mol fraction: 0.03

**PREPARED BY:**

C.L. Young

**EXPERIMENTAL VALUES:**

| $T$/K | Mole fraction** of Dioxane, $x_{C_4H_8O_2}$ | Ostwald coefficient,* $L$ |
|---|---|---|
| 283.15 | 0.03 | 0.07729 |
| 288.15 | | 0.06877 |
| 293.15 | | 0.06207 |
| 298.15 | | 0.05678 |
| 303.15 | | 0.05262 |

* Smoothed values obtained from

$kT \ln L = 13{,}529.1 - 86.676\ (T/K) + 0.11940\ (T/K)^2$ cal mol$^{-1}$
where k is in units of cal mol$^{-1}$ K$^{-1}$

** Mole fraction before saturation with ethane which is virtually the same as after saturation.

**AUXILIARY INFORMATION**

**METHOD/APPARATUS/PROCEDURE:**

The apparatus was similar to that described by Ben-Naim and Baer (1) and Wen and Hung (2). It consists of three main parts, a dissolution cell of 300 to 600 cm$^3$ capacity, a gas volume measuring column, and a manometer. The solvent is degassed in the dissolution cell, the gas is introduced and dissolved while the liquid is kept stirred by a magnetic stirrer immersed in the water bath. Dissolution of the gas results in the change in the height of a column of mercury which is measured by a cathetometer.

**SOURCE AND PURITY OF MATERIALS:**

1. Matheson sample, purity 99.9 mol per cent.
2. AR grade
3. Deionised, doubly distilled.

**ESTIMATED ERROR:**

$\delta T$/K = ±0.01; $\delta L/L$ = ±0.005; (estimated by compiler)

**REFERENCES:**

1. Ben-Naim, A.; Baer, S. *Trans. Faraday Soc.* 1963, *59* 2735.

2. Wen, W.-Y.; Hung, J.H. *J. Phys. Chem.* 1970, *74*, 170.

COMPONENTS:

(1) Ethane; $C_2H_6$; [74-84-0]

(2) Sulfinybismethane, (Dimethylsulfoxide, DMSO); $C_2H_6OS$; [67-68-5]

(3) Water; $H_2O$; [7732-18-5]

ORIGINAL MEASUREMENTS:

Ben-Naim, A.; Yaacobi, M.

*J. Phys. Chem.* 1974, *78*, 170-5.

VARIABLES:

$T$/K: 283.15-303.15
$P$/KPa: 101.325 (1 atm)
$x_2$/mol fraction: 0.03

PREPARED BY:

C.L. Young

EXPERIMENTAL VALUES:

| $T$/K | Mole fraction** of DMSO, $x_{DMSO}$ | Ostwald coefficient,* $L$ |
|---|---|---|
| 283.15 | 0.03 | 0.07095 |
| 288.15 | | 0.06233 |
| 293.15 | | 0.05544 |
| 298.15 | | 0.04989 |
| 303.15 | | 0.04539 |

* Smoothed values obtained from

$kT \ln L = 11{,}689.2 - 72{,}532\ (T/K) + 0.09180\ (T/K)^2$ cal mol$^{-1}$
where k is in units of cal mol$^{-1}$ K$^{-1}$

** Mole fraction before saturation with ethane which is virtually the same as the mole fraction after saturation.

AUXILIARY INFORMATION

METHOD/APPARATUS/PROCEDURE:

The apparatus was similar to that described by Ben-Naim and Baer (1) and Wen and Hung (2). It consists of three main parts, a dissolution cell of 300 to 600 cm$^3$ capacity, a gas volume measuring column, and a manometer. The solvent is degassed in the dissolution cell, the gas is introduced and dissolved while the liquid is kept stirred by a magnetic stirrer immersed in the water bath. Dissolution of the gas results in the change in the height of a column of mercury which is measured by a cathetometer.

SOURCE AND PURITY OF MATERIALS:

1. Matheson sample, purity 99.9 mol per cent.
2. CP grade
3. Deionised, doubly distilled.

ESTIMATED ERROR:

$\delta T$/K = ±0.01; $\delta L/L$ = ±0.005; (estimated by compiler).

REFERENCES:


1. Ben-Naim, A.; Baer, S. *Trans. Faraday Soc.* 1963, *59*, 2735.
2. Wen, W.-Y.; Hung, J.H. *J. Phys. Chem.* 1970, *74*, 170

**COMPONENTS:**

(1) Ethane; $C_2H_2$; [74-84-0]

(2) β-D-Fructofuranasyl -α-D-glucopyranoside, (Sucrose); $C_{12}H_{22}O_{11}$; [57-50-1]

(3) Water; $H_2O$; [7732-18-5]

**ORIGINAL MEASUREMENTS:**

Ben-Naim, A.; Yaacobi, M.

*J. Phys. Chem.* 1974, *78*, 170-5.

**VARIABLES:**

$P$/KPa: 101.325 (1 atm)
$T$/K: 283.15-303.15
$c_2$/mol dm$^{-3}$: 0.5

**PREPARED BY:**

C.L. Young

**EXPERIMENTAL VALUES:**

| $T$/K | Conc. of sucrose $c_2$/mol dm$^{-3}$ | Ostwald coefficient,* $L$ |
|---|---|---|
| 283.15 | 0.5 | 0.05609 |
| 288.15 | | 0.04880 |
| 293.15 | | 0.04293 |
| 298.15 | | 0.03818 |
| 303.15 | | 0.03430 |

* Smoothed values obtained from

$kT \ln L = 10{,}670 - 64.765\ (T/K) + 0.07542\ (T/K)^2$ cal mol$^{-1}$
where k is in units of cal mol$^{-1}$ K$^{-1}$

AUXILIARY INFORMATION

**METHOD/APPARATUS/PROCEDURE:**

The apparatus was similar to that described by Ben-Naim and Baer (1) and Wen and Hung (2). It consists of three main parts, a dissolution cell of 300 to 600 cm$^3$ capacity, a gas volume measuring column, and a manometer. The solvent is degassed in the dissolution cell, the gas is introduced and dissolved while the liquid is kept stirred by a magnetic stirrer immersed in the water bath. Dissolution of the gas results in the change in the height of a column of mercury which is measured by a cathetometer.

**SOURCE AND PURITY OF MATERIALS:**

1. Matheson sample, purity 99.9 mol per cent.
2. AR grade.
3. Deionised, doubly distilled.

**ESTIMATED ERROR:**

$\delta T$/K = ±0.01; $\delta L/L$ = ±0.005. (estimated by compiler).

**REFERENCES:**


1. Ben-Naim, A.; Baer, S. *Trans. Faraday Soc.* 1963, *59*, 2735.
2. Wen, W.-Y.; Hung, J.H. *J. Phys. Chem.* 1970, *74*, 170.

**COMPONENTS:**

(1) Ethane; $C_2H_6$; [74-84-0]

(2) Ethanol; $C_2H_6O$; [74-17-5]

(3) Water; $H_2O$; [7732-18-5]

**ORIGINAL MEASUREMENTS:**

Yaacobi, M.; Ben-Naim, A.

*J. Solution Chem.* 1973, *2*, 425-443.

**VARIABLES:**

$T$/K: 283.15 - 303.15

$P$/kPa: 101.325 (1 atm)

$C_2H_6O/x_2$: 0 - 1.0

**PREPARED BY:**

W. Hayduk

**EXPERIMENTAL VALUES:**

| $t$/°C | $T$/K | Mole fraction Ethanol[1] / $x_2$ | Ostwald Coefficient[1] $L/cm^3 cm^{-3}$ | Mole fraction[2] $/10^4 x_1$ |
|---|---|---|---|---|
| 10 | 283.15 | 0.00[3] | 0.06905[3] | 0.536 |
| | | 0.02 | 0.07477 | 0.603 |
| | | 0.03 | 0.07583 | 0.623 |
| | | 0.045 | 0.07683 | 0.648 |
| | | 0.06 | 0.07691 | 0.666 |
| | | 0.09 | 0.07675 | 0.698 |
| | | 0.12 | 0.08243 | 0.786 |
| | | 0.15 | 0.09794 | 0.978 |
| | | 0.20 | 0.1563 | 1.683 |
| | | 0.40 | 0.659 | 9.28 |
| | | 0.60 | 1.392 | 24.39 |
| | | 0.80 | 2.242 | 47.18 |
| | | 1.00[4] | 3.361[4] | 82.80 |

[1] Original data for solubility in ethanol-water mixed solvent solution.

[2] Calculated by compiler using density data for ethanol-water solutions, and assuming ideal gas behavior for ethane.

[3] Also reported by Ben-Naim, Wilf and Yaacobi in *J. Phys. Chem.* 1973, *77*, 95.

[4] Also reported by Ben-Naim and Yaacobi in *J. Phys. Chem.* 1974, *73*, 175.

**AUXILIARY INFORMATION**

**METHOD/APPARATUS/PROCEDURE:**

The method is volumetric utilizing an all-glass apparatus consisting of a dissolution cell of 300 to 600 $cm^3$ capacity, a gas volume measuring column, and a manometer. The solvent is degassed in the dissolution cell. Gas dissolves while the liquid is stirred using a magnetic stirrer. The volume of gas confined over mercury is read initially and after equilibration, by means of a cathetometer.

The apparatus is described by Ben-Naim and Baer (1) but it includes the modification introduced by Wen and Hung (2) of replacing the stopcocks with Teflon needle valves.

**SOURCE AND PURITY OF MATERIALS:**

1. Matheson; purity 99.9 mole per cent.

2. Absolute alcohol for alcohol-rich solutions. Analytical alcohol for water-rich solutions; purity not specified.

3. Purified by ion exchange and double distillation.

**ESTIMATED ERROR:**

$\delta L/L = 0.01$

$\delta x_2 = 0.01$

$\delta T/K = 0.05$

Estimated by compiler.

**REFERENCES:**


1. Ben-Naim, A.; Baer, S. *Trans. Faraday Soc.* 1963, *59*,

2. Wen, W.-Y.; Hung, J.H. *J. Phys. Chem.* 1970, *74*, 170.

| COMPONENTS: | ORIGINAL MEASUREMENTS: |
|---|---|
| (1) Ethane; $C_2H_6$; [74-84-0]<br>(2) Ethanol; $C_2H_6O$; [74-17-5]<br>(3) Water; $H_2O$; [7732-18-5] | Yaacobi, M.; Ben-Naim, A.<br>*J. Solution Chem.* 1973, *2*, 425-443. |

EXPERIMENTAL VALUES: continued

| $t$/°C | $T$/K | Mole fraction Ethanol[1] / $x_2$ | Ostwald Coefficient[1] $L$/cm$^3$cm$^{-3}$ | Mole fraction[2] /$10^4 x_1$ |
|---|---|---|---|---|
| 15 | 288.15 | 0.00[3] | 0.05912[3] | 0.451 |
| | | 0.02 | 0.06488 | 0.515 |
| | | 0.03 | 0.06577 | 0.532 |
| | | 0.045 | 0.06738 | 0.559 |
| | | 0.06 | 0.06878 | 0.586 |
| | | 0.09 | 0.07072 | 0.633 |
| | | 0.12 | 0.07871 | 0.739 |
| | | 0.15 | 0.09540 | 0.939 |
| | | 0.20 | 0.1537 | 1.63 |
| | | 0.40 | 0.638 | 8.87 |
| | | 0.60 | 1.314 | 22.74 |
| | | 0.80 | 2.111 | 43.89 |
| | | 1.00[4] | 3.117[4] | 75.93 |
| 20 | 293.15 | 0.00[3] | 0.05139[3] | |
| | | 0.02 | 0.05696 | 0.445 |
| | | 0.03 | 0.05830 | 0.464 |
| | | 0.045 | 0.06034 | 0.493 |
| | | 0.06 | 0.06247 | 0.524 |
| | | 0.09 | 0.06640 | 0.585 |
| | | 0.12 | 0.07617 | 0.705 |
| | | 0.15 | 0.09432 | 0.915 |
| | | 0.20 | 0.1528 | 1.60 |
| | | 0.40 | 0.617 | 8.47 |
| | | 0.60 | 1.245 | 21.29 |
| | | 0.80 | 1.990 | 40.90 |
| | | 1.00[4] | 2.910 | 70.12 |
| 25 | 298.15 | 0.00[3] | 0.04533[3] | |
| | | 0.02 | 0.05057 | 0.389 |
| | | 0.03 | 0.05274 | 0.413 |
| | | 0.045 | 0.05512 | 0.444 |
| | | 0.06 | 0.05760 | 0.476 |
| | | 0.09 | 0.06343 | 0.551 |
| | | 0.12 | 0.07466 | 0.681 |
| | | 0.15 | 0.09463 | 0.906 |
| | | 0.20 | 0.1534 | 1.59 |
| | | 0.40 | 0.596 | 8.09 |
| | | 0.60 | 1.183 | 20.00 |
| | | 0.80 | 1.874 | 38.08 |
| | | 1.00[4] | 2.730[4] | 65.05 |
| 30 | 303.15 | 0.00[3] | 0.04054 | 0.295 |
| | | 0.02 | 0.04538 | 0.343 |
| | | 0.03 | 0.04866 | 0.375 |
| | | 0.045 | 0.05130 | 0.407 |
| | | 0.06 | 0.05386 | 0.438 |
| | | 0.09 | 0.06163 | 0.528 |
| | | 0.12 | 0.07407 | 0.667 |
| | | 0.15 | 0.09624 | 0.909 |
| | | 0.20 | 0.1556 | 1.59 |
| | | 0.40 | 0.576 | 7.72 |
| | | 0.60 | 1.127 | 18.84 |
| | | 0.80 | 1.767 | 35.52 |
| | | 1.00[4] | 2.580 | 60.80 |

**COMPONENTS:**

(1) Ethane; $C_2H_6$; [74-84-0]

(2) Urea; $CH_4N_2O$; [57-13-6]

(3) Water; $H_2O$; [7732-18-5]

**ORIGINAL MEASUREMENTS:**

Ben-Naim, A.; Yaacobi, M.

*J. Phys. Chem.* 1974, *78*, 170-5

**VARIABLES:**

$T$/K: 283.15-303.15
$P$/KPa: 101.325 (1 atm)
$c_2$/mol dm$^{-3}$: 1.0-7.0

**PREPARED BY:**

C.L. Young

**EXPERIMENTAL VALUES:**

| $T$/K | Conc. of urea/mol dm$^{-3}$ | Ostwald coefficient, $^*L$ |
|---|---|---|
| 283.15 | 1.0 | 0.06401 |
| 288.15 | | 0.05603 |
| 293.15 | | 0.04942 |
| 298.15 | | 0.04391 |
| 303.15 | | 0.03928 |
| 283.15 | 2.0 | 0.06042 |
| 288.15 | | 0.05325 |
| 293.15 | | 0.04750 |
| 298.15 | | 0.04285 |
| 303.15 | | 0.03908 |
| 283.15 | 4.0 | 0.05420 |
| 288.15 | | 0.04879 |
| 293.15 | | 0.04417 |
| 298.15 | | 0.04018 |
| 303.15 | | 0.03674 |
| 283.15 | 7.0 | 0.04897 |
| 288.15 | | 0.04421 |
| 293.15 | | 0.04063 |
| 298.15 | | 0.03798 |
| 303.15 | | 0.03607 |

* Smoothed values obtained from
$kT \ln L = 7{,}273.5 - 41.403\ (T/K) + 0.03621\ (T/K)^2$ cal mol$^{-1}$;
$kT \ln L = 11{,}371.2 - 70.985\ (T/K) + 0.08917\ (T/K)^2$ cal mol$^{-1}$;
$kT \ln L = 5{,}153.0 - 30.054\ (T/K) + 0.02140\ (T/K)^2$ cal mol$^{-1}$;
$kT \ln L = 16{,}909.2 - 112.90\ (T/K) + 0.16662\ (T/K)^2$ cal mol$^{-1}$;
where k is in units of cal mol$^{-1}$ K$^{-1}$, for concentration of 1.0, 2.0, 4.0, 7.0 mol l$^{-1}$ respectively.

**AUXILIARY INFORMATION**

**METHOD/APPARATUS/PROCEDURE:**

The apparatus was similar to that described by Ben-Naim and Baer (1) and Wen and Hung (2). It consists of three main parts, a dissolution cell of 300 to 600 cm$^3$ capacity, a gas volume measuring column, and a manometer. The solvent is degassed in the dissolution cell, the gas is introduced and dissolved while the liquid is kept stirred by a magnetic stirrer immersed in the water bath. Dissolution of the gas results in the change in the height of a column of mercury which is measured by a cathetometer.

**SOURCE AND PURITY OF MATERIALS:**

1. Matheson sample purity 99.9 mol per cent.
2. AR grade.
3. Deionised, doubly distilled.

**ESTIMATED ERROR:**

$\delta T$/K = ±0.01; $\delta L/L$ = ±0.005.
(estimated by compiler).

**REFERENCES:**


1. Ben-Naim, A.; Baer, S. *Trans. Faraday Soc.* 1963, *59*, 2735.
2. Wen, W.-Y.; Hung, J.H. *J. Phys. Chem.* 1970, *74*, 170.

**COMPONENTS:**

(1) Ethane; $C_2H_6$; [74-84-0]

(2) Urea; $CH_4N_2O$; [57-13-6]

(3) Water; $H_2O$; [7732-18-5]

**ORIGINAL MEASUREMENTS:**

Wen, W.-Y.; Hung, J. H.

*J. Phys. Chem.* 1970, *74*, 170 - 180.

**VARIABLES:**

$T$/K: 278.15 - 308.15

$P$/kPa: 101.325 (1 atm)

$m_2$/mol kg$^{-1}$: 0, 0.495

**PREPARED BY:**

H. L. Clever

**EXPERIMENTAL VALUES:**

| $T$/K | Urea Molality $m_2$/mol kg$^{-1}$ | Ethane Solubility $S_1$/cm$^3$ (STP) kg$^{-1}$ | Setschenow Constant[1] k/kg mol$^{-1}$ |
|---|---|---|---|
| 278.15 | 0 | 80.19 ± 0.23 | |
| | 0.495 | 78.89 | +0.013 |
| 288.15 | 0 | 55.55 ± 0.15 | |
| | 0.495 | 55.55 | -0.002 |
| 298.15 | 0 | 41.20 ± 0.12 | |
| | 0.495 | 41.60 | -0.010 |
| 308.15 | 0 | 32.27 ± 0.10 | |
| | 0.495 | 32.72 | -0.012 |

[1] Setschenow constant, k/kg mol$^{-1}$ = $(1/(m_2/\text{mol kg}^{-1}))\log(S_1^\circ/S_1)$

The authors specify the value of the constant for $m_2$/mol kg$^{-1}$ = 0.1.

## AUXILIARY INFORMATION

**METHOD/APPARATUS/PROCEDURE:**

The apparatus was similar to that described by Ben-Naim and Baer (1). Teflon needle valves were used in place of stopcocks.

The apparatus consists of three main parts, a dissolution cell of 300 to 600 cm$^3$ capacity, a gas volume measuring column, and a manometer.

The solvent is degassed in the dissolution cell, the gas is introduced and dissolved while the liquid is kept stirred by a magnetic stirrer immersed in the water bath. Dissolution of the gas results in the change in the height of a column of mercury which is measured by a cathetometer.

**SOURCE AND PURITY OF MATERIALS:**

(1) Ethane. Matheson Co. Stated to be better than 99.9 per cent pure.

(2) Urea. No information.

(3) Water. Distilled from an all Pyrex apparatus. Specific conductivity $1.5 \times 10^{-6}$ (ohm cm)$^{-1}$.

**ESTIMATED ERROR:**

$\delta T$/K = ±0.005

$\delta S_1/S_1$ = ±0.003

**REFERENCES:**

1. Ben-Naim, A.; Baer, S. *Trans. Faraday Soc.* 1963, *59*, 2735.

COMPONENTS:

(1) Ethane; $C_2H_6$; [74-84-0]
(2) Urea; $CH_4N_2O$; [57-13-6]
(3) Water; $H_2O$; [7732-18-5]

ORIGINAL MEASUREMENTS:

Wetlaufer, D. B.; Malik, S. K.; Stoller, L.; Coffin, R. L.
*J. Am. Chem. Soc.*
1964, *86*, 508-514.

VARIABLES:

$T$/K: 278.2-318.2
$c_2$/mol dm$^{-3}$: 6.96

PREPARED BY:

C. L. Young

EXPERIMENTAL VALUES:

| $T$/K | Conc. of urea in soln. $c_2$/mol dm$^{-3}$ | $10^3$ Conc. of ethane† in soln. $c_1$/mol dm$^{-3}$ | Mole fraction* of ethane $x_{C_2H_6}$ |
|---|---|---|---|
| 278.2 | 6.96 | 0.00233 | 0.0000518 |
| 298.2 | 6.96 | 0.00161 | 0.0000358 |
| 318.2 | 6.96 | 0.00123 | 0.0000274 |

† at a partial pressure of 101.3 kPa.

* calculated by compiler.

AUXILIARY INFORMATION

METHOD/APPARATUS/PROCEDURE:

Modified Van Slyke-Neill apparatus fitted with a magnetic stirrer. Solution was saturated with gas and then sample transferred to the Van Slyke extraction chamber.

SOURCE AND PURITY OF MATERIALS:

1. Matheson c.p. grade, purity 99 mole per cent or better.
2. Distilled.
3. Commercial sample, purified by two recrystallizations from 65% ethanol.

ESTIMATED ERROR:

$\delta T$/K = ±0.05; $\delta x_{C_2H_6}$ = ±2%.

REFERENCES:

COMPONENTS:

(1) Ethane; $C_2H_6$; [74-84-0]

(2) Water; $H_2O$; [7732-18-5]

(3) Aqueous organic solutions at elevated pressures.

EVALUATOR:

Walter Hayduk
Department of Chemical Engineering
University of Ottawa
Ottawa, Canada
K1N 9B4

CRITICAL EVALUATION:

There are no comparable data for the solubilities of ethane in the aqueous ethanolamine solutions, 2-aminoethanol (monoethanolamine) and 2,2′-iminobisethanol (diethanolamine) solutions at two pressures considerably above atmospheric, at two temperatures, and also at two concentrations as reported by Lawson and Garst (1). These data appear erratic whcn compared with the ethane solubility in water. Nor are they self-consistent in that they do not show comparable effects with increasing temperature, pressure or concentration. Essentially identical values are listed for 5% diethanolamine for both temperatures, 310.9 and 338.7 K. This was checked and found in the original reference, but is considered most unlikely.

Because of these inconsistencies these data are classified as doubtful.

REFERENCES

1. Lawson, J.D.; Garst, A.W. *J. Chem. Eng. Data* 1976, *21*, 30-32.

COMPONENTS:

(1) Ethane; $C_2H_6$; [74-84-0]

(2) 2-Aminoethanol, (Monoethanolamine); $C_2H_7NO$; [141-43-5]

(3) Water; $H_2O$; [7732-18-5]

ORIGINAL MEASUREMENTS:

Lawson, J.D.; Garst, A.W.

*J. Chem. Eng. Data* 1976, *21*, 30-2.

VARIABLES:

$T$/K: 310.93, 338.71
$P$/MPa: 3.39-6.43
$c_2$/Wt.%: 15,40

PREPARED BY:

C.L. Young

EXPERIMENTAL VALUES:

| $T$/K | $P$/MPa | Conc. wt. % amine | Mole fraction of ethane in liquid, $x_{C_2H_6}$ | $10^5$x Solubility / mol $g^{-1}$ (soln) |
|---|---|---|---|---|
| 310.93 | 3.385 | 15 | 0.000785 | 3.90 |
| | 5.985 | | 0.00101 | 5.04 |
| 338.71 | 3.454 | 15 | 0.000626 | 3.11 |
| | 6.584 | | 0.00100 | 4.98 |
| | 3.468 | 40 | 0.00107 | 4.25 |
| | 6.426 | | 0.00160 | 6.38 |

AUXILIARY INFORMATION

METHOD/APPARATUS/PROCEDURE:

Rocking equilibrium cell fitted with liquid sampling valve. Pressure measured with Bourdon gauge. Cell charged with amine then ethane added. Liquid phase samples analysed volumetrically.

SOURCE AND PURITY OF MATERIALS:

1. Purity 99 mole per cent minimum.

2. Commercial sample, purity better than 99 mole per cent as determined by acid titration.

3. Distilled.

ESTIMATED ERROR:

$\delta T$/K = ±0.15; $\delta P$/MPa = ±0.5%
$\delta x_{C_2H_6}$ = ±3%.

REFERENCES:

**COMPONENTS:**

(1) Ethane; $C_2H_6$; [74-84-0]

(2) 2,2'-Iminobisethanol, (Diethanolamine); $C_4H_{11}NO$; [111-42-2]

(3) Water; $H_2O$; [7732-18-5]

**ORIGINAL MEASUREMENTS:**

Lawson, J.D.; Garst, A.W.

*J. Chem. Eng Data* 1976, *21*, 30-2.

**VARIABLES:**

$T$/K: 310.93, 338.71
$P$/MPa: 3.45-6.70
$c_2$/Wt.%: 5,25

**PREPARED BY:**

C.L. Young

**EXPERIMENTAL VALUES:**

| $T$/K | $P$/MPa | Conc. Wt % $C_4H_{11}NO$ | Mole fraction of ethane in liquid, $x_{C_2H_6}$ | $10^5$x Solubility /mol g$^{-1}$ (soln.) |
|---|---|---|---|---|
| 310.93 | 3.454 | 5 | 0.000650 | 2.65 |
| | 6.598 | | 0.000727 | 3.94 |
| | 3.309 | 25 | 0.000870 | 3.83 |
| | 5.985 | | 0.00114 | 5.02 |
| 338.71 | 3.454 | 5 | 0.000650 | 2.65 |
| | 6.598 | | 0.000727 | 3.94 |
| | 3.434 | 25 | 0.000713 | 3.14 |
| | 6.701 | | 0.00109 | 4.78 |

AUXILIARY INFORMATION

**METHOD/APPARATUS/PROCEDURE:**

Rocking equilibrium cell fitted with liquid sampling valve. Pressure measured with Bourdon gauge. Cell charged with amine and then ethane added. Liquid phase samples analysed volumetrically.

**SOURCE AND PURITY OF MATERIALS:**

1. Purity 99 mole per cent minimum.

2. Commercial sample, purity better than 99 mole per cent as determined by acid titration.

3. Distilled.

**ESTIMATED ERROR:**

$\delta T$/K = ±0.15; $\delta P$/MPa = ±0.5%
$\delta x_{C_2H_6}$ = ±3%.

**REFERENCES:**

| COMPONENTS: | EVALUATOR: |
|---|---|
| (1) Ethane; $C_2H_6$; [74-84-0]<br>(2) Paraffin solvents | Walter Hayduk<br>Department of Chemical Engineering<br>University of Ottawa<br>Ottawa, Canada K1N 9B4 |

CRITICAL EVALUATION:

The solubility of ethane in the n-alkanes which are normally liquids has been studied by at least nine groups of workers with only partially consistent results as can be observed in the figure below (which shows solubilities at 298.15 K and a partial pressure of 101.325 kPa). Although all the workers show a general increase in solubility with carbon number, the extent of the increase is variable. Whereas several groups (1,6,7,9) used chromatographic techniques, the other workers used volumetric methods. The solubility in pentane(1) appears significantly lower than the likely solubility and that in dodecane(6) significantly higher than the likely solubility. These values are classed as doubtful. Only for three of the solvents, hexane, heptane and hexadecane are the solubilities consistent among at least three workers within an experimental error of ±2%. The data of Monfort and Arriaga(6) and Cukor and Prausnitz(8) for solubilities in hexadecane were linearly extrapolated on a ln $x_1$ versus 1/T plot to 298.15 K. These latter three sets of data are classed as tentative.

More accurate solubility determinations in most of the n-alkanes are required.

An approximate equation for the solubilities of ethane at 298.15 K at a partial pressure of 101.325 kPa in the n-alkanes from pentane to hexadecane is as follows:

$$x_1 = 0.02835 + 0.000569\ C_n$$

This equation is shown as a dotted line in the figure but because of the paucity and lack of consistency of data on which it is based,

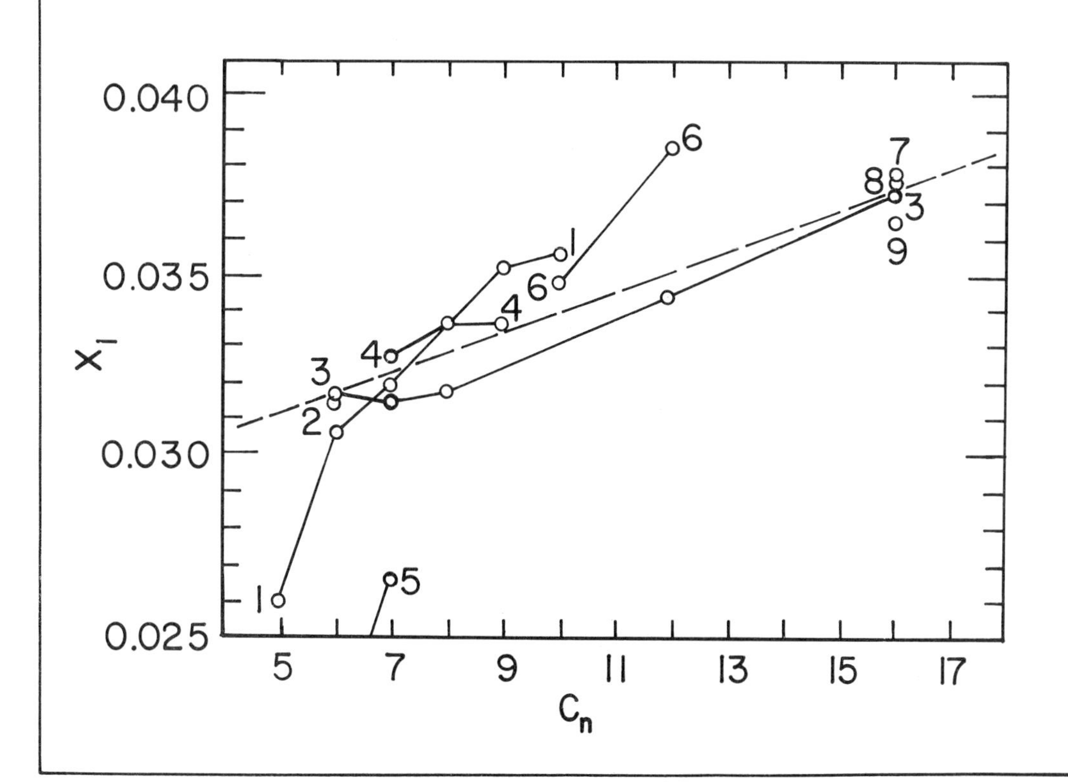

COMPONENTS:

(1) Ethane; $C_2H_6$; [74-84-0]

(2) Paraffin solvents

EVALUATOR:

Walter Hayduk
Department of Chemical Engineering
University of Ottawa
Ottawa, Canada K1N 9B4

CRITICAL EVALUATION:

...continued

it must be considered of doubtful reliability. The references for ethane solubilities in normal paraffin solvents are: pentane (1), hexane (1,2,3,5,10,14), heptane (1,3,4,5,14), octane (1,3,4), nonane (1,4), decane (1,6), dodecane (3,6) and hexadecane (3,7,8,9).

The solubility of ethane in neohexane (10) is approximately twice that in n-hexane at the same temperature. It is considered tentative. The solubilities in isooctane (11) and heptamethyl nonane (9) are likewise considered tentative.

Solubilities in the higher molecular weight solvents heptadecane (7), octadecane (9,12), eicosane (12,13), docosane (12) and squalane (13) are also available at temperatures ranging from about 300 to 475 K. The two sets of data for solubilities in eicosane check within 5%. All these data are considered tentative.

It is noted that solubilities in the branch-chained isomers, neohexane, isooctane etc., are consistently somewhat higher than in their straight-chained counterparts of the same carbon number.

The early solubilities of McDaniel (5) in hexane and heptane are considered doubtful and are rejected.

Solubilities are also available at 298.15 K and 101.325 kPa in the two-component solvent solutions composed of hexane and hexadecane (3). These data are classified as tentative.

References

1. Jadot, R. *J. Chem. Phys.* 1972, *69*, 1036-1040.
2. Waters, J.A.; Mortimer, G.A. *J. Chem. Eng. Data* 1972, *17*, 156-157.
3. Hayduk, W.; Cheng, S.C. *Can. J. Chem. Eng.* 1970, *48*, 93-99.
4. Thomsen, E.S.; Gjaldbaek, J.C. *Acta Chem. Scand.* 1963, *17*, 127-133.
5. McDaniel, A.S. *J. Phys. Chem.* 1911, *15*, 587-610.
6. Monfort, J.P.; Arriaga, J.L. *Chem. Eng. Commun.* 1980, *7*, 17-25.
7. Lenoir, J-Y.; Renault, P.; Renon, H. *J. Chem. Eng. Data* 1971, *16*, 340-342.
8. Cukor, P.M.; Prausnitz, J.M. *J. Phys. Chem.* 1972, *76*, 598-601.
9. Richon, D.; Renon, H. *J. Chem. Eng. Data* 1980, *25*, 59-60.
10. Tilquin, B.; Decanniere, L.; Fontaine, R.; Claes, P. *Ann. Soc. Sc. Bruxelles (Belgium)* 1967, *81*, 191-199.
11. Kobatake, Y.; Hildebrand, J.H. *J. Phys. Chem.* 1961, *65*, 331-335.
12. Ng, S.; Harris, H.G.; Prausnitz, J.M. *J. Chem. Eng. Data* 1969, *14*, 482-483.
13. Chappelow, C.C.; Prausnitz, J.M. *Am. Inst. Chem. Engnrs. J.* 1974, *20*, 1097-1104.
14. Malik, V.K.; Hayduk, W. *Can. J. Chem. Eng.* 1970, *46*, 462-466.

**COMPONENTS:**

(1) Ethane; $C_2H_6$; [74-84-0]

(2) Pentane; $C_5H_{12}$; [109-66-0]
or
Hexane; $C_6H_{14}$; [110-54-3]

**ORIGINAL MEASUREMENTS:**

Jadot, R.

*J. Chim. Phys.* 1972, *69*, 1036-40

**VARIABLES:**

$T$/K: 298.15

**PREPARED BY:**

C.L. Young

**EXPERIMENTAL VALUES:**

| $T$/K | Henry's Law Constant, $H$/atm | Mole fraction$^+$ at partial pressure of 101.3 kPa, $x_{C_2H_6}$ | #$\Delta H^\infty$ /cal mol$^{-1}$ (/J mol$^{-1}$) |
|---|---|---|---|
| | Pentane; $C_5H_{12}$; [109-66-0] | | |
| 298.15 | 38.48 | 0.02599 | - |
| | Hexane; $C_6H_{14}$; [110-54-3] | | |
| 298.15 | 32.54 | 0.03073 | 220 (920) |

+ Calculated by compiler assuming $x_{C_2H_6} = 1/H$

# Excess partial molar enthalpy of solution at infinite dilution.

AUXILIARY INFORMATION

**METHOD/APPARATUS/PROCEDURE:**

The conventional gas chromatographic technique was used. The carrier gas was helium. The value of Henry's law constant was calculated from the retention time. The value applies to very low partial pressures of gas and there may be a substantial difference from that measured at 1 atm. pressure. There is also considerable uncertainty in the value of Henry's constant since no allowance was made for surface adsorption.

**SOURCE AND PURITY OF MATERIALS:**

No details given.

**ESTIMATED ERROR:**

$\delta T$/K = ±0.05; $\delta H$ = ±2%

**REFERENCES:**

**COMPONENTS:**

(1) Ethane; $C_2H_6$; [74-84-0]

(2) Hexane; $C_6H_{14}$; [110-54-3]

**ORIGINAL MEASUREMENTS:**

Waters, J. A.; Mortimer, G. A.

*J. Chem. Eng. Data* 1972, *17*, 156 - 157.

**VARIABLES:**

$T$/K: 273.15 - 303.15
$p$/kPa: 101.325 (1 atm)

**PREPARED BY:**

H. L. Clever

**EXPERIMENTAL VALUES:**

| Temperature $t/^0C$ | $T$/K | Ethane $c_1$/ mol dm$^{-3}$atm$^{-1}$ | Mol Fraction $10^2x_1$ | Bunsen Coefficient $\alpha$/cm$^3$(STP)cm$^{-3}$atm$^{-1}$ | | Ostwald Coefficient $L$/cm$^3$cm$^{-3}$ |
|---|---|---|---|---|---|---|
| 0 | 273.15 | 0.381 ± 0.006 | 4.62 | 8.54 ± 0.13 | 8.68[1] | 8.54 |
| 10 | 283.15 | 0.317 ± 0.010 | 3.92 | 7.10 ± 0.22 | 7.14[1] | 7.36 |
| 20 | 293.15 | 0.264 ± 0.005 | 3.32 | 5.89 ± 0.11 | 5.68[1] | 6.32 |
| 25 | 298.15 | 0.247 ± 0.005 | 3.15 | 5.54 ± 0.11 | - | 6.05 |
| 30 | 303.15 | 0.225 ± 0.012 | 2.89 | 5.04 ± 0.27 | 4.96[1] | 5.59 |

[1] Measured by gas chromatography.

The Ostwald coefficient and mole fraction values were calculated by the compiler assuming ideal gas behavior.

Smoothed Data: For use between 273.15 and 303.15 K

$$\ln x_1 = -7.7775 + 12.8452/(T/100\ \text{K})$$

The standard error about the regression line is 2.86 x $10^{-4}$.

| $T$/K | Mol Fraction $10^2x_1$ |
|---|---|
| 273.15 | 4.62 |
| 283.15 | 3.91 |
| 293.15 | 3.35 |
| 298.15 | 3.11 |
| 303.15 | 2.90 |

**AUXILIARY INFORMATION**

**METHOD/APPARATUS/PROCEDURE:**

The solubility measurements were carried out in duplicate by technique B of Waters, Mortimer and Clements (1). The amount of gas required to saturate a known volume of solvent at a known temperature and partial pressure of the gas is determined. The change of pressure of the gas in a reservoir is determined as the previously evacuated absorption is filled and the liquid in it is saturated with gas.

The solvent is degassed by pumping out at $10^{-4}$ mmHg at liquid nitrogen temperature.

The volumes of the gas reservoir and the absorption vessel are known.

**SOURCE AND PURITY OF MATERIALS:**

(1) Ethane. Matheson Co., Inc. Research grade. Stated to be 99.9 mole per cent.

(2) Hexane. Phillips Petroleum Co. Maximum impurites stated to be 0.5 per cent benzene and 0.5 per cent methylcyclopentane. The hexane was passed through 3A Molecular Sieves and dried over sodium before use.

**ESTIMATED ERROR:**

$\delta p$/psia = ± 0.002
$\delta T$/K = ± 0.1
$\delta\alpha/\alpha$ = ± 0.02

**REFERENCES:**

1. Waters, J. A.; Mortimer, G. A.; Clements, H. E. *J. Chem. Eng. Data* 1970, *15*, 174.

**COMPONENTS:**

(1) Ethane; $C_2H_6$; [74-84-0]

(2) Hexane; $C_6H_{14}$; [110-54-3]

**ORIGINAL MEASUREMENTS:**

McDaniel, A.S.

*J. Phys. Chem.*, 1911, *15*, 587-610.

**VARIABLES:**

$T$/K: 295.25 - 328.15
$P$/kPa: 101.325 (1 atm)

**PREPARED BY:**

W. Hayduk

**EXPERIMENTAL VALUES:**

| $t$/°C | $T$/K | Ostwald coefficient[1] $L$/cm$^3$cm$^{-3}$ | Bunsen coefficient[3] $\alpha$/cm$^3$(STP)cm$^{-3}$atm$^{-1}$ | Mole fraction[3] $10^4x_1$ |
|---|---|---|---|---|
| 22.1 | 295.25 | 3.35 | 3.10 | 178 |
| 25.0 | 298.15 | 3.29[2] | 3.01 | 174 |
| 30.0 | 303.15 | 3.18 | 2.87 | 167 |
| 55.0 | 328.15 | 2.88 | 2.40 | 144 |

[1] Original data listed as Absorption coefficient interpreted by compiler to be equivalent to Ostwald coefficient as listed here.

[2] Ostwald coefficient (Absorption coefficient) as estimated at 298.15 K by author.

[3] Bunsen coefficient and mole fraction solubility calculated by compiler assuming ideal gas behavior.

[4] McDaniel's results are consistently from 20 to 80 per cent too low when compared with more reliable data.

**AUXILIARY INFORMATION**

**METHOD/APPARATUS/PROCEDURE:**

Glass apparatus consisting of a gas burette connected to a solvent contacting chamber. Gas pressure or volume adjusted using mercury displacement. Equilibration achieved at atmospheric pressure by hand shaking apparatus, and incrementally adding gas to contacting chamber. Solubility measured by obtaining total uptake of gas by known volume of solvent.

**SOURCE AND PURITY OF MATERIALS:**

1. Prepared by reaction of ethyl iodide with zinc-copper. Purity not measured.

2. Source not given; purity specified as 99 per cent.

**ESTIMATED ERROR:**

$\delta L/L = -0.20$
(estimated by compiler; see note [4] above)

**REFERENCES:**

**COMPONENTS:**

(1) Ethane; $C_2H_6$; [74-84-0]

(2) Hexane; $C_6H_{14}$; [110-54-3]

**ORIGINAL MEASUREMENTS:**

Tilquin, B.; Decannière, L.; Fontaine, R.; Claes, P.

*Ann. Soc. Sc. Bruxelles (Belgium)* 1967, *81*, 191-199.

**VARIABLES:**

$T$/K: 288.15
$P$/kPa: 4.11 - 8.13

**PREPARED BY:**

W. Hayduk

**EXPERIMENTAL VALUES:**

| $t$/°C | $T$/K | Ostwald coefficient[1] $L$/cm$^3$cm$^{-3}$ | Mole fraction[2] / $x_1$ | Henry's constant[2] $H$/atm |
|---|---|---|---|---|
| 15.0 | 288.15 | 6.655 | 0.03525 | 28.37 |

[1] Original data at low pressure reported as distribution coefficient; but if Henry's law and ideal gas law apply, distribution coefficient is equivalent to Ostwald coefficient as shown here.

[2] Calculated by compiler for a gas partial pressure of 101.325 kPa assuming that Henry's law and ideal gas law apply.

AUXILIARY INFORMATION

**METHOD/APPARATUS/PROCEDURE:**

All glass apparatus used at very low gas partial pressures, containing a replaceable degassed solvent ampule equipped with a breakable point which could be broken by means of a magnetically activated plunger. Quantity of gas fed into system determined by measuring the pressure change in a known volume. Quantity of liquid measured by weight. Pressure change observed after solvent released. Experimental details described by Rzad and Claes[1].

**SOURCE AND PURITY OF MATERIALS:**

1. Source not given; minimum purity specified as 99.0 mole per cent.

2. Fluka pure grade; minimum purity specified as 99.0 mole per cent.

**ESTIMATED ERROR:**

$\delta T$/K = 0.05
$\delta x_1/x_1$ = 0.01
(estimated by compiler)

**REFERENCES:**

1. Rzad, S.; Claes, P.

*Bull. Soc. Chim. Belges*, 1964, *73*, 689.

COMPONENTS:

(1) Ethane; $C_2H_6$; [74-84-0]

(2) Hexane; $C_6H_{14}$; [110-54-3]

ORIGINAL MEASUREMENTS:

Hayduk, W.; Cheng, S.C.

*Can. J. Chem. Eng.* 1970, *48*, 93-99.

VARIABLES:

$T$/K: 298.15 - 303.15

$P$/kPa: 101.325

PREPARED BY:

W. Hayduk

EXPERIMENTAL VALUES:

| $T$/K | Ostwald Coefficient[1] $L$/cm$^3$cm$^{-3}$ | Bunsen Coefficient[2] $\alpha$/cm$^3$(STP)cm$^{-3}$atm$^{-1}$ | Mole fraction[2] $10^4x_1$ |
|---|---|---|---|
| 298.15 | 6.09 | 5.58 | 317 |
| 303.15 | 5.74 | 5.17 | 297 |

[1] Original data.

[2] Bunsen coefficient and mole fraction calculated by compiler assuming ideal gas behavior.

AUXILIARY INFORMATION

METHOD/APPARATUS/PROCEDURE:

Volumetric method using glass apparatus. Degassed solvent contacted the gas while flowing in a thin film through an absorption spiral into a solution buret. Dry gas was maintained at atmospheric pressure in a gas buret by raising the mercury level. Volumes of solution and residual gas were read at regular intervals.

For some experiments the solvent flow was controlled with a stopcock; for the rest, a calibrated syringe pump was used.

Degassing was accomplished using a two-stage vacuum process described by Clever et al. (1).

SOURCE AND PURITY OF MATERIALS:

1. Matheson C.P. grade. Purity 99.5 mole per cent minimum.

2. Canadian Laboratory Supplies. Chromatoquality grade. Purity 99.0 mole per cent minimum.

ESTIMATED ERROR:

$\delta T$/K = 0.05

$\delta P$/kPa = 0.05

$\delta x_1/x_1$ = 0.01

REFERENCES:

1. Clever, H.L.; Battino, R.; Saylor, J.H.; Gross, P.M. *J. Phys. Chem.*, 1957, *61*, 1078.

**COMPONENTS:**

(1) Ethane; $C_2H_6$; [74-84-0]

(2) Hexane; $C_6H_{14}$; [110-54-3]

**ORIGINAL MEASUREMENTS:**

Malik, V.K.; Hayduk, W.

*Can. J. Chem. Eng.* 1970, *46*, 462-466.

**VARIABLES:**

$T$/K: 303.15

$P$/kPa: 101.325

**PREPARED BY:**

W. Hayduk

**EXPERIMENTAL VALUES:**

| $t$/°C | $T$/K | Ostwald coefficient[1] $L$/cm$^3$cm$^{-3}$ | Bunsen coefficient[2] $\alpha$/cm$^3$(STP)cm$^{-3}$atm$^{-1}$ | Mole fraction[2] $10^4x_1$ |
|---|---|---|---|---|
| 30.0 | 303.15 | 5.72 | 5.15 | 296 |

[1] Original data.

[2] Bunsen coefficient and mole fraction calculated by compiler assuming ideal gas behavior.

**AUXILIARY INFORMATION**

**METHOD/APPARATUS/PROCEDURE:**

Volumetric method using glass apparatus. Degassed solvent contacted the gas while flowing in a thin film through an absorption spiral into a solution buret. Dry gas was maintained at atmospheric pressure in a gas buret by raising the mercury level. Volumes of solution and residual gas were read at regular intervals.

**SOURCE AND PURITY OF MATERIALS:**

1. Matheson C.P. grade. Purity 99.5 mole per cent minimum.

2. Canadian Laboratory Supplies. Chromatoquality grade. Purity 99.0 mole per cent minimum.

**ESTIMATED ERROR:**

$\delta T$/K = 0.05

$\delta P$/kPa = 0.05

$\delta x_1/x_1$ = 0.01

**REFERENCES:**

**COMPONENTS:**

(1) Ethane; $C_2H_6$; [74-84-0]

(2) Heptane; $C_7H_{16}$; [142-82-5]

**ORIGINAL MEASUREMENTS:**

Hayduk, W.; Cheng, S.C.

*Can. J. Chem. Eng.* 1970, *48*, 93-99.

**VARIABLES:**

$T$/K: 293.15 - 313.15

$P$/kPa: 101.325

**PREPARED BY:**

W. Hayduk

**EXPERIMENTAL VALUES:**

| $T$/K | Ostwald Coefficient[1] $L$/cm$^3$cm$^{-3}$ | Bunsen Coefficient[2] $\alpha$/cm$^3$(STP)cm$^{-3}$atm$^{-1}$ | Mole fraction[2] $10^4 x_1$ | Mole fraction[3] $10^4 x_1$ |
|---|---|---|---|---|
| 293.15 | 5.743 | 5.351 | 338 | 338 |
| 298.15 | 5.393 | 4.941 | 315 | 315 |
| 303.15 | 5.090 | 4.586 | 295 | 294 |
| 313.15 | 4.420 | 3.857 | 258 | 258 |

[1] Original data.

[2] Bunsen coefficient and mole fraction calculated by compiler assuming ideal gas behavior.

[3] Smoothed data based on equation applicable between 293.15 and 313.15 K as derived by compiler:

$$\ln x_1 = -7.6106 + 1238.31/T$$

AUXILIARY INFORMATION

**METHOD/APPARATUS/PROCEDURE:**

Volumetric method using glass apparatus. Degassed solvent contacted the gas while flowing in a thin film through an absorption spiral into a solution buret. Dry gas was maintained at atmospheric pressure in a gas buret by raising the mercury level. Volumes of solution and residual gas were read at regular intervals.

For some experiments the solvent flow was controlled with a stopcock; for the rest, a calibrated syringe pump was used.

Degassing was accomplished using a two-stage vacuum process described by Clever et al. (1).

**SOURCE AND PURITY OF MATERIALS:**

1. Matheson C.P. grade. Purity 99.5 mole per cent minimum.

2. Canadian Laboratory Supplies. Chromatoquality grade. Purity 99.0 mole per cent minimum.

**ESTIMATED ERROR:**

$\delta T$/K = 0.05

$\delta P$/kPa = 0.05

$\delta x_1/x_1$ = 0.01

**REFERENCES:**

1. Clever, H.L.; Battino, R.; Saylor, J.H.; Gross, P.M. *J. Phys. Chem.*, 1957, *61*, 1078.

**COMPONENTS:**

(1) Ethane; $C_2H_6$; [74-84-0]

(2) Heptane; $C_7H_{16}$; [142-82-5]

**ORIGINAL MEASUREMENTS:**

Malik, V.K.; Hayduk, W.

*Can. J. Chem. Eng.* 1968, *46*, 462-466.

**VARIABLES:**

$T$/K: 303.15 - 313.15

**PREPARED BY:**

W. Hayduk

**EXPERIMENTAL VALUES:**

| $t$/°C | $T$/K | Ostwald coefficient[1] $L$/cm$^3$cm$^{-3}$ | Bunsen coefficient[2] $\alpha$/cm$^3$(STP)cm$^{-3}$atm$^{-1}$ | Mole fraction[2] $10^4x_1$ |
|---|---|---|---|---|
| 30.0 | 303.15 | 5.08 | 4.58 | 294 |
| 40.0 | 313.15 | 4.48 | 3.91 | 261 |

[1] Original data.

[2] Bunsen coefficient and mole fraction calculated by compiler assuming ideal gas behavior.

**AUXILIARY INFORMATION**

**METHOD/APPARATUS/PROCEDURE:**

Volumetric method using glass apparatus. Degassed solvent contacting the gas while flowing in a thin film through an absorption spiral into a solution buret. Dry gas was maintained at atmospheric pressure in a gas buret by raising the mercury level. Volumes of solution and residual gas were read at regular intervals.

**SOURCE AND PURITY OF MATERIALS:**

1. Matheson C.P. grade. Purity 99.5 mole per cent minimum.

2. Canadian Laboratory Supplies. Chromatoquality grade. Purity 99.0 mole per cent minimum.

**ESTIMATED ERROR:**

$\delta T$/K = 0.05
$\delta P$/kPa = 0.05
$\delta x_1/x_1$ = 0.01

**REFERENCES:**

**COMPONENTS:**

(1) Ethane; $C_2H_6$; [74-84-0]

(2) Heptane; $C_7H_{16}$; [142-82-5]

**ORIGINAL MEASUREMENTS:**

Thomsen, E. S.; Gjaldbaek, J. C.

*Acta Chem. Scand.* 1963, *17*, 127 - 133.

**VARIABLES:**

$T$/K: 298.15

$p_1$/kPa: 101.325 (1 atm)

**PREPARED BY:**

E. S. Thomsen

**EXPERIMENTAL VALUES:**

| $T$/K | Mol Fraction $10^2 x_1$ | Bunsen Coefficient $\alpha$/cm$^3$ (STP) cm$^{-3}$ atm$^{-1}$ | Ostwald Coefficient $L$/cm$^3$ cm$^{-3}$ |
|---|---|---|---|
| 298.05 | 3.26 | 5.12 | 5.59 |
| 298.15 | 3.31 | 5.20 | 5.68 |

The mole fraction and Ostwald solubility values were calculated by the compiler.

**AUXILIARY INFORMATION**

**METHOD/APPARATUS/PROCEDURE:**

A calibrated all-glass combined manometer and bulb was enclosed in an air thermostat and shaken until equilibrium. Mercury was used for calibration and as the confining liquid. The solvents were degassed in the apparatus. Details are in references 1 and 2.

The absorbed volume of gas was calculated from the initial and final amounts, both saturated with solvent vapor. The amount of solvent was determined by the weight of displaced mercury.

The saturation of the liquid with the gas was carried out close to atmospheric pressure. The solubility values were reported for one atmosphere gas pressure assuming Henry's law is obeyed.

**SOURCE AND PURITY OF MATERIALS:**

(1) Ethane. Phillips Petroleum Co. Research grade. Contained 0.2 per cent air and 0.1 per cent unidentified impurity.

(2) Heptane. Merck. Distillation range 0.09 K.

**ESTIMATED ERROR:**

$\delta T$/K = 0.05

$\delta x_1/x_1$ = 0.015

**REFERENCES:**

1. Lannung, A. *J. Am. Chem. Soc.* 1930, *52*, 68.

2. Gjaldbaek, J. C. *Acta Chem. Scand.* 1952, *6*, 623.

**COMPONENTS:**

(1) Ethane; $C_2H_6$; [74-84-0]

(2) Heptane; $C_7H_{16}$; [142-82-5]

**ORIGINAL MEASUREMENTS:**

McDaniel, A.S.

*J. Phys. Chem.* 1911, *15*, 587-610.

**VARIABLES:**

$T$/K: 298.15 - 313.15
$P$/kPa: 101.325 (1 atm)

**PREPARED BY:**

W. Hayduk

**EXPERIMENTAL VALUES:**

| $t$/°C | $T$/K | Ostwald Coefficient[1] $L$/cm$^3$cm$^{-3}$ | Bunsen Coefficient[2] $\alpha$/cm$^3$(STP)cm$^{-3}$atm$^{-1}$ | Mole fraction[2] $10^4 x_1$ |
|---|---|---|---|---|
| 25.0 | 298.15 | 4.50 | 4.12 | 266 |
| 30.0 | 303.15 | 4.42 | 3.98 | 257 |
| 40.0 | 313.15 | 4.26 | 3.72 | 243 |

[1] Original data listed as Absorption coefficient interpreted by compiler to be equivalent to Ostwald coefficient as listed here.

[2] Bunsen coefficient and mole fraction solubility calculated by compiler assuming ideal gas behavior.

[3] McDaniels results are consistently from 20 to 80 per cent too low when compared with more reliable data.

## AUXILIARY INFORMATION

**METHOD/APPARATUS/PROCEDURE:**

Glass apparatus consisting of a gas burette connected to a solvent contacting chamber. Gas pressure or volume adjusted using mercury displacement. Equilibration achieved at atmospheric pressure by hand shaking apparatus, and incrementally adding gas to contacting chamber. Solubility measured by obtaining total uptake of gas by known volume of solvent.

**SOURCE AND PURITY OF MATERIALS:**

1. Prepared by reaction of ethyl iodide with zinc-copper. Purity not measured.

2. Source not given; purity specified as 99 per cent.

**ESTIMATED ERROR:**

$\delta L/L = -0.20$
(estimated by compiler; see note [3] above.)

**REFERENCES:**

| COMPONENTS: | ORIGINAL MEASUREMENTS: |
|---|---|
| (1) Ethane; $C_2H_6$; [74-84-0]<br>(2) Heptane; $C_7H_{16}$; [142-82-5]<br>or<br>Octane; $C_8H_{18}$; [111-65-9] | Jadot, R.<br>*J. Chim. Phys.* 1972, *69*, 1036-40. |
| **VARIABLES:** | **PREPARED BY:** |
| $T$/K: 298.15 | C.L. Young |

EXPERIMENTAL VALUES:

| $T$/K | Henry's Law Constant, $H$/atm | Mole fraction$^+$ at partial pressure of 101.3kPa, $x_{C_2H_6}$ | #$\Delta H^\infty$ /cal mol$^{-1}$ (/J mol$^{-1}$) |
|---|---|---|---|
| | Heptane; $C_7H_{14}$; [142-82-5] | | |
| 298.15 | 31.209 | 0.032042 | 260 (1088) |
| | Octane; $C_8H_{18}$; [111-65-9] | | |
| 298.15 | 29.69 | 0.03368 | 290 (1213) |

\+ Calculated by compiler assuming $x_{C_2H_6} = 1/H$

\# Excess partial molar enthalpy of solution at infinite dilution.

AUXILIARY INFORMATION

METHOD/APPARATUS/PROCEDURE:

The conventional gas chromatographic technique was used. The carrier gas was helium. The value of Henry's law constant was calculated from the retention time. The value applies to very low partial pressures of gas and there may be a substantial difference from that measured at 1 atm. pressure. There is also considerable uncertainty in the value of Henry's constant since no allowance was made for surface adsorption.

SOURCE AND PURITY OF MATERIALS:

No details given.

ESTIMATED ERROR:

$\delta T$/K = ±0.05; $\delta H$ = ±2%.

REFERENCES:

**COMPONENTS:**

(1) Ethane; $C_2H_6$; [74-84-0]

(2) Octane; $C_8H_{18}$; [111-65-9]

**ORIGINAL MEASUREMENTS:**

Thomsen, E. S.; Gjaldbaek, J. C.

*Acta Chem. Scand.* 1963, *17*, 127 - 133.

**VARIABLES:**

$T$/K: 298.15-298.45

$p_1$/kPa: 101.325 (1 atm)

**PREPARED BY:**

E. S. Thomsen

**EXPERIMENTAL VALUES:**

| $T$/K | Mol Fraction $10^2 x_1$ | Bunsen Coefficient $\alpha/cm^3(STP)cm^{-3}atm^{-1}$ | Ostwald Coefficient $L/cm^3 cm^{-3}$ |
|---|---|---|---|
| 298.15 | 3.37 | 4.78 | 5.22 |
| 298.25 | 3.43 | 4.87 | 5.32 |
| 298.25 | 3.36 | 4.76 | 5.20 |
| 298.45 | 3.32 | 4.71 | 5.15 |

The mole fraction and Ostwald solubility values were calculated by the compiler.

**AUXILIARY INFORMATION**

**METHOD/APPARATUS/PROCEDURE:**

A calibrated all-glass combined manometer and bulb was enclosed in an air thermostat and shaken until equilibrium. Mercury was used for calibration and as the confining liquid. The solvents were degassed in the apparatus. Details are in references 1 and 2.

The absorbed volume of gas was calculated from the initial and final amounts, both saturated with solvent vapor. The amount of solvent was determined by the weight of displaced mercury.

The saturation of the liquid with the gas was carried out close to atmospheric pressure. The solubility values were reported for one atmosphere gas pressure assuming Henry's law is obeyed.

**SOURCE AND PURITY OF MATERIALS:**

(1) Ethane. Phillips Petroleum Co. Research grade. Contained 0.2 per cent air and 0.1 per cent unidentified impurity.

(2) Octane. British Drug House. Fractionated; about 1 per cent impurity; distillation range 0.02 K for the sample.

**ESTIMATED ERROR:**

$\delta T/K = 0.05$

$\delta x_1/x_1 = 0.015$

**REFERENCES:**

1. Lannung, A. *J. Am. Chem. Soc.* 1930, *52*, 68.

2. Gjaldbaek, J. C. *Acta Chem. Scand.* 1952, *6*, 623.

**COMPONENTS:**

(1) Ethane; $C_2H_6$; [74-84-0]

(2) Octane; $C_8H_{18}$; [111-65-9]

**ORIGINAL MEASUREMENTS:**

Hayduk, W.; Cheng, S.C.

*Can. J. Chem. Eng.* 1970, *48*, 93-99.

**VARIABLES:**

$T$/K: 298.15
$P$/kPa: 101.325

**PREPARED BY:**

W. Hayduk

**EXPERIMENTAL VALUES:**

| $T$/K | Ostwald Coefficient[1] $L$/cm$^3$cm$^{-3}$ | Bunsen Coefficient[2] $\alpha$/cm$^3$(STP)cm$^{-3}$atm$^{-1}$ | Mole fraction[2] $10^4x_1$ |
|---|---|---|---|
| 298.15 | 4.917 | 4.505 | 318 |

[1] Original data.

[2] Bunsen coefficient and mole fraction calculated by compiler assuming ideal gas behavior.

**AUXILIARY INFORMATION**

**METHOD/APPARATUS/PROCEDURE:**

Volumetric method using glass apparatus. Degassed solvent contacted the gas while flowing in a thin film through an absorption spiral into a solution buret. Dry gas was maintained at atmospheric pressure in a gas buret by raising the mercury level. Volumes of solution and residual gas were read at regular intervals.

For some experiments the solvent flow was controlled with a stopcock; for the rest, a calibrated syringe pump was used.

Degassing was accomplished using a two-stage vacuum process described by Clever et al. (1).

**SOURCE AND PURITY OF MATERIALS:**

1. Matheson C.P. grade. Purity 99.5 mole per cent minimum.

2. Canadian Laboratory Supplies. Chromatoquality grade. Purity 99.0 mole per cent minimum.

**ESTIMATED ERROR:**

$\delta T$/K = 0.05
$\delta P$/kPa = 0.05
$\delta x_1/x_1$ = 0.01

**REFERENCES:**

1. Clever, H.L.; Battino, R.; Saylor, J.H.; Gross, P.M. *J. Phys. Chem.*, 1957, *61*, 1078.

**COMPONENTS:**

(1) Ethane; $C_2H_6$; [74-84-0]

(2) Nonane; $C_9H_{20}$; [111-84-2]

**ORIGINAL MEASUREMENTS:**

Thomsen, E. S.; Gjaldbaek, J. C.

*Acta Chem. Scand.* 1963, *17*, 127 - 133.

**VARIABLES:**

$T$/K: 298.15

$p_1$/kPa: 101.325 (1 atm)

**PREPARED BY:**

E. S. Thomsen

**EXPERIMENTAL VALUES:**

| $T$/K | Mol Fraction $10^2 x_1$ | Bunsen Coefficient $\alpha/cm^3(STP)cm^{-3}atm^{-1}$ | Ostwald Coefficient $L/cm^3 cm^{-3}$ |
|---|---|---|---|
| 298.15 | 3.32 | 4.28 | 4.67 |
| 298.15 | 3.38 | 4.36 | 4.76 |
| 298.15 | 3.38 | 4.36 | 4.76 |

The mole fraction and Ostwald solubility values were calculated by the compiler.

**AUXILIARY INFORMATION**

**METHOD/APPARATUS/PROCEDURE:**

A calibrated all-glass combined manometer and bulb was enclosed in an air thermostat and shaken until equilibrium. Mercury was used for calibration and as the confining liquid. The solvents were degassed in the apparatus. Details are in references 1 and 2.

The absorbed volume of gas was calculated from the initial and final amounts, both saturated with solvent vapor. The amount of solvent was determined by the weight of displaced mercury.

The saturation of the liquid with the gas was carried out close to atmospheric pressure. The solubility values were reported for one atmosphere gas pressure assuming Henry's law is obeyed.

**SOURCE AND PURITY OF MATERIALS:**

(1) Ethane. Phillips Petroleum Co. Research grade. Contained 0.2 per cent air and 0.1 per cent unidentified impurity.

(2) Nonane. Fluka. "purum" grade. Fractionated; distillation range 0.08 K.

**ESTIMATED ERROR:**

$\delta T$/K = 0.05

$\delta x_1/x_1$ = 0.015

**REFERENCES:**

1. Lannung, A. *J. Am. Chem. Soc.* 1930, *52*, 68.

2. Gjaldbaek, J. C. *Acta Chem. Scand.* 1952, *6*, 623.

**COMPONENTS:**

(1) Ethane; $C_2H_6$; [74-84-0]

(2) Nonane; $C_9H_{20}$; [111-84-2]
or
Decane; $C_{10}H_{22}$; [124-18-5]

**ORIGINAL MEASUREMENTS:**

Jadot, R.

*J. Chim. Phys.* 1972, *69*, 1036-40.

**VARIABLES:**

$T$/K: 298.15

**PREPARED BY:**

C.L. Young

**EXPERIMENTAL VALUES:**

| $T$/K | Henry's Law Constant, $H$/atm | Mole fraction$^{+}$ at partial pressure of 101.3kPa, $x_{C_2H_6}$ | #$\Delta H^{\infty}$ /cal mol$^{-1}$ (/J mol$^{-1}$) |
|---|---|---|---|
| | | Nonane; $C_9H_{20}$; [111-84-2] | |
| 298.15 | 28.30 | 0.03534 | 325(1360) |
| | | Decane; $C_{10}H_{22}$; [124-18-5] | |
| 298.15 | 27.90 | 0.03584 | 360(1506) |

\+ Calculated by compiler assuming $x_{C_2H_6} = 1/H$

\# Excess partial molar enthalpy of solution at infinite dilution.

**AUXILIARY INFORMATION**

**METHOD/APPARATUS/PROCEDURE:**

The conventional gas chromatographic technique was used. The carrier gas was helium. The value of Henry's law constant was calculated from the retention time. The value applies to very low partial pressures of gas and there may be a substantial difference from that measured at 1 atm. pressure. There is also considerable uncertainty in the value of Henry's constant since no allowance was made for surface adsorption.

**SOURCE AND PURITY OF MATERIALS:**

No details given

**ESTIMATED ERROR:**

$\delta T$/K = ±0.05; $\delta H$ = ±2%

**REFERENCES:**

**COMPONENTS:**

(1) Ethane; $C_2H_6$; [74-84-0]

(2) Decane; $C_{10}H_{22}$; [124-18-5]

**ORIGINAL MEASUREMENTS:**

Monfort, J. P.; Arriaga, J. L.

*Chem. Eng. Commun.*

1980, *7*, 17-25.

**VARIABLES:**

$T$/K: 278.15-323.15

**PREPARED BY:**

C. L. Young

**EXPERIMENTAL VALUES:**

| $T$/K | Henry's Law Constant, $H$/atm | Mole fraction* of ethane at 101.3 kPa partial pressure $x_{C_2H_6}$ |
|---|---|---|
| 278.15 | 19.30 | 0.05181 (0.0509)† |
| 293.15 | 26.61 | 0.03758 (0.0385) |
| 303.15 | 30.98 | 0.03228 (0.0324) |
| 323.15 | 41.57 | 0.02406 (0.0238) |

* Calculated by compiler assuming a linear relationship between $p_{C_2H_6}$ and $x_{C_2H_6}$ (i.e., $x_{C_2H_6}$(1 atm) = $1/H_{C_2H_6}$).

† From equation of smoothed data between 278.15 and 323.15 K:

$\ln x_1 = 1521.2/T - 8.4462$

Correlation coefficient = 0.9983

**AUXILIARY INFORMATION**

**METHOD/APPARATUS/PROCEDURE:**

Chromatographic determination with exponential dilutor. Solvent saturated with gas. A stripping gas was slowly passed through the solution and the concentration of dissolved gas measured using gas chromatography. Details in source.

**SOURCE AND PURITY OF MATERIALS:**

1. Matheson sample, purity 99 per cent by mass.

2. Merck spectroscopic grade.

**ESTIMATED ERROR:**

$\delta T$/K = ±0.03; $\delta H$ = ±3%.

**REFERENCES:**

COMPONENTS:

(1) Ethane; $C_2H_6$; [74-84-0]

(2) Dodecane; $C_{12}H_{26}$; [112-40-3]

ORIGINAL MEASUREMENTS:

Monfort, J. P.; Arriaga, J. L.
*Chem. Eng. Commun.*
1980, *7*, 17-25.

VARIABLES:

$T$/K: 278.15-323.15

PREPARED BY:

C. L. Young

EXPERIMENTAL VALUES:

| $T$/K | Henry's Law Constant, $H$/atm | Mole fraction* of ethane at 101.3 kPa partial pressure $x_{C_2H_6}$ |
|---|---|---|
| 278.15 | 18.45 | 0.05420 (0.0542)† |
| 298.15 | 25.89 | 0.03862 (0.0389) |
| 323.15 | 36.46 | 0.02743 (0.0273) |

* Calculated by compiler assuming a linear relationship between $p_{C_2H_6}$ and $x_{C_2H_6}$ (i.e., $x_{C_2H_6}(1\ atm) = 1/H_{C_2H_6}$).

† From equation of smoothed data between 278.15 and 323.15 K:

$\ln x_1 = 1359.8/T - 7.8076$

Correlation coefficient = 0.998

AUXILIARY INFORMATION

METHOD/APPARATUS/PROCEDURE:

Chromatographic determination with exponential dilutor. Solvent saturated with gas. A stripping gas was slowly passed through the solution and the concentration of dissolved gas measured using gas chromatography. Details in source.

SOURCE AND PURITY OF MATERIALS:

1. Matheson sample, purity 99 per cent by mass.

2. Merck spectroscopic grade.

ESTIMATED ERROR:

$\delta T$/K = ±0.03; $\delta H$ = ±3%.

REFERENCES:

**COMPONENTS:**

(1) Ethane; $C_2H_6$; [74-84-0]

(2) Dodecane; $C_{12}H_{26}$; [112-40-3]

**ORIGINAL MEASUREMENTS:**

Hayduk, W.; Cheng, S.C.

*Can. J. Chem. Eng.* 1970, *48*, 93-99.

**VARIABLES:**

$T$/K: 298.15
$P$/kPa: 101.325

**PREPARED BY:**

W. Hayduk

**EXPERIMENTAL VALUES:**

| $T$/K | Ostwald Coefficient[1] $L/cm^3cm^{-3}$ | Bunsen Coefficient[2] $\alpha/cm^3(STP)cm^{-3}atm^{-1}$ | Mole fraction[2] $10^4x_1$ |
|---|---|---|---|
| 298.15 | 3.827 | 3.506 | 345 |

[1] Original data.

[2] Bunsen coefficient and mole fraction calculated by compiler assuming ideal gas behavior.

**AUXILIARY INFORMATION**

**METHOD/APPARATUS/PROCEDURE:**

Volumetric method using glass apparatus. Degassed solvent contacted the gas while flowing in a thin film through an absorption spiral into a solution buret. Dry gas was maintained at atmospheric pressure in a gas buret by raising the mercury level. Volumes of solution and residual gas were read at regular intervals.

For some experiments the solvent flow was controlled with a stopcock; for the rest, a calibrated syringe pump was used.

Degassing was accomplished using a two-stage vacuum process described by Clever et al. (1).

**SOURCE AND PURITY OF MATERIALS:**

1. Matheson C.P. grade. Purity 99.5 mole per cent minimum.

2. Canadian Laboratory Supplies. Chromatoquality grade. Purity 99.0 mole per cent minimum.

**ESTIMATED ERROR:**

$\delta T$/K = 0.05
$\delta P$/kPa = 0.05
$\delta x_1/x_1$ = 0.01

**REFERENCES:**

1. Clever, H.L.; Battino, R.; Saylor, J.H.; Gross, P.M. *J. Phys. Chem.*, 1957, *61*, 1078.

**COMPONENTS:**

(1) Ethane; $C_2H_6$; [74-84-0]

(2) Hexadecane; $C_{16}H_{34}$; [544-76-3]

**ORIGINAL MEASUREMENTS:**

Hayduk, W.; Cheng, S.C.

*Can. J. Chem. Eng.* 1970, *48*, 93-99.

**VARIABLES:**

$T$/K: 298.15
$P$/kPa: 101.325

**PREPARED BY:**

W. Hayduk

**EXPERIMENTAL VALUES:**

| $T$/K | Ostwald Coefficient[1] $L/cm^3cm^{-3}$ | Bunsen Coefficient[2] $\alpha/cm^3(STP)cm^{-3}atm^{-1}$ | Mole fraction[2] $10^4x_1$ |
|---|---|---|---|
| 298.15 | 3.223 | 2.953 | 373 |

[1] Original data.

[2] Bunsen coefficient and mole fraction calculated by compiler assuming ideal gas behavior.

**AUXILIARY INFORMATION**

**METHOD/APPARATUS/PROCEDURE:**

Volumetric method using glass apparatus. Degassed solvent contacted the gas while flowing in a thin film through an absorption spiral into a solution buret. Dry gas was maintained at atmospheric pressure in a gas buret by raising the mercury level. Volumes of solution and residual gas were read at regular intervals.

For some experiments the solvent flow was controlled with a stopcock; for the rest, a calibrated syringe pump was used.

Degassing was accomplished using a two-stage vacuum process described by Clever et al. (1).

**SOURCE AND PURITY OF MATERIALS:**

1. Matheson C.P. grade. Purity 99.5 mole per cent minimum.

2. Canadian Laboratory Supplies. Olefin-free. Purity 99.0 mole per cent minimum.

**ESTIMATED ERROR:**

$\delta T$/K = 0.05
$\delta P$/kPa = 0.05
$\delta x_1/x_1$ = 0.01

**REFERENCES:**

1. Clever, H.L.; Battino, R.; Saylor, J.H.; Gross, P.M. *J. Phys. Chem.*, 1957, *61*, 1078.

**COMPONENTS:**

(1) Ethane; $C_2H_6$; [74-84-0]

(2) Hexadecane; $C_{16}H_{34}$; [544-76-3] or Heptadecane; $C_{17}H_{36}$; [629-78-7]

**ORIGINAL MEASUREMENTS:**

Lenoir, J-Y; Renault, P.; Renon, H.

*J. Chem. Eng. Data* 1971, *16*, 340-2

**VARIABLES:**

$T$/K: 298.15, 323.15

**PREPARED BY:**

C.L. Young

**EXPERIMENTAL VALUES:**

| $T$/K | Henry's Constant $H_{C_2H_6}$/atm | Mole fraction at 1 atm* $x_{C_2H_6}$ |
|---|---|---|
| | Hexadecane; $C_{16}H_{34}$; [544-76-3] | |
| 298.15 | 26.4 | 0.0379 |
| | Heptadecane; $C_{17}H_{36}$; [629-78-7] | |
| 323.15 | 35.1 | 0.0285 |

* Calculated by compiler assuming a linear function of $p_{C_2H_6}$ vs $x_{C_2H_6}$, i.e. $x_{C_2H_6}$(1 atm) = $1/H_{C_2H_6}$

**AUXILIARY INFORMATION**

**METHOD/APPARATUS/PROCEDURE:**

A conventional gas-liquid chromatographic unit fitted with a thermal conductivity detector was used. The carrier gas was helium. The value of Henry's law constant was calculated from the retention time. The value applies to very low partial pressures of gas and there may be a substantial difference from that measured at 1 atm. pressure. There is also considerable uncertainty in the value of Henry's constant since surface adsorption was not allowed for although its possible existence was noted.

**SOURCE AND PURITY OF MATERIALS:**

(1) L'Air liquide sample, minimum purity 99.9 mole per cent.

(2) Touzart and Matignon or Serlabo sample, purity 99 mole per cent.

**ESTIMATED ERROR:**

$\delta T$/K = ±0.1; $\delta H$/atm = ±6% (estimated by compiler).

**REFERENCES:**

**COMPONENTS:**

(1) Ethane; $C_2H_6$; [74-84-0]

(2) Hexadecane; $C_{16}H_{34}$; [544-76-3]
or
Octadecane; $C_{18}H_{38}$; [593-45-3]

**ORIGINAL MEASUREMENTS:**

Richon, D.; Renon, H.

*J. Chem. Eng. Data* 1980, *25*, 59-60.

**VARIABLES:**

$T$/K: 298.15, 323.15

**PREPARED BY:**

C. L. Young

**EXPERIMENTAL VALUES:**

| $T$/K | Limiting value of Henry's constant, $H^{\infty}$ /atm | Mole fraction of ethane, * $x_{C_2H_6}$ |
|---|---|---|
| | Hexadecane | |
| 298.15 | 27.3 | 0.0366 |
| | Octadecane | |
| 323.15 | 35.0 | 0.0286 |

* Calculated by compiler assuming mole fraction is a linear function of pressure up to 1 atm.

**AUXILIARY INFORMATION**

**METHOD/APPARATUS/PROCEDURE:**

Inert gas stripping plus gas chromatographic method. Details given in ref. (1). Method based on passing constant stream of inert gas through dissolved gas-solvent mixture and periodically injecting mixture into gas chromatograph. Henry's law constant determined from variation of gas peak area with time.

**SOURCE AND PURITY OF MATERIALS:**

1. L'Air Liquide sample, purity 99.9 mole per cent.
2. Hexadecane was a Merck sample, Octadecane was a Fluka sample, both had purities of not less than 99 mole per cent.

**ESTIMATED ERROR:**

$\delta T$/K = ±0.05; $\delta x_{C_2H_6}$ = ±4%
(estimated by compiler).

**REFERENCES:**

1. Leroi, J. C.; Masson, J. C.; Renon, H.; Fabries, J. F.; Sannier, H. *Ind. Eng. Chem. Process. Des. Develop.* 1977, *16*, 139.

**COMPONENTS:**

(1) Ethane; $C_2H_6$; [74-84-0]

(2) Hexadecane; $C_{16}H_{34}$; [544-76-3]

**ORIGINAL MEASUREMENTS:**

Cukor, P.M.; Prausnitz, J.M.

*J. Phys. Chem.* 1972, *76*, 598-601

**VARIABLES:**

$T$/K: 300-475

**PREPARED BY:**

C.L. Young

**EXPERIMENTAL VALUES:**

| $T$/K | Henry's Constant[a] /atm | Mole fraction[b] of ethane in liquid, $/x_{C_2H_6}$ |
|---|---|---|
| 300 | 27.1 | 0.0369 (0.0358)[c] |
| 325 | 36.8 | 0.0272 (0.0269) |
| 350 | 48.0 | 0.0208 (0.0211) |
| 375 | 60.2 | 0.0166 (0.0171) |
| 400 | 72.8 | 0.0137 (0.0142) |
| 425 | 84.6 | 0.0118 (0.0121) |
| 450 | 94.6 | 0.0105 (0.0105) |
| 475 | 102.2 | 0.00978 (0.00919) |

[a] Quoted in supplemenatry material for original paper.

[b] Calculated by compiler for a partial pressure of 1 atmosphere.

[c] From equation of smoothed data:

$$\ln x_1 = 1105.9/T - 7.0174$$

Correlation coefficient = 0.9974

**AUXILIARY INFORMATION**

**METHOD/APPARATUS/PROCEDURE:**

Volumetric apparatus similar to that described by Dymond and Hildebrand (1). Pressure measured with a null detector and precision gauge. Details in ref. (2).

**SOURCE AND PURITY OF MATERIALS:**

No details given.

**ESTIMATED ERROR:**

$\delta T/K = \pm 0.05$; $\delta x_{C_2H_6} = \pm 2\%$

**REFERENCES:**

1. Dymond, J.; Hildebrand, J.H. *Ind. Eng. Chem. Fundam.* 1967, *6*, 130.

2. Cukor, P.M.; Prausnitz, J.M. *Ind.Eng.Chem.Fundam.* 1971, *10*, 638.

COMPONENTS:

(1) Ethane; $C_2H_6$; [74-84-0]
(2) Octadecane; $C_{18}H_{38}$; [593-45-3]

ORIGINAL MEASUREMENTS:

Ng. S.; Harris, H.G.; Prausnitz, J.M.

*J. Chem. Eng. Data* 1969, *14*, 482-3.

VARIABLES:

$T$/K: 308.2-423.2

PREPARED BY:

C.L. Young

EXPERIMENTAL VALUES:

| $T$/K | Henry's Constant, $H$ / atm | Mole fraction of ethane in liquid, $x_{C_2H_6}$ + |
|---|---|---|
| 308.2 | 27.3 | 0.0366 (0.0361)* |
| 323.2 | 33.5 | 0.0299 (0.0296) |
| 343.2 | 43.4 | 0.0230 (0.0234) |
| 363.2 | 54.0 | 0.0185 (0.0190) |
| 373.2 | 58.1 | 0.0172 (0.0172) |
| 423.2 | 86.4 | 0.0116 (0.0114) |

+ at 1 atmosphere partial pressure, calculated by compiler assuming mole fraction equals $1/H$.

* from equation of smoothed data for temperatures between 308.9 and 423.2 K:

$$\ln x_1 = 1308.9/T - 7.5689$$

Correlation coefficient = 0.9991

AUXILIARY INFORMATION

METHOD/APPARATUS/PROCEDURE:

Gas chromatographic method. Solvent supported on Chromosorb P in 6 m column. Gas injected as sample, helium used as carrier gas. Henry's law constant calculated from knowledge of retention time and flow rate

SOURCE AND PURITY OF MATERIALS:

1. Matheson sample, purity greater than 99 mole per cent.

2. Matheson, Coleman and Bell sample, m.pt. 27-28.5°C.

ESTIMATED ERROR:

$\delta T$/K = ±0.1; $\delta H$/atm = ±5%

REFERENCES:

COMPONENTS:

(1) Ethane; $C_2H_6$; [74-84-0]

(2) Eicosane; $C_{20}H_{42}$; [112-95-8]

ORIGINAL MEASUREMENTS:

Ng. S.; Harris, H.G.; Prausnitz, J.M.

*J. Chem. Eng. Data* 1969, *14*, 482-3.

VARIABLES:

$T$/K: 323.2-413.2

PREPARED BY:

C. L. Young

EXPERIMENTAL VALUES:

| $T$/K | Henry's Constant, $H$ /atm | Mole fraction* of ethane in liquid, $x_{C_2H_6}$ |
|---|---|---|
| 323.2 | 33.1 | 0.0302 (0.0301)† |
| 343.2 | 42.2 | 0.0237 (0.0238) |
| 373.2 | 56.9 | 0.0176 (0.0175) |
| 393.2 | 68.3 | 0.0146 (0.0147) |
| 413.2 | 79.8 | 0.0125 (0.0125) |

† From the equation of smoothed data between 323.2 and 413.2 K :

$$\ln x_1 = 1308.2/T - 7.5498$$

Correlation coefficient = 0.9999

* At 1 atmosphere partial pressure, calculated by compiler assuming mole fraction equals $1/H$

AUXILIARY INFORMATION

METHOD/APPARATUS/PROCEDURE:

Gas chromatographic method. Solvent supported on Chromosorb P in 6 m column. Gas injected as sample, helium used as carrier gas. Henry's law constant calculated from knowledge of retention time and flow rate.

SOURCE AND PURITY OF MATERIALS:

1. Matheson sample, purity greater than 99 mole per cent.

2. Matheson, Coleman and Bell sample, m.pt. 35-36.5°C.

ESTIMATED ERROR:

$\delta T$/K = $\pm 0.1$; $\delta H$/atm = $\pm 5\%$

REFERENCES:

**COMPONENTS:**

(1) Ethane; $C_2H_6$; [74-84-0]

(2) Eicosane; $C_{20}H_{42}$; [112-95-8]

**ORIGINAL MEASUREMENTS:**

Chappelow, C.C.; Prausnitz, J.M.

*Am. Inst. Chem. Engnrs. J.* 1974, *20*, 1097-1104.

**VARIABLES:**

$T$/K: 325-475

**PREPARED BY:**

C.L. Young

**EXPERIMENTAL VALUES:**

| $T$/K | Henry's Constant[a] /atm. | Mole fraction[b] of ethane at 1 atm partial pressure, $x_{C_2H_6}$ |
|---|---|---|
| 325 | 35.7 | 0.0280 (0.0273)[c] |
| 350 | 47.3 | 0.0211 (0.0213) |
| 375 | 59.1 | 0.0169 (0.0172) |
| 400 | 71.0 | 0.0141 (0.0143) |
| 425 | 83.1 | 0.0120 (0.0121) |
| 450 | 95.1 | 0.0105 (0.0105) |
| 475 | 107 | 0.00935 (0.00916) |

[a] Authors stated measurements were made at several pressures and values of solubility used were all within the Henry's Law region.

[b] Calculated by compiler assuming linear relationship between mole fraction and pressure.

[c] From: $\ln x_1 = 1123.1/T - 7.0567$

AUXILIARY INFORMATION

**METHOD/APPARATUS/PROCEDURE:**

Volumetric apparatus similar to that described by Dymond and Hildebrand (1). Pressure measured with a null detector and precision gauge. Details in ref. (2).

**SOURCE AND PURITY OF MATERIALS:**

Solvent degassed, no other details given.

**ESTIMATED ERROR:**

$\delta T/\mathrm{K} = \pm 0.1$; $\delta x_{C_2H_6} = \pm 1\%$

**REFERENCES:**

1. Dymond, J.; Hildebrand, J.H. *Ind.Eng.Chem.Fundam.* 1967, *6*, 130.
2. Cukor, P.M.; Prausnitz, J.M. *Ind.Eng.Chem.Fundam.* 1971, *10*, 638.

COMPONENTS:

(1) Ethane; $C_2H_6$; [74-84-0]

(2) Docosane; $C_{22}H_{46}$; [629-97-0]

ORIGINAL MEASUREMENTS:

Ng. S.; Harris, H.G.; Prausnitz, J.M.

*J. Chem. Eng. Data*, 1969, *14*, 482-3.

VARIABLES:

$T$/K: 333.2-473.2

PREPARED BY:

C. L. Young

EXPERIMENTAL VALUES:

| $T$/K | Henry's Constant, $H$ /atm | Mole fraction* of ethane in liquid, $x_{C_2H_6}$ |
|---|---|---|
| 333.2 | 36.1 | 0.0277 (0.0275)† |
| 383.2 | 60.2 | 0.0166 (0.0167) |
| 408.2 | 75.1 | 0.0133 (0.0136) |
| 433.2 | 86.2 | 0.0116 (0.0114) |
| 453.2 | 98.5 | 0.0102 (0.00998) |
| 473.2 | 115.0 | 0.00870 (0.00886) |

† From equation of smoothed data between 333.2 and 473.2 K:

$$\ln x_1 = 1277.64/T - 7.4264$$

Correlation coefficient = 0.9989

* At 1 atmosphere partial pressure, calculated by compiler assuming mole fraction equals $1/H$

AUXILIARY INFORMATION

METHOD/APPARATUS/PROCEDURE:

Gas chromatographic method. Solvent supported on Chromosorb P in 6 m column. Gas injected as sample, helium used as carrier gas. Henry's law constant calculated from knowledge of retention time and flow rate.

SOURCE AND PURITY OF MATERIALS:

1. Matheson sample, purity greater than 99 mole per cent.

2. Matheson, Coleman and Bell sample, m.pt. 43-45°C.

ESTIMATED ERROR:

$\delta T$/K = $\pm$0.1; $\delta H$/atm = $\pm$5%

REFERENCES:

**COMPONENTS:**

(1) Ethane; $C_2H_6$; [74-84-0]

(2) 2,2'-Dimethylbutane (Neo-hexane); $C_6H_{14}$; [75-83-2]

**ORIGINAL MEASUREMENTS:**

Tilquin, B.; Decannière, L.; Fontaine, R.; Claes, P.

*Ann. Soc. Sc. Bruxelles (Belgium)* 1967, *81*, 191-199.

**VARIABLES:**

$T$/K: 288.15
$P$/kPa: 2.05 - 2.11

**PREPARED BY:**

W. Hayduk

**EXPERIMENTAL VALUES:**

| $t$/°C | $T$/K | Ostwald coefficent[1] $L$ | Mole fraction[2] / $x_1$ | Henry's constant[2] $H$/atm |
|---|---|---|---|---|
| 15.0 | 288.15 | 14.61 | 0.07530 | 13.28 |

[1] Original data at low pressure reported as distribution coefficient; but if Henry's law and ideal gas law apply, distribution coefficient is equivalent to Ostwald coefficient as shown here.

[2] Calculated by compiler for a gas partial pressure of 101.325 kPa assuming that Henry's law and ideal gas law apply.

**AUXILIARY INFORMATION**

**METHOD/APPARATUS/PROCEDURE:**

All glass apparatus used at very low gas partial pressures, containing a replaceable degassed solvent ampule equipped with a breakable point which could be broken by means of a magnetically activated plunger. Quantity of gas fed into system determined by measuring the pressure change in a known volume. Quantity of liquid measured by weight. Pressure change observed after solvent released. Experimental details described by Rzad and Claes[1].

**SOURCE AND PURITY OF MATERIALS:**

1. Source not given; minimum purity specified as 99.0 mole per cent.

2. Fluka pure grade; minimum purity specified as 99.0 mole per cent.

**ESTIMATED ERROR:**

$T$/K = 0.05
$\delta x_1/x_1$ = 0.01
(estimated by compiler)

**REFERENCES:**

1. Rzad, S.; Claes, P.

*Bull. Soc. Chim. Belges*, 1964, *73*, 689.

**COMPONENTS:**

(1) Ethane; $C_2H_6$; [74-84-0]

(2) 2,2,4-Trimethylpentane or isooctane; $C_8H_{18}$; [540-84-1]

**ORIGINAL MEASUREMENTS:**

Kobatake, Y.; Hildebrand, J. H.

*J. Phys. Chem.* 1961, *65*, 331 - 335.

**VARIABLES:**

$T$/K: 287.16 - 304.95
$p_1$/kPa: 101.325 (1 atm)

**PREPARED BY:**

M. E. Derrick
H. L. Clever

**EXPERIMENTAL VALUES:**

| Temperature $t/^{\circ}C$ | $T$/K | Mol Fraction $10^2x_1$ | Bunsen Coefficient $\alpha/cm^3(STP)\ cm^{-3}\ atm^{-1}$ | Ostwald Coefficient $L/cm^3\ cm^{-3}$ |
|---|---|---|---|---|
| 14.01 | 287.16 | 3.5327 | 5.00 | 5.26 |
| 20.50 | 293.65 | 3.1934 | 4.48 | 4.82 |
| 25.00 | 298.15 | 2.938 [1] | 4.08 | 4.45 |
| 31.80 | 304.95 | 2.6034 | 3.58 | 4.00 |

[1] The value is enclosed in ( ) in the original paper, it may be a value that was smoothed by the authors.

The Bunsen and Ostwald coefficients were calculated by the compiler assuming that the gas was ideal.

Smoothed Data: For use between 287.16 and 304.95 K.

$$\ln x_1 = -8.5920 + 15.0913/(T/100\ K)$$

The standard error about the regression line is $2.78 \times 10^{-4}$

| $T$/K | Mol Fraction $10^2x_1$ |
|---|---|
| 288.15 | 3.492 |
| 293.15 | 3.194 |
| 298.15 | 2.929 |
| 303.15 | 2.695 |

AUXILIARY INFORMATION

**METHOD/APPARATUS/PROCEDURE:**

The apparatus consists of a gas measuring buret, an absorption pipet, and a reservoir for the solvent. The buret is thermostated at 25°C, the pipet at any temperature from 5 to 30°C. The pipet contains an iron bar in glass for magnetic stirring. The pure solvent is degassed by freezing with liquid nitrogen, evacuating, then boiling with a heat lamp. The degassing process is repeated three times. The solvent is flowed into the pipet where it is again boiled for final degassing. Manipulation of the apparatus is such that the solvent never comes in contact with stopcock grease. The liquid in the pipet is sealed off by mercury. Its volume is the difference between the capacity of the pipet and the volume of mercury that confines it. Gas is admitted into the pipet. Its exact amount is determined by P-V measurements in the buret before and after introduction of the gas into the pipet. The stirrer is set in motion. Equilibrium is attained within 24 hours.

**SOURCE AND PURITY OF MATERIALS:**

(1) Ethane. Matheson Co., Inc. Research grade. Dried by passage over $P_2O_5$ followed by multiple trap vaporization and evacuation at liquid $N_2$ temperature.

(2) Isooctane. Phillips Petroleum Co. Pure grade. Dried over $Mg(ClO_4)_2$ and fractionated through a 15 plate column at a reflux ratio of 20:1. B.p. t/°C 99.1.

**ESTIMATED ERROR:**

$\delta T/K = 0.02$
$\delta x_1/x_1 = 0.003$

**REFERENCES:**

COMPONENTS:

(1) Ethane; $C_2H_6$; [74-84-0]

(2) 2,2,4,4,6,8,8-Heptamethyl nonane; $C_{16}H_{34}$; [4390-04-9]

ORIGINAL MEASUREMENTS:

Richon, D.; Renon, H.

*J. Chem. Eng. Data* 1980, *25*, 59-60.

VARIABLES:

$T$/K: 298.15

PREPARED BY:

C. L. Young

EXPERIMENTAL VALUES:

| $T$/K | Limiting value of Henry's constant, $H^{\infty}$/atm | Mole fraction of ethane, * $x_{C_2H_6}$ |
|---|---|---|
| 298.15 | 6.71 | 0.149 |

* Calculated by compiler assuming mole fraction is a linear function of pressure up to 1 atm.

AUXILIARY INFORMATION

METHOD/APPARATUS/PROCEDURE:

Inert gas stripping plus gas chromatographic method. Details given in ref. (1). Method based on passing constant stream of inert gas through dissolved gas-solvent mixture and periodically injecting mixture into gas chromatograph. Henry's law constant determined from variation of gas peak area with time.

SOURCE AND PURITY OF MATERIALS:

1. L'Air Liquide sample, purity 99.9 mole per cent.
2. Sigma sample, purity not less than 99 mole per cent.

ESTIMATED ERROR:

$\delta T$/K = ±0.05; $\delta x_{C_2H_6}$ = ±4% (estimated by compiler).

REFERENCES:

1. Leroi, J. C.; Masson, J. C.; Renon, H.; Fabries, J. F.; Sannier, H. *Ind. Eng. Chem. Process. Des. Develop.* 1977, *16*, 139.

**COMPONENTS:**

1. Ethane; $C_2H_6$; [74-84-0]
2. 2,6,10,15,19,23-Hexamethyltetracosane (Squalane); $C_{30}H_{62}$; [111-01-3]

**ORIGINAL MEASUREMENTS:**

Chappelow, C.C.; Prausnitz, J.M.

*Am. Inst. Chem. Engnrs. J.* 1974, *20*, 1097-1104.

**VARIABLES:**

$T$/K: 300-475

**PREPARED BY:**

C.L. Young

**EXPERIMENTAL VALUES:**

| $T$/K | Henry's Constant[a] /atm | Mole fraction[b] of ethane at 1 atm partial pressure, $x_{C_2H_6}$ | |
|---|---|---|---|
| 300 | 22.2 | 0.0450 | (0.0450)[c] |
| 325 | 28.6 | 0.0350 | (0.0351) |
| 350 | 35.2 | 0.0284 | (0.0284) |
| 375 | 42.2 | 0.0237 | (0.0236) |
| 400 | 49.7 | 0.0201 | (0.0200) |
| 425 | 57.8 | 0.0173 | (0.0174) |
| 450 | 65.9 | 0.0152 | (0.0153) |
| 475 | 73.2 | 0.0137 | (0.0137) |

[a] Authors stated measurements were made at several pressures and values of solubility used were all within the Henry's Law region.

[b] Calculated by compiler assuming linear relationship between mole fraction and pressure.

[c] From equation of smoothed data developed by compiler:

$$\ln x_1 = 972.01/T - 6.3402$$

correlation coefficient = 0.9999

**AUXILIARY INFORMATION**

**METHOD/APPARATUS/PROCEDURE:**

Volumetric apparatus similar to that described by Dymond and Hildebrand (1). Pressure measured with a null detector and precision gauge. Details in ref. (2).

**SOURCE AND PURITY OF MATERIALS:**

Solvent degassed, no other details given.

**ESTIMATED ERROR:**

$\delta T$/K = ±0.1; $\delta x_{C_2H_6}$ = ±1%

**REFERENCES:**

1. Dymond, J.; Hildebrand, J.H. *Ind.Chem.Eng.Fundam.* 1967, *6*, 130.
2. Cukor, P.M.; Prausnitz, J.M. *Ind.Chem.Eng.Fundam.* 1971, *10*, 638.

**COMPONENTS:**

(1) Ethane; $C_2H_6$; [74-84-0]

(2) Hexane; $C_6H_{14}$; [110-54-3]

(3) Hexadecane; $C_{16}H_{34}$; [544-76-3]

**ORIGINAL MEASUREMENTS:**

Hayduk, W.; Cheng, S.C.

*Can J. Chem. Eng.* 1970, *48*, 93-99.

**VARIABLES:**

$T$/K: 298.15

$P$/kPa: 101.325

$x_3$/Mol fraction: 0-1

**PREPARED BY:**

W. Hayduk

**EXPERIMENTAL VALUES:**

| Mole fraction[1] Hexadecane in solvent, $x_3$ | Ostwald Coefficient[2] $L$/cm$^3$cm$^{-3}$ | Bunsen Coefficient[3] $\alpha$/cm$^3$(STP)cm$^{-3}$atm$^{-1}$ | Mole fraction[3] $10^4x_1$ |
|---|---|---|---|
| 0.0 | 6.09 | 5.58 | 317 |
| 0.096 | 5.42 | 4.97 | 316 |
| 0.180 | 5.10 | 4.67 | 324 |
| 0.275 | 4.58 | 4.20 | 320 |
| 0.373 | 4.32 | 3.96 | 328 |
| 0.420 | 4.18 | 3.83 | 330 |
| 0.650 | 3.68 | 3.37 | 344 |
| 0.777 | 3.47 | 3.18 | 353 |
| 1.0 | 3.22 | 2.95 | 373 |

[1] $x_3$ reported on a solute-free basis in solutions containing components 2 and 3.

[2] Original data.

[3] Bunsen coefficient and mole fraction solubility calculated by compiler assuming ideal gas behavior and assuming also that there is no volume change of mixing of the solvent components.

AUXILIARY INFORMATION

**METHOD/APPARATUS/PROCEDURE:**

Volumetric method using glass apparatus. Degassed solvent contacted the gas while flowing in a thin film through an absorption spiral into a solution buret. Dry gas was maintained at atmospheric pressure in a gas buret by raising the mercury level. Volumes of solution and residual gas were read at regular intervals.

For some experiments the solvent flow was controlled with a stopcock; for the rest, a calibrated syringe pump was used.

Degassing was accomplished using a two-stage vacuum process described by Clever et al. (1). The two-component solvent solutions were analyzed after degassing.

**SOURCE AND PURITY OF MATERIALS:**

1. Matheson C.P. grade. Purity 99.5 mole per cent minimum.
2. Canadian Laboratory Supplies. Chromatoquality grade. Purity 99.0 mole per cent minimum.
3. Canadian Laboratory Supplies. Olefin-free. Purity 99.0 mole per cent minimum.

**ESTIMATED ERROR:**

$\delta T$/K = 0.05

$\delta P$/kPa = 0.05

$\delta x_1/x_1$ = 0.01

$\delta x_3$ = 0.01

**REFERENCES:**

1. Clever, H.L.; Battino, R.; Saylor, J.H.; Gross, P.M. *J. Phys. Chem.*, 1957, *61*, 1078.

COMPONENTS:

(1) Ethane; $C_2H_6$; [74-84-0]

(2) Paraffin solvents at elevated pressures

EVALUATOR:

Walter Hayduk
Department of Chemical Engineering
University of Ottawa
Ottawa, Canada K1N 9B4

CRITICAL EVALUATION:

Ethane solubilities at elevated pressures are available in propane (1,2), butane (3), 2-methylpropane (4), pentane (5), hexane (6,16), heptane (7), octane (8), decane (9), dodecane (10,11,12) and eicosane (13). Except for propane, hexane and dodecane which have two sources of data, there is only one source of data for each solvent. It is possible to check the consistency of the high pressure solubility data by plotting the logarithm of the solubility versus the logarithm of the pressure. A consistent plot may be obtained if the mole fraction solubility is corrected to a gas partial pressure equivalent to the total pressure using Henry's law when the gas contains a significant quantity of the solvent vapor. Since such a plot is approximately linear for moderate pressures, it is possible to extrapolate it to obtain a solubility at atmospheric pressure. It is also possible to extrapolate the high pressure data to a composition corresponding to pure ethane. At the pure ethane composition one would expect that the pressure would correspond to the vapor pressure of ethane at the particular temperature. Estimates of hypothetical ethane vapor pressures could be made even for temperatures above the critical. The data were tested for consistency by comparison with the solubility at atmospheric pressure and with extrapolated ethane vapor pressures at high pressures. The data of Legret et al. (10)

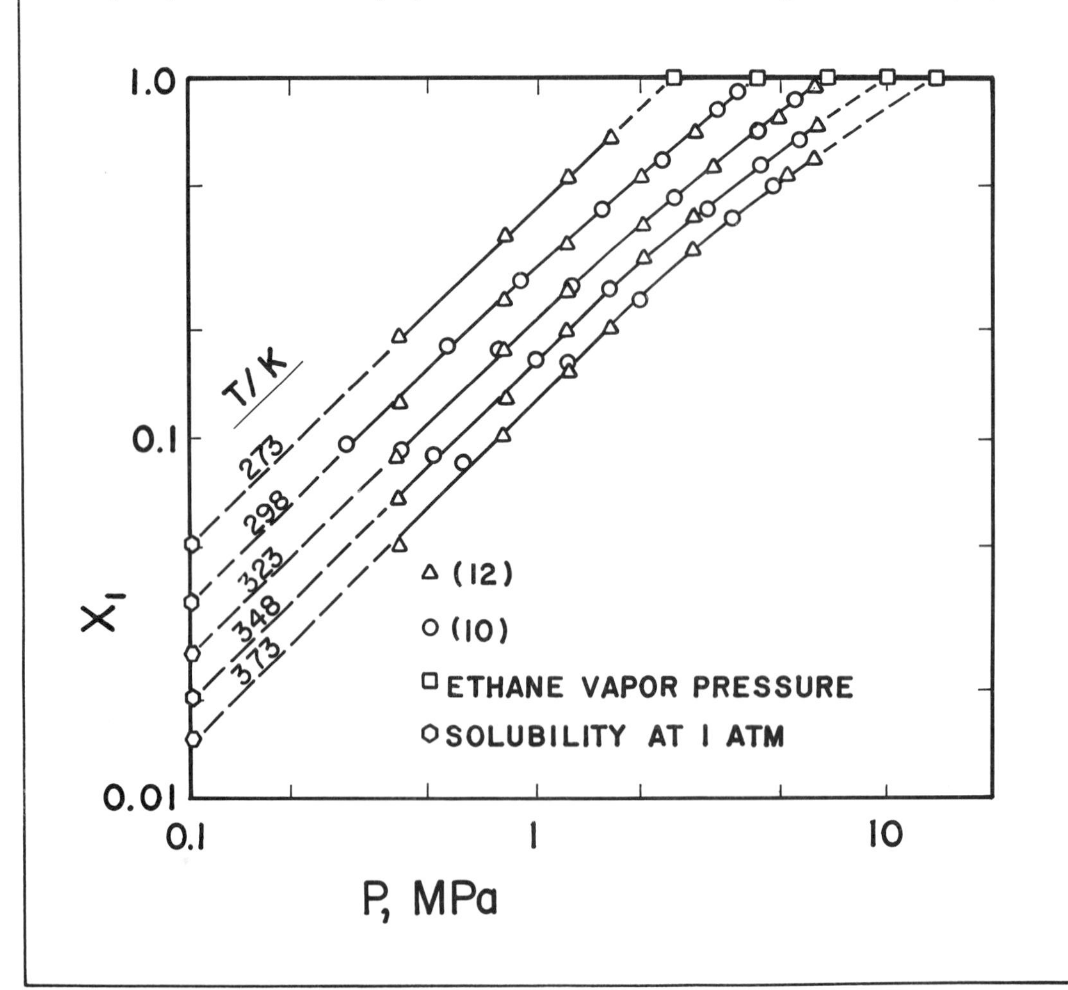

COMPONENTS:

(1) Ethane; $C_2H_6$; [74-84-0]

(2) Paraffin solvents at elevated pressures

EVALUATOR:

Walter Hayduk
Department of Chemical Engineering
University of Ottawa
Ottawa, Canada K1N 9B4

CRITICAL EVALUATION:

...continued

and Lee and Kohn (12) for the solubilities in dodecane are shown in the figure above. The data check one another closely although the more recent data of Legret et al. (10) tend to be several percent higher in some instances for the lower pressure measurements, and are considered more accurate. The data of Meskel-Lesavre et al. (11) and of Legret et al. (10) are from the same laboratory, and check closely with one another when the measurements are reported for the same temperature. The former data are not shown on the figure. All these data are also consistent with both the extrapolated values of solubility at atmospheric pressure (14) and ethane vapor pressures. They are classified as tentative.

The consistency test is not expected to apply at high temperatures when extrapolation for ethane vapor pressure becomes uncertain, or for the highly volatile solvents propane, butane and 2-methylpropane (isobutane).

The two sources of data for propane (1,2) and hexane (6,16) are for different temperature ranges so that a direct comparison of the values are impossible. Both sets of data for propane appear consistent and are classified as tentative. The solubilities in hexane at 298.15 K of Ohgaki et al. (16) appear entirely consistent with the solubility at atmospheric pressure and also with the ethane vapor pressure. On the other hand, the solubilities of Zais and Silberberg (6) satisfy neither of the above tests. Nor do they appear self-consistent. Whereas the former data (16) for solubilities in hexane are classified as tentative, the latter (6) are classified as doubtful. Some doubt is also cast on the accuracy of the solubilities in heptane of Mehra and Thodos (7) which do not appear consistent. They are classified as tentative since no other data are available. It is noted that for comparison, the data in pentane and octane solvents are highly consistent.

All the remaining data are classified as tentative. Solubilities are available in the two-component, mixed solvent solutions of butane and pentane at 338.71 K and elevated pressures. These data, by Herlihy and Thodos (15) are consistent with solubilities in butane (3) and pentane (5) and are classified as tentative.

References

1. Miksovsky, I.; Wichterle, I. *Coll. Czech. Comm.* 1975, *40*, 365-370.

2. Kahre, L.C. *J. Chem. Eng. Data* 1973, *18*, 267-270.

3. Lhotak, V.; Wichterle, I. *Fluid Phase Equilib.* 1981, *6*, 229-235.

4. Besserer, G.J.; Robinson, D.B. *J. Chem. Eng. Data* 1973, *18*, 301-304.

5. Reamer, H.H.; Sate, B.H.; Lacey, W.N. *J. Chem. Eng. Data* 1960, *5*, 44-50.

6. Zais, E.J.; Silberberg, I.H. *J. Chem. Eng. Data* 1970, *15*, 253-256.

COMPONENTS:

(1) Ethane; $C_2H_6$; [74-84-0]

(2) Paraffin solvents at elevated pressures

EVALUATOR:

Walter Hayduk
Department of Chemical Engineering
University of Ottawa
Ottawa, Canada K1N 9B4

CRITICAL EVALUATION:

...continued

7. Mehra, V.S.; Thodos, G. *J. Chem. Eng. Data* 1965, *10*, 211-214.

8. Rodrigues, A.B.J.; McCaffrey, D.S.; Kohn, J.P. *J. Chem. Eng. Data* 1968, *13*, 164-168.

9. Reamer, H.H.; Sage, B.H. *J. Chem. Eng. Data* 1962, *7*, 161-168.

10. Legret, D.; Richon, D.; Renon, H. *Ind. Eng. Chem. Fundam.* 1980, *19*, 122-126.

11. Meskel-Lesavre, M.; Richon, D.; Renon, H. *Ind. Eng. Chem. Fundam.* 1981, *20*, 284-289.

12. Lee, K.H.; Kohn, J.P. *J. Chem. Eng. Data* 1969, *14*, 292-295.

13. Puri, S.; Kohn, J.P. *J. Chem. Eng. Data* 1970, *15*, 372-374.

14. Hayduk, W.; Buckley, W.D.; *Can. J. Chem. Eng.* 1971, *49*, 667-671.

15. Herlihy, J.C.; Thodos, G. *J. Chem. Eng. Data* 1962, *7*, 346-351.

16. Ohgaki, K.; Sano, R.; Katayama, T. *J. Chem. Eng. Data* 1976, *21*, 55-58.

COMPONENTS:

(1) Ethane; $C_2H_6$; [74-84-0]

(2) Propane; $C_3H_8$; [74-98-6]

ORIGINAL MEASUREMENTS:

Kahre, L. C.

*J. Chem. Eng. Data* 1973, *18*, 267-270.

VARIABLES:

$T$/K: 288.76

$P$/MPa: 0.966-2.920

PREPARED BY:

C. L. Young

EXPERIMENTAL VALUES:

| $T$/K | $P$/MPa | Mole fraction of ethane in liquid, $x_{C_2H_6}$ | in gas, $y_{C_2H_6}$ |
|---|---|---|---|
| 288.76 | 0.966 | 0.1078 | 0.2795 |
| | 1.168 | 0.2003 | 0.4376 |
| | 1.422 | 0.3094 | 0.5750 |
| | 1.652 | 0.4037 | 0.6659 |
| | 1.861 | 0.4855 | 0.7290 |
| | 1.979 | 0.5346 | 0.7621 |
| | 2.200 | 0.6209 | 0.8134 |
| | 2.362 | 0.6810 | 0.8463 |
| | 2.682 | 0.7867 | 0.9009 |
| | 2.920 | 0.8635 | 0.9365 |

AUXILIARY INFORMATION

METHOD/APPARATUS/PROCEDURE:

Windowed cell fitted with stirrer which could be used to recirculate cell contents through by-pass. Phases analysed by gas chromatography. Temperature measured with thermocouple and pressure measured with Bourdon gauge. Density data given in source.

SOURCE AND PURITY OF MATERIALS:

1. and 2. Phillips Petroleum research grade sample.

ESTIMATED ERROR:

$\delta T$/K = ±0.06; $\delta P$/MPa = ±0.003; $\delta x_{C_2H_6}$ = ±0.0005 (estimated by compiler).

REFERENCES:

COMPONENTS:

(1) Ethane; $C_2H_6$; [74-84-0]

(2) Propane; $C_3H_8$; [74-98-6]

ORIGINAL MEASUREMENTS:

Miksovsky, I.; Wichterle, I.
*Coll. Czech. Chem. Comm.* 1975, *40*, 365-370.

VARIABLES:

$T$/K: 303.15-369.15

$P$/MPa: 1.159-5.087

PREPARED BY:

C. L. Young

EXPERIMENTAL VALUES:

| $T$/K | $P$/MPa | Mole fraction of ethane in liquid, $x_{C_2H_6}$ | Mole fraction of ethane in vapor, $y_{C_2H_6}$ | $T$/K | $P$/MPa | Mole fraction of ethane in liquid, $x_{C_2H_6}$ | Mole fraction of ethane in vapor, $y_{C_2H_6}$ |
|---|---|---|---|---|---|---|---|
| 303.15 | 1.159 | 0.0245 | 0.0689 | 323.15 | 2.654 | 0.2363 | 0.3877 |
| | 1.226 | 0.0538 | 0.1409 | | 2.768 | 0.2660 | 0.4208 |
| | 1.398 | 0.1244 | 0.2890 | | 3.007 | 0.3217 | 0.4848 |
| | 1.594 | 0.1748 | 0.3848 | | 3.432 | 0.4278 | 0.5870 |
| | 1.989 | 0.3217 | 0.5602 | | 3.756 | 0.5015 | 0.6460 |
| | 2.377 | 0.4392 | 0.6641 | | 4.114 | 0.5653 | 0.6967 |
| | 3.023 | 0.6180 | 0.7837 | | 4.594 | 0.6666 | 0.7576 |
| | 3.379 | 0.7085 | 0.8332 | | 4.790 | 0.7035 | 0.7715 |
| | 3.763 | 0.8074 | 0.8881 | | 4.938 | 0.7332 | 0.7786 |
| | 4.105 | 0.8813 | 0.9280 | | 5.016 | 0.7504 | 0.7823 |
| | 4.242 | 0.9124 | 0.9459 | | 5.059 | 0.7586 | 0.7810 |
| | 4.355 | 0.9374 | 0.9597 | | 5.078 | 0.7621 | 0.7765 |
| | 4.455 | 0.9527 | 0.9686 | | 5.087 | 0.7755 | 0.7755 |
| | 4.546 | 0.97590 | 0.98375 | 343.17 | 2.833 | 0.0475 | 0.0894 |
| | 4.552 | 0.97632 | 0.98377 | | 3.247 | 0.1319 | 0.2109 |
| | 4.620 | 0.99080 | 0.99353 | | 3.726 | 0.2224 | 0.3143 |
| | 4.642 | 0.99473 | 0.99615 | | 4.251 | 0.3269 | 0.4197 |
| | 4.644 | 0.99476 | 0.99621 | | 4.724 | 0.4039 | 0.4807 |
| 323.15 | 1.812 | 0.0271 | 0.0600 | | 4.830 | 0.4208 | 0.4882 |
| | 1.951 | 0.0665 | 0.1385 | | 4.887 | 0.4335 | 0.4897 |
| | 2.003 | 0.0778 | 0.1599 | | 4.935 | 0.4450 | 0.4890 |
| | 2.133 | 0.1097 | 0.2168 | | 4.943 | 0.4475 | 0.4890 |
| | 2.279 | 0.1466 | 0.2661 | | 4.947 | 0.4486 | 0.4889 |
| | 2.608 | 0.2287 | 0.3779 | | | | |

(cont.)

AUXILIARY INFORMATION

METHOD/APPARATUS/PROCEDURE:

Glass windowed cell. Vapor recirculated by magnetic pump. Pressure measured by Bourdon gauge isolated from cell contents by a null membrane pressure transducer. Temperature measured with platinum resistance thermometer. Samples analysed by gas chromatography. Details in ref. (1).

SOURCE AND PURITY OF MATERIALS:

1. Ethane, Fluka research grade; purity 99.97 mole per cent; main impurity acetylene.

2. Propane, Fluka research grade; purity 99.9 mole per cent; main impurity isobutane.

ESTIMATED ERROR:

$\delta T$/K = ±0.01; $\delta P$/MPa = ±0.005;

$\delta x_{C_2H_6}$, $\delta y_{C_2H_6}$ = ±3.0%.

REFERENCES:

1. Miksovsky, I.; Wichterle, I. *Coll. Czech. Chem. Comm.* 1975, *40*, 360.

| COMPONENTS: | ORIGINAL MEASUREMENTS: |
|---|---|
| (1) Ethane; $C_2H_6$; [74-84-0] | Miksovsky, I.; Wichterle, I. |
| (2) Propane; $C_3H_8$; [74-98-6] | *Coll. Czech. Chem. Comm.* 1975, *40*, 365-370. |

EXPERIMENTAL VALUES:

| $T$/K | $P$/MPa | Mole fraction of ethane in liquid, $x_{C_2H_6}$ | Mole fraction of ethane in vapor, $y_{C_2H_6}$ | $T$/K | $P$/MPa | Mole fraction of ethane in liquid, $x_{C_2H_6}$ | Mole fraction of ethane in vapor, $y_{C_2H_6}$ |
|---|---|---|---|---|---|---|---|
| 343.17 | 4.965 | 0.4542 | 0.4874 | 363.16 | 4.534 | 0.14017 | 0.14017 |
| | 4.972 | 0.4567 | 0.4867 | 367.66 | 4.114 | 0.00555 | 0.00756 |
| | 4.973 | 0.4589 | 0.4865 | | 4.186 | 0.01785 | 0.02327 |
| | 4.996 | 0.4783 | 0.4783 | | 4.246 | 0.02790 | 0.03463 |
| 363.16 | 3.840 | 0.01161 | 0.01725 | | 4.319 | 0.04130 | 0.04721 |
| | 3.899 | 0.02137 | 0.03071 | | 4.356 | 0.04924 | 0.05308 |
| | 3.997 | 0.03954 | 0.05433 | | 4.371 | 0.05521 | 0.05521 |
| | 4.102 | 0.05657 | 0.07462 | 369.15 | 4.221 | 0.00473 | 0.00567 |
| | 4.188 | 0.07161 | 0.09123 | | 4.281 | 0.01504 | 0.01690 |
| | 4.368 | 0.10173 | 0.12157 | | 4.296 | 0.01793 | 0.01983 |
| | 4.473 | 0.12241 | 0.13672 | | 4.307 | 0.02111 | 0.02111 |
| | 4.526 | 0.13333 | 0.14048 | | | | |

Vapor-liquid equilibrium data near the critical temperature of ethane is given in original.

**COMPONENTS:**

(1) Ethane; $C_2H_6$; [74-84-0]

(2) Butane; $C_4H_{10}$; [106-97-8]

**ORIGINAL MEASUREMENTS:**

Lhotak, V.; Wichterle, I.

*Fluid Phase Equilib.*

1981, *6*, 229-235.

**VARIABLES:**

$T$/K: 303.15-363.40

$P$/MPa: 0.441-5.326

**PREPARED BY:**

C. L. Young

**EXPERIMENTAL VALUES:**

| $T$/K | $P$/MPa | Mole fraction of ethane in liquid, $x_{C_2H_6}$ | in gas, $y_{C_2H_6}$ |
|---|---|---|---|
| 303.15 | $0.441_4$ | 0.044 | 0.251 |
| | $0.473_0$ | 0.050 | 0.316 |
| | $1.131_5$ | 0.257 | 0.750 |
| | $1.237_6$ | 0.288 | 0.787 |
| | $1.644_2$ | 0.397 | 0.835 |
| | $1.705_2$ | 0.403 | 0.848 |
| | $2.109_0$ | 0.512 | 0.883 |
| | $2.811_3$ | 0.670 | 0.927 |
| | $3.313_6$ | 0.791 | 0.944 |
| | $3.533_1$ | 0.837 | 0.951 |
| | $4.126_7$ | 0.932 | 0.970 |
| 323.15 | $0.691_8$ | 0.048 | 0.222 |
| | $0.893_3$ | 0.097 | 0.405 |
| | $1.076_1$ | 0.133 | 0.525 |
| | $1.081_2$ | 0.132 | 0.533 |
| | $1.224_1$ | 0.164 | 0.576 |
| | $1.533_8$ | 0.232 | 0.670 |
| | $1.730_6$ | 0.269 | 0.704 |
| | $2.212_6$ | 0.358 | 0.777 |
| | $2.713_8$ | 0.460 | 0.816 |
| | $2.831_0$ | 0.486 | 0.830 |
| | $3.424_9$ | 0.600 | 0.861 |

| $T$/K | $P$/MPa | Mole fraction of ethane in liquid, $x_{C_2H_6}$ | in gas, $y_{C_2H_6}$ |
|---|---|---|---|
| 323.15 | $3.792_7$ | 0.651 | 0.873 |
| | $4.587_2$ | 0.763 | 0.898 |
| | $4.863_6$ | 0.833 | 0.912 |
| 343.17 | $1.180_4$ | 0.079 | 0.273 |
| | $1.533_0$ | 0.130 | 0.450 |
| | $2.124_3$ | 0.233 | 0.600 |
| | $2.750_7$ | 0.341 | 0.681 |
| | $3.654_5$ | 0.477 | 0.747 |
| | $4.485_3$ | 0.603 | 0.790 |
| | $4.877_3$ | 0.643 | 0.801 |
| 363.40 | $1.544_9$ | 0.048 | 0.169 |
| | $1.731_9$ | 0.073 | 0.247 |
| | $2.103_8$ | 0.120 | 0.347 |
| | $2.537_1$ | 0.188 | 0.439 |
| | $2.670_4$ | 0.201 | 0.457 |
| | $3.078_4$ | 0.261 | 0.529 |
| | $3.704_4$ | 0.346 | 0.590 |
| | $4.499_6$ | 0.445 | 0.641 |
| | $4.579_1$ | 0.450 | 0.646 |
| | $4.935_1$ | 0.500 | 0.666 |
| | $5.326_5$ | 0.553 | 0.692 |

**AUXILIARY INFORMATION**

**METHOD/APPARATUS/PROCEDURE:**

Static equilibrium cell, made of glass and stainless steel and fitted with a magnetically driven stirer. Details of cell and procedure in source. Temperature measured with a platinum resistance thermometer and pressure with a transducer and dead weight gauge combination. Samples of both phases analysed by gas chromatography.

**SOURCE AND PURITY OF MATERIALS:**

1. Fluka sample, purity 99.97 mole per cent.
2. Fluka sample, purity 99.96 mole per cent, 2-methylpropane and 2,2-dimethyl propane being major impurities.

**ESTIMATED ERROR:**

$\delta T$/K = ±0.01; $\delta P$/MPa = ±0.001;

$\delta x_{C_2H_6}$, $\delta y_{C_2H_6}$ = ±0.001.

**REFERENCES:**

**COMPONENTS:**

(1) Ethane; $C_2H_6$; [74-84-0]

(2) 2-Methylpropane; $C_4H_{10}$; [75-28-5]

**ORIGINAL MEASUREMENTS:**

Besserer, G.J.; Robinson, D.B.

*J. Chem. Eng. Data* 1973, *18*, 301-4.

**VARIABLES:**

$T$/K: 311.3-394.0

$P$/MPa: 1.07-5.37

**PREPARED BY:**

C.L. Young

**EXPERIMENTAL VALUES:**

| $T$/K | $P$/MPa | Mole fraction of ethane in liquid, $x_{C_2H_6}$ | in gas, $y_{C_2H_6}$ |
|---|---|---|---|
| 311.3 | 1.07 | 0.1782 | 0.5524 |
| | 1.43 | 0.2742 | 0.6862 |
| | 1.51 | 0.2951 | - |
| | 2.25 | 0.4841 | 0.8277 |
| | 2.30 | 0.4978 | - |
| | 2.76 | 0.5955 | 0.8639 |
| | 3.10 | 0.6648 | 0.8879 |
| | 3.56 | 0.7536 | 0.9152 |
| | 4.03 | 0.8314 | 0.9267 |
| | 4.03 | 0.8318 | 0.9370 |
| | 4.39 | 0.8858 | 0.9481 |
| | 4.41 | 0.8875 | 0.9524 |
| | 4.58 | 0.9135 | 0.9588 |
| | 4.91 | 0.9541 | 0.9788 |
| | 4.98 | 0.9626 | - |
| 344.5 | 1.32 | 0.0367 | 0.1771 |
| | 1.54 | 0.0867 | 0.2999 |
| | 1.99 | 0.1697 | 0.4513 |
| | 2.32 | 0.2328 | 0.5330 |
| | 2.88 | 0.3285 | 0.6201 |
| | 3.54 | 0.4333 | 0.6962 |
| | 3.92 | 0.4931 | 0.7194 |
| | 4.19 | 0.5366 | 0.7382 |
| | 4.78 | 0.6240 | 0.7766 |

cont...

**AUXILIARY INFORMATION**

**METHOD/APPARATUS/PROCEDURE:**

Cell fitted with two moveable pistons which enabled cell contents to be circulated in external line. Fitted with optical system which allowed measurement of refractive index. Temperature measured with iron-constantan thermocouple and pressure with strain gauge transducer. Components charged into cell mixed by piston movement. Samples withdrawn and analysed by G.C. Details in ref. (1).

**SOURCE AND PURITY OF MATERIALS:**

1. Phillips Petroleum research grade sample, purity better than 99.9 mole per cent.

2. Matheson Co. instrument grade sample purity better than 99.9 mole per cent.

**ESTIMATED ERROR:**

$\delta T$/K = ±0.05; $\delta P/10^5$Pa = ±0.2; $\delta x_{C_2H_6}, \delta y_{C_2H_6}$ = ±0.003

**REFERENCES:**


1. Besserer, G.J.; Robinson, D.B. *Can. J. Chem. Eng.* 1971, *49*, 651

COMPONENTS:

(1) Ethane; $C_2H_6$; [74-84-0]

(2) 2-Methylpropane; $C_4H_{10}$; [75-28-5]

ORIGINAL MEASUREMENTS:

Besserer, G.J.; Robinson, D.B.

*J. Chem. Eng. Data* 1973, *18*, 301-4.

EXPERIMENTAL VALUES:

| $T$/K | $P$/MPa | Mole fraction of ethane in liquid, $x_{C_2H_6}$ | in gas, $y_{C_2H_6}$ |
|---|---|---|---|
| 344.5 | 5.37 | 0.7118 | 0.7792 |
| 377.4 | 2.34 | 0.0226 | 0.0593 |
| | 2.72 | 0.0812 | 0.1817 |
| | 3.04 | 0.1345 | 0.2657 |
| | 3.37 | 0.1638 | 0.3342 |
| | 3.67 | 0.2269 | 0.3854 |
| | 4.15 | 0.2742 | 0.4048 |
| | 4.53 | 0.3421 | 0.4554 |
| | 4.83 | 0.3811 | 0.4694 |
| | 4.96 | 0.4169 | 0.4690 |
| 394.0 | 3.20 | 0.0211 | 0.0431 |
| | 3.40 | 0.0583 | 0.1153 |
| | 3.76 | 0.1183 | 0.2031 |
| | 3.96 | 0.1372 | 0.2082 |
| | 4.19 | 0.1672 | 0.2197 |

COMPONENTS:

(1) Ethane; $C_2H_6$; [74-84-0]

(2) Pentane; $C_5H_{12}$; [109-66-0]

ORIGINAL MEASUREMENTS:

Reamer, H. H.; Sage, B. H.; Lacey, W. N.
*J. Chem. Eng. Data*
1960, *5*, 44-50.

VARIABLES:

$T$/K: 277.59-444.26

$P$/MPa: 0.344-6.83

PREPARED BY:

C. L. Young

EXPERIMENTAL VALUES:

| $T$/K | $P$/MPa | Mole fraction of ethane in liquid, $x_{C_2H_6}$ | Mole fraction of ethane in gas, $y_{C_2H_6}$ | $T$/K | $P$/MPa | Mole fraction of ethane in liquid, $x_{C_2H_6}$ | Mole fraction of ethane in gas, $y_{C_2H_6}$ |
|---|---|---|---|---|---|---|---|
| 277.59 | 0.344 | 0.1432 | 0.9158 | 344.26 | 1.72 | 0.2443 | 0.8100 |
| | 0.689 | 0.2891 | 0.9504 | | 2.07 | 0.2982 | 0.8391 |
| | 1.03 | 0.4316 | 0.9659 | | 2.41 | 0.3500 | 0.8592 |
| | 1.38 | 0.5659 | 0.9763 | | 2.75 | 0.3991 | 0.8722 |
| | 1.72 | 0.6950 | 0.9838 | | 3.10 | 0.4471 | 0.8823 |
| | 2.07 | 0.8141 | 0.9901 | | 3.45 | 0.4940 | 0.8909 |
| | 2.41 | 0.9235 | 0.9960 | | 4.14 | 0.5804 | 0.9032 |
| 310.93 | 0.344 | 0.0624 | 0.6808 | | 4.83 | 0.6579 | 0.9091 |
| | 0.689 | 0.1519 | 0.8448 | | 5.52 | 0.7295 | 0.9100 |
| | 1.03 | 0.2371 | 0.8897 | | 6.21 | 0.8028 | 0.8908 |
| | 1.38 | 0.3201 | 0.9134 | | 6.37 | 0.8502 | 0.8502 |
| | 1.72 | 0.4002 | 0.9284 | 377.59 | 0.689 | 0.0048 | 0.0462 |
| | 2.07 | 0.4774 | 0.9389 | | 1.03 | 0.0506 | 0.3350 |
| | 2.41 | 0.5511 | 0.9472 | | 1.38 | 0.0947 | 0.4820 |
| | 2.75 | 0.6219 | 0.9548 | | 1.72 | 0.1367 | 0.5698 |
| | 3.10 | 0.6879 | 0.9609 | | 2.07 | 0.1796 | 0.6358 |
| | 3.45 | 0.7465 | 0.9673 | | 2.41 | 0.2213 | 0.6837 |
| | 4.14 | 0.8503 | 0.9782 | | 2.75 | 0.2630 | 0.7188 |
| | 4.83 | 0.9274 | 0.9854 | | 3.10 | 0.3032 | 0.7456 |
| | 5.21 | 0.9778 | 0.9778 | | 3.45 | 0.3430 | 0.7661 |
| 344.26 | 0.689 | 0.0755 | 0.5855 | | 4.14 | 0.4188 | 0.7938 |
| | 1.03 | 0.1323 | 0.7018 | | 4.83 | 0.4886 | 0.8002 |
| | 1.38 | 0.1900 | 0.7692 | | | | |

(cont.)

AUXILIARY INFORMATION

METHOD/APPARATUS/PROCEDURE:

Static PVT cell fitted with dead weight pressure balance and platinum resistance thermometer. Bubble point determined from discontinuity in slope of pressure-volume isotherm. Gas phase composition determined by analysis using partial condensation techniques.
Details in source and ref. (1).

SOURCE AND PURITY OF MATERIALS:

1. Phillips Petroleum Co. research grade sample, purity at least 99.94 mole per cent.

2. Phillips Petroleum Co. research grade sample, purity better than 99.84 mole per cent.

ESTIMATED ERROR:

$\delta T$/K = ±0.02; $\delta P$/MPa = ±0.07;
$\delta x_{C_2H_6}$, $\delta y_{C_2H_6}$ = ±0.005.

REFERENCES:

1. Sage, B. H.; Lacey, W. N.
*Trans. Am. Inst. Mining Met. Engnrs.* 1940, *136*, 136.

COMPONENTS:

(1) Ethane; $C_2H_6$; [74-84-0]

(2) Pentane; $C_5H_{12}$; [109-66-0]

ORIGINAL MEASUREMENTS:

Reamer, H. H.; Sage, B. H.; Lacey, W. N.

*J. Chem. Eng. Data* 1960, *5*, 44-50.

EXPERIMENTAL VALUES:

| $T$/K | $P$/MPa | Mole fraction of ethane in liquid, $x_{C_2H_6}$ | Mole fraction of ethane in gas, $y_{C_2H_6}$ | $T$/K | $P$/MPa | Mole fraction of ethane in liquid, $x_{C_2H_6}$ | Mole fraction of ethane in gas, $y_{C_2H_6}$ |
|---|---|---|---|---|---|---|---|
| 377.59 | 5.52 | 0.5567 | 0.8019 | 410.93 | 4.83 | 0.3426 | 0.6107 |
| | 6.21 | 0.6243 | 0.7971 | | 5.52 | 0.3988 | 0.6165 |
| | 6.83 | 0.7189 | 0.7189 | | 6.21 | 0.4702 | 0.6139 |
| 410.93 | 1.38 | 0.0084 | 0.0481 | | 6.58 | 0.5630 | 0.5630 |
| | 1.72 | 0.0436 | 0.2042 | 444.26 | 2.41 | 0.0132 | 0.0385 |
| | 2.07 | 0.0821 | 0.3257 | | 2.75 | 0.0451 | 0.1274 |
| | 2.41 | 0.1172 | 0.4050 | | 3.10 | 0.0762 | 0.2062 |
| | 2.75 | 0.1520 | 0.4612 | | 3.45 | 0.1070 | 0.2698 |
| | 3.10 | 0.1859 | 0.5046 | | 4.14 | 0.1606 | 0.3299 |
| | 3.45 | 0.2197 | 0.5402 | | 4.83 | 0.2216 | 0.3359 |
| | 4.14 | 0.2842 | 0.5874 | | 5.16 | 0.2946 | 0.2946 |

**COMPONENTS:**

(1) Ethane; $C_2H_6$; [74-84-0]

(2) Hexane; $C_6H_{14}$; [110-54-3]

**ORIGINAL MEASUREMENTS:**

Zais, E.J.; Silberberg, I.H.

*J. Chem. Eng. Data* 1970, *15*, 253-256.

**VARIABLES:**

$T$/K: 339 - 450

$P$/MPa: 0.09-7.90

**PREPARED BY:**

W. Hayduk

**EXPERIMENTAL VALUES:**

| $t$/°F | $T$/K | Pressure, pounds per square inch /psia | Pressure[3] $P$/MPa | Equilibrium compositions in mole fraction ethane: Vapor phase / $y_1$ | Liquid phase / $x_1$ |
|---|---|---|---|---|---|
| 150 | 338.71 | 13.6[1] | 0.0934 | 0.0 | 0.0 |
| | | 59.4 | 0.4080 | 0.807 | - |
| | | 127.6 | 0.8765 | 0.902 | - |
| | | 224.0 | 1.539 | 0.941 | 0.297 |
| | | 290.0 | 1.992 | 0.948 | 0.357 |
| | | 394.0 | 2.706 | 0.965 | 0.427 |
| | | 502.7 | 3.453 | 0.965 | 0.517 |
| | | 594.7 | 4.085 | 0.970 | - |
| | | 660.8 | 4.539 | - | 0.640 |
| | | 754.3 | 5.181 | 0.974 | - |
| | | 837.8 | 5.755 | - | 0.762 |
| | | 917.8 | 6.304 | 0.959 | 0.853 |
| | | 926.9 | 6.367 | 0.958 | 0.859 |
| | | 932.2 | 6.403 | 0.964 | 0.907 |
| | | 938[2] | 6.443 | 0.920[2] | 0.920[2] |

1 Vapor pressure of n-hexane.
2 Critical pressure, composition; extrapolated by authors.
3 Pressure calculated by compiler.

(continued)

**AUXILIARY INFORMATION**

**METHOD/APPARATUS/PROCEDURE:**

Stainless steel apparatus with gas recirculation to bottom of cell by magnetic pump. Equilibration for 2 to 4 h followed by phase separation for 3 to 6 h. Sampling at constant pressure by injecting mercury while samples withdrawn. Pressures measured by bourdon gauge. Analysis by gas chromatography.

**SOURCE AND PURITY OF MATERIALS:**

1. Phillips Petroleum research grade with stated minimum purity of 99.98 mole per cent.

2. Phillips Petroleum pure grade with stated minimum purity of 99.0 mole per cent; further purified by solidification under vacuum.

**ESTIMATED ERROR:**

$\delta T$/K: 0.2

$\delta P$/MPa: 0.01 for $P$<7 MPa

$\delta P$/MPa: 0.05 for $P$>7 MPa

$\delta x_1, \delta y_1$: 0.010 (authors)

**REFERENCES:**

**COMPONENTS:**

(1) Ethane; $C_2H_6$; [74-84-0]

(2) Hexane; $C_6H_{14}$; [110-54-3]

**ORIGINAL MEASUREMENTS:**

Zais, E.J.; Silberberg, I.H.

*J. Chem. Eng. Data* 1970, *15*, 253-256.

**VARIABLES:**

$T$/K: 339 - 450
$P$/MPa: 0.41 - 7.88

**PREPARED BY:**

W. Hayduk

**EXPERIMENTAL VALUES:**

| $t$/°F | $T$/K | Pressure, pounds per square inch /psia | Pressure[3] $P$/MPa | Equilibrium composition in mole fraction ethane: Vapor phase / $y_1$ | Liquid phase / $x_1$ |
|---|---|---|---|---|---|
| 250 | 394.26 | 59.7[1] | 0.4116 | 0.0 | 0.0 |
| | | 97.3 | 0.6709 | 0.396 | - |
| | | 194.0 | 1.338 | - | 0.137 |
| | | 203.0 | 1.400 | 0.700 | 0.140 |
| | | 405.5 | 2.796 | 0.799 | 0.275 |
| | | 587.8 | 4.053 | 0.845 | - |
| | | 807.3 | 5.566 | 0.865 | - |
| | | 822.0 | 5.668 | - | 0.505 |
| | | 936.4 | 6.456 | 0.875 | - |
| | | 940.0 | 6.481 | 0.878 | - |
| | | 1015.0 | 6.998 | 0.860 | 0.553 |
| | | 1110.0 | 7.653 | 0.837 | 0.612 |
| | | 1120.0 | 7.722 | - | 0.648 |
| | | 1143.0 | 7.881 | - | 0.700 |
| | | 1146.0[2] | 7.902 | 0.735[2] | 0.735[2] |

[1] Vapor pressure of hexane.
[2] Critical pressure, composition by extrapolation by authors.
[3] Pressure calculated by compiler.

(continued)

**AUXILIARY INFORMATION**

**METHOD/APPARATUS/PROCEDURE:**

Stainless steel apparatus with gas recirculation to bottom of cell by magnetic pump. Equilibration for 2 to 4 h followed by phase separation for 3 to 6 h. Sampling at constant pressure by injecting mercury while samples withdrawn. Pressures measured by bourdon gauge. Analysis by gas chromatography.

**SOURCE AND PURITY OF MATERIALS:**

1. Phillips Petroleum research grade with stated minimum purity of 99.98 mole per cent.

2. Phillips Petroleum pure grade with stated minimum purity of 99.0 mole per cent; further purified by solidification under vacuum.

**ESTIMATED ERROR:**

$\delta T$/K: 0.2
$\delta P$/MPa: 0.01 for $P$<7 MPa
$\delta P$/MPa: 0.05 for $P$>7 MPa
$\delta x_1, \delta y_1$: 0.010 (authors)

**REFERENCES:**

COMPONENTS:

(1) Ethane; $C_2H_6$; [74-84-0]

(2) Hexane; $C_6H_{14}$; [110-54-3]

ORIGINAL MEASUREMENTS:

Zais, E.J.; Silberberg, I.H.

*J. Chem. Eng. Data* 1970, *15*, 253-256.

VARIABLES:

$T$/K: 339 - 450
$P$/MPa: 0.41 - 7.88

PREPARED BY:

W. Hayduk

EXPERIMENTAL VALUES:

| $t$/°F | $T$/K | Pressure, pounds per square inch /psia | Pressure[3] $P$/MPa | Equilibrium compositions in mole fraction ethane: Vapor phase / $y_1$ | Liquid phase / $x_1$ |
|---|---|---|---|---|---|
| 350 | 449.82 | 179.5[1] | 1.238 | 0 | 0 |
| | | 263.6 | 1.818 | - | 0.075 |
| | | 447.2 | 3.083 | 0.515 | 0.150 |
| | | 620.9 | 4.281 | 0.603 | 0.210 |
| | | 722.6 | 4.982 | - | 0.252 |
| | | 825.5 | 5.692 | 0.623 | - |
| | | 930.8 | 6.418 | 0.601 | 0.397 |
| | | 980.0[2] | 6.757 | 0.525[2] | 0.525[2] |

[1] Vapor pressure of hexane.
[2] Critical pressure, composition by extrapolation by authors.
[3] Pressure calculated by compiler.

Smoothed data also listed by authors.

AUXILIARY INFORMATION

METHOD/APPARATUS/PROCEDURE:

Stainless steel apparatus with gas recirculation to bottom of cell by magnetic pump. Equilibration for 2 to 4 h followed by phase separation for 3 to 6 h. Sampling at constant pressure by injecting mercury while samples withdrawn. Pressures measured by bourdon gauge. Analysis by gas chromatography.

SOURCE AND PURITY OF MATERIALS:

1. Phillips Petroleum research grade with stated minimum purity of 99.98 mole per cent.

2. Phillips Petroleum pure grade with stated minimum purity of 99.0 mole per cent; further purified by solidification under vacuum.

ESTIMATED ERROR:

$\delta T$/K: 0.2
$\delta P$/MPa: 0.01 for $P$<7 MPa
$\delta P$/MPa: 0.05 for $P$>7 MPa
$\delta x_1$, $\delta y_1$: 0.010 (authors)

REFERENCES:

**COMPONENTS:**

(1) Ethane; $C_2H_6$; [74-84-0]

(2) Hexane; $C_6H_{14}$; [110-54-3]

**ORIGINAL MEASUREMENTS:**

Ohgaki, K.; Sano, R.; Katayama. T.

*J. Chem. Eng. Data* 1976, *21*, 55-58.

**VARIABLES:**

$T$/K: 298.15

$P/10^5$Pa: 5.08-35.49

**PREPARED BY:**

C.L. Young

**EXPERIMENTAL VALUES:**

| $T$/K | $P/10^5$Pa | Mole fraction of ethane in liquid, $x_{C_2H_6}$ | in vapor, $y_{C_2H_6}$ |
|---|---|---|---|
| 298.15 | 5.078 | 0.1497 | 0.9544 |
| | 9.015 | 0.2698 | 0.9722 |
| | 11.110 | 0.3306 | 0.9776 |
| | 15.192 | 0.4502 | 0.9825 |
| | 19.975 | 0.5833 | 0.9854 |
| | 29.698 | 0.8097 | 0.9869 |
| | 35.494 | 0.9135 | 0.9886 |

AUXILIARY INFORMATION

**METHOD/APPARATUS/PROCEDURE:**

Static equilibrium cell fitted with windows and magnetic stirrer. Temperature of thermostatic liquid measured with platinum resistance thermometer. Pressure measured using dead weight gauge and differential pressure transducer. Samples of vapor and liquid analysed by gas chromatography. Details in source and ref. (1).

**SOURCE AND PURITY OF MATERIALS:**

1. Takachiho Kagakukogyo Co. sample, purity better than 99.7 mole per cent.

2. Merck sample purity about 99.95 mole per cent.

**ESTIMATED ERROR:** $\delta T$/K = ±0.01; $\delta P/10^5$Pa = ±0.01; $\delta x_{C_2H_6}$ (for $x_{C_2H_6} < 0.5$) = ±1%. $\delta(1-x_{C_2H_6})$ (for $x_{C_2H_6} > 0.5$) = ±1% (similarly for vapor composition $y$).

**REFERENCES:**

1. Ohgaki, K.; Katayama, T.; *J. Chem. Eng. Data* 1975, *20*, 264.

COMPONENTS:

(1) Ethane; $C_2H_6$; [74-84-0]

(2) Heptane; $C_7H_{16}$; [142-82-5]

ORIGINAL MEASUREMENTS:

Mehra, V.S.; Thodos, G.

*J. Chem. Eng. Data* 1965, *10*, 211-4.

VARIABLES:

$T$/K: 338.71-449.82

$P/10^5$Pa: 31.4-85.2

PREPARED BY:

C.L. Young

EXPERIMENTAL VALUES:

| $T$/K | $P/10^5$Pa | Mole fraction of ethane in liquid, $x_{C_2H_6}$ | in vapor, $y_{C_2H_6}$ |
|---|---|---|---|
| 338.71 | 31.4 | 0.517 | 0.982 |
| | 39.2 | 0.616 | 0.983 |
| | 46.1 | 0.699 | 0.983 |
| | 54.0 | 0.776 | 0.982 |
| | 61.2 | 0.848 | 0.977 |
| | 65.3 | 0.887 | 0.972 |
| | 66.7 | 0.903 | 0.967 |
| 366.48 | 36.1 | 0.452 | 0.961 |
| | 43.0 | 0.517 | 0.960 |
| | 50.0 | 0.580 | 0.961 |
| | 59.8 | 0.662 | 0.959 |
| | 67.2 | 0.738 | 0.953 |
| | 74.7 | 0.798 | 0.944 |
| | 77.7 | 0.829 | 0.938 |
| 394.26 | 39.8 | 0.410 | 0.919 |
| | 49.5 | 0.476 | 0.925 |
| | 60.3 | 0.563 | 0.923 |
| | 70.3 | 0.631 | 0.917 |
| | 78.7 | 0.700 | 0.904 |
| | 83.8 | 0.738 | 0.888 |
| 422.04 | 40.7 | 0.340 | 0.859 |

cont...

AUXILIARY INFORMATION

METHOD/APPARATUS/PROCEDURE:

Variable volume vapor-liquid equilibrium cell with moveable piston. Details given in ref. Pressure measured with Bourdon gauge. Ethane introduced into cell and then Heptane added. Samples of vapor and liquid with-drawn. frozen in liquid nitrogen and then completely vaporized and analysed by gas chromatography with thermal conductivity detector.

SOURCE AND PURITY OF MATERIALS:

1. Phillips Petroleum Co. sample purity 99.91 mole per cent.

2. Phillips Petroleum Co. sample, purity 99.78 mole per cent.

ESTIMATED ERROR:

$\delta T$/K = ±0.2; $\delta P/10^5$Pa = ±0.2; $\delta x_{C_2H_6}$, $\delta y_{C_2H_6}$ = ±1%.

REFERENCES:

COMPONENTS:

(1) Ethane; $C_2H_6$; [74-84-0]

(2) Heptane; $C_7H_{16}$; [142-82-5]

ORIGINAL MEASUREMENTS:

Mehra, V.S. Thodos, G.

*J. Chem. Eng. Data* 1965, *10*, 211-4.

EXPERIMENTAL VALUES:

| $T$/K | $P/10^5$Pa | Mole fraction of ethane in liquid, $x_{C_2H_6}$ | in vapor, $y_{C_2H_6}$ |
|---|---|---|---|
| 422.04 | 51.4 | 0.425 | 0.869 |
| | 63.4 | 0.512 | 0.862 |
| | 74.5 | 0.593 | 0.859 |
| | 81.5 | 0.658 | 0.840 |
| | 85.2 | 0.690 | 0.821 |
| 449.82 | 40.4 | 0.296 | 0.767 |
| | 49.4 | 0.333 | 0.783 |
| | 59.0 | 0.407 | 0.792 |
| | 68.5 | 0.473 | 0.783 |
| | 76.0 | 0.536 | 0.772 |
| | 79.7 | 0.569 | 0.756 |

COMPONENTS:

(1) Ethane; $C_2H_6$; [74-84-0]

(2) Octane; $C_8H_{18}$; [111-65-9]

ORIGINAL MEASUREMENTS:

Rodrigues, A.B.J.; McCaffrey, D.S.; Kohn, J.P.

*J. Chem. Eng. Data* 1968, *13*, 164-8

VARIABLES:

$T$/K: 273.15-373.15

$P$/MPa: 0.405-5.269

PREPARED BY:

C.L. Young

EXPERIMENTAL VALUES:

| $T$/K | $P$/atm | $P$/MPa | Mole fraction of ethane in liquid, $x_{C_2H_6}$ | in vapor, $y_{C_2H_6}$ |
|---|---|---|---|---|
| 273.15 | 4.0 | 0.405 | 0.178 | - |
| | 8.0 | 0.811 | 0.350 | - |
| | 12.0 | 1.216 | 0.525 | - |
| | 16.0 | 1.621 | 0.697 | - |
| | 20.0 | 2.027 | 0.869 | - |
| | 22.0 | 2.229 | 0.952 | - |
| 298.15 | 4.0 | 0.405 | 0.112 | - |
| | 8.0 | 0.811 | 0.223 | - |
| | 12.0 | 1.216 | 0.334 | - |
| | 16.0 | 1.621 | 0.447 | - |
| | 20.0 | 2.027 | 0.547 | - |
| | 24.0 | 2.432 | 0.643 | - |
| | 28.0 | 2.837 | 0.735 | - |
| | 32.0 | 3.242 | 0.828 | - |
| | 36.0 | 3.648 | 0.922 | - |
| | 40.0 | 4.053 | 0.984 | - |
| 313.15 | 4.0 | 0.405 | 0.093 | 0.9704 |
| | 8.0 | 0.811 | 0.186 | 0.9798 |
| | 12.0 | 1.216 | 0.277 | 0.9868 |
| | 16.0 | 1.621 | 0.363 | 0.9916 |
| | 20.0 | 2.027 | 0.439 | 0.9947 |
| | 24.0 | 2.432 | 0.514 | 0.9967 |
| | 28.0 | 2.837 | 0.583 | 0.9977 |
| | 32.0 | 3.242 | 0.652 | 0.9979 |

cont...

AUXILIARY INFORMATION

METHOD/APPARATUS/PROCEDURE:

Borosilicate glass cell. Temperature measured with Platinum resistance thermometer. Pressure measured on Bourdon gauge. Samples of ethane added to n-octane and equilibrated. Vapor phase composition calculated assuming ideal gas behaviour. Liquid phase composition estimated from known overall composition and volumes of both phases. Details in source and ref. (1).

SOURCE AND PURITY OF MATERIALS:

1. Matheson Co. sample fractionated, purity 99.7 mole %.

2. Phillips Petroleum Co. "pure" grade purity > 99 mole %.

ESTIMATED ERROR:

$\delta T$/K = ±0.07; $\delta P$/MPa = 0.007; $\delta x_{C_2H_6} = \delta y_{C_2H_6} \leqslant 0.003$.

REFERENCES:

1. Kohn, J.P.; and Kurata, F.; *Petrol Process*, 1956, *11*, 57.

COMPONENTS:

(1) Ethane; $C_2H_6$; [74-84-0]

(2) Octane, $C_8H_{18}$; [111-65-9]

ORIGINAL MEASUREMENTS:

Rodrigues, A.B.J.; McCaffrey, D.S. Kohn, J.P.

*J. Chem. Eng. Data* 1968, *13*, 164-8.

EXPERIMENTAL VALUES:

| $T$/K | $P$/atm | $P$/MPa | Mole fraction of ethane in liquid, $x_{C_2H_6}$ | in vapor, $y_{C_2H_6}$ |
|---|---|---|---|---|
| 313.15 | 36.0 | 3.648 | 0.721 | 0.9979 |
| | 40.0 | 4.053 | 0.792 | 0.9981 |
| | 44.0 | 4.458 | 0.859 | 0.9984 |
| | 48.0 | 4.864 | 0.921 | 0.9989 |
| | 52.0 | 5.269 | 0.973 | 0.9994 |
| 323.15 | 4.0 | 0.405 | 0.084 | 0.9464 |
| | 8.0 | 0.811 | 0.162 | 0.9657 |
| | 12.0 | 1.216 | 0.248 | 0.9761 |
| | 16.0 | 1.621 | 0.322 | 0.9827 |
| | 20.0 | 2.027 | 0.392 | 0.9870 |
| | 24.0 | 2.432 | 0.458 | 0.9900 |
| | 28.0 | 2.837 | 0.517 | 0.9918 |
| | 32.0 | 3.242 | 0.577 | 0.9929 |
| | 36.0 | 3.648 | 0.636 | 0.9933 |
| | 40.0 | 4.053 | 0.693 | 0.9934 |
| | 44.0 | 4.458 | 0.749 | 0.9931 |
| | 48.0 | 4.864 | 0.807 | 0.9924 |
| | 52.0 | 5.269 | 0.863 | - |
| 348.15 | 4.0 | 0.405 | 0.057 | 0.9361 |
| | 8.0 | 0.811 | 0.126 | 0.9518 |
| | 12.0 | 1.216 | 0.173 | 0.9636 |
| | 16.0 | 1.621 | 0.231 | 0.9721 |
| | 20.0 | 2.027 | 0.288 | 0.9787 |
| | 24.0 | 2.432 | 0.346 | 0.9823 |
| | 28.0 | 2.837 | 0.399 | 0.9855 |
| | 32.0 | 3.242 | 0.449 | 0.9875 |
| | 36.0 | 3.648 | 0.493 | 0.9882 |
| | 40.0 | 4.053 | 0.537 | 0.9881 |
| | 44.0 | 4.458 | 0.578 | 0.9876 |
| | 48.0 | 4.864 | 0.622 | 0.9863 |
| | 52.0 | 5.269 | 0.663 | - |
| 373.15 | 4.0 | 0.405 | 0.047 | 0.8987 |
| | 8.0 | 0.811 | 0.093 | 0.9146 |
| | 12.0 | 1.216 | 0.139 | 0.9278 |
| | 16.0 | 1.621 | 0.186 | 0.9400 |
| | 20.0 | 2.027 | 0.232 | 0.9507 |
| | 24.0 | 2.432 | 0.278 | 0.9596 |
| | 28.0 | 2.837 | 0.324 | 0.9660 |
| | 32.0 | 3.242 | 0.367 | 0.9702 |
| | 36.0 | 3.648 | 0.405 | 0.9726 |

COMPONENTS:

(1) Ethane; $C_2H_6$; [74-84-0]

(2) Decane; $C_{10}H_{22}$; [124-18-5]

ORIGINAL MEASUREMENTS:

Reamer, H.H.; Sage, C.H.

*J. Chem. Eng. Data* 1962, *7*, 161-8.

VARIABLES:

$T$/K: 277.6-510.9

$P$/MPa: 3.45-118.25

PREPARED BY:

C.L. Young

EXPERIMENTAL VALUES:

| $T$/K | $P/10^5$Pa | Mole fraction of ethane in liquid, $x_{C_2H_6}$ | Mole fraction of ethane in vapor, $y_{C_2H_6}$ |
|---|---|---|---|
| 277.6 | 3.45 | 0.1284 | 0.9992 |
| | 6.89 | 0.2576 | 0.9994 |
| | 10.34 | 0.3874 | 0.9995 |
| | 13.79 | 0.5166 | 0.9996 |
| | 17.24 | 0.6458 | 0.9997 |
| | 20.68 | 0.7769 | 0.9998 |
| | 24.13 | 0.9065 | 0.9999 |
| 310.9 | 6.89 | 0.1565 | 0.9985 |
| | 13.79 | 0.3060 | 0.9988 |
| | 20.68 | 0.4474 | 0.9988 |
| | 27.58 | 0.5801 | 0.9988 |
| | 34.47 | 0.7035 | 0.9988 |
| | 41.37 | 0.8165 | 0.9988 |
| | 48.26 | 0.9190 | 0.9988 |
| | 53.64 | 0.995 | 0.995 |
| 344.3 | 6.89 | 0.1114 | 0.9947 |
| | 13.79 | 0.2164 | 0.9964 |
| | 20.68 | 0.3144 | 0.9968 |
| | 27.58 | 0.4056 | 0.9970 |
| | 34.47 | 0.4897 | 0.9970 |
| | 41.37 | 0.5687 | 0.9970 |
| | 48.26 | 0.6432 | 0.9970 |
| | 55.16 | 0.7127 | 0.9970 |

cont...

AUXILIARY INFORMATION

METHOD/APPARATUS/PROCEDURE:

PVT cell charged with mixture of known composition. Pressure measured with pressure balance and temperature measured using resistance thermometer. Bubble point determined from discontinuity in PV isotherm. Details in source and ref. (1).

SOURCE AND PURITY OF MATERIALS:

1. Phillips Petroleum Co. sample purity 99.9 mole per cent.

2. Phillips Petroleum Co. sample purity 99.35 mole per cent. Major impurities being isomers.

ESTIMATED ERROR:

$\delta T$/K = ±0.018; $\delta P/10^5$Pa = ±0.05% or ±0.06 (whichever is greater) $\delta x_{C_2H_6}$, $\delta y_{C_2H_6}$ = ±0.003.

REFERENCES:

1. Sage, B.H.; Lacey, W.N.; *Trans. Am. Inst. Mining and Met. Engrs.*, 1940, *136*, 136.

COMPONENTS:

(1) Ethane; $C_2H_6$; [74-84-0]

(2) Decane; $C_{10}H_{22}$; [124-18-5]

ORIGINAL MEASUREMENTS

Reamer, H.H.; Sage, C.H.

*J. Chem. Eng. Data* 1962, *7*, 161-8.

EXPERIMENTAL VALUES:

| $T$/K | $P/10^5$Pa | Mole fraction of ethane in liquid, $x_{C_2H_6}$ | in vapor, $y_{C_2H_6}$ |
|---|---|---|---|
| 344.3 | 62.05 | 0.7777 | 0.9970 |
| | 68.95 | 0.8364 | 0.9970 |
| | 75.84 | 0.8933 | 0.9960 |
| | 81.63 | 0.964 | 0.964 |
| 377.6 | 6.89 | 0.0835 | 0.9817 |
| | 13.79 | 0.1648 | 0.9896 |
| | 20.68 | 0.2421 | 0.9919 |
| | 27.58 | 0.3144 | 0.9930 |
| | 34.47 | 0.3821 | 0.9934 |
| | 41.37 | 0.4466 | 0.9936 |
| | 48.26 | 0.5058 | 0.9935 |
| | 55.15 | 0.5609 | 0.9934 |
| | 62.05 | 0.6134 | 0.9930 |
| | 68.95 | 0.6626 | 0.9919 |
| | 75.84 | 0.7098 | 0.9902 |
| | 82.74 | 0.7551 | 0.9877 |
| | 89.63 | 0.7987 | 0.9846 |
| | 96.53 | 0.8445 | 0.9804 |
| | 103.42 | 0.8986 | 0.9589 |
| | 104.73 | 0.927 | 0.927 |
| 410.9 | 6.89 | 0.0675 | 0.9450 |
| | 13.79 | 0.1357 | 0.9702 |
| | 20.68 | 0.2001 | 0.9784 |
| | 27.58 | 0.2604 | 0.9817 |
| | 34.47 | 0.3164 | 0.9833 |
| | 41.37 | 0.3693 | 9.9835 |
| | 48.26 | 0.4188 | 0.9838 |
| | 55.15 | 0.4674 | 0.9835 |
| | 62.05 | 0.5126 | 0.9827 |
| | 68.95 | 0.5567 | 0.9810 |
| | 75.84 | 0.5985 | 0.9785 |
| | 82.74 | 0.6380 | 0.9755 |
| | 89.63 | 0.6765 | 0.9715 |
| | 96.53 | 0.7140 | 0.9660 |
| | 103.42 | 0.7522 | 0.9560 |
| | 110.32 | 0.7987 | 0.9353 |
| | 116.31 | 0.888 | 0.888 |
| 444.3 | 6.89 | 0.0522 | 0.8673 |
| | 13.79 | 0.1107 | 0.9306 |
| | 20.68 | 0.1673 | 0.9498 |
| | 27.58 | 0.2219 | 0.9574 |
| | 34.47 | 0.2738 | 0.9610 |
| | 41.37 | 0.3231 | 0.9619 |
| | 48.26 | 0.3698 | 0.9625 |
| | 55.15 | 0.4133 | 0.9625 |
| | 62.05 | 0.4540 | 0.9620 |
| | 68.95 | 0.4936 | 0.9610 |
| | 75.84 | 0.5309 | 0.9598 |
| | 82.74 | 0.5678 | 0.9579 |
| | 89.63 | 0.6060 | 0.9550 |
| | 96.53 | 0.6452 | 0.9490 |
| | 103.42 | 0.6849 | 0.9390 |
| | 110.32 | 0.7292 | 0.9181 |
| | 117.21 | 0.7902 | 0.8645 |
| | 118.25 | 0.835 | 0.835 |

COMPONENTS:

(1) Ethane; $C_2H_6$; [74-84-0]

(2) Decane; $C_{10}H_{22}$; [124-18-5]

ORIGINAL MEASUREMENTS:

Reamer, H.H.; Sage, C.H.

*J. Chem. Eng. Data.* 1962, *7*, 161-8.

EXPERIMENTAL VALUES:

| $T$/K | $P/10^5$Pa | Mole fraction of ethane in liquid, $x_{C_2H_6}$ | in vapor, $y_{C_2H_6}$ |
|---|---|---|---|
| 477.6 | 6.89 | 0.0358 | 0.6830 |
| | 13.79 | 0.0874 | 0.8362 |
| | 20.68 | 0.1383 | 0.8845 |
| | 27.58 | 0.1879 | 0.9038 |
| | 34.47 | 0.2357 | 0.9097 |
| | 41.37 | 0.2817 | 0.9118 |
| | 48.26 | 0.3277 | 0.9129 |
| | 55.15 | 0.3716 | 0.9138 |
| | 62.05 | 0.4139 | 0.9140 |
| | 68.95 | 0.4529 | 0.9138 |
| | 75.84 | 0.4911 | 0.9130 |
| | 82.74 | 0.5276 | 0.9118 |
| | 89.63 | 0.5679 | 0.9097 |
| | 96.53 | 0.6040 | 0.9032 |
| | 103.42 | 0.6451 | 0.8850 |
| | 110.32 | 0.6967 | 0.8389 |
| | 113.07 | 0.778 | 0.778 |
| 510.9 | 6.89 | 0.0170 | 0.3347 |
| | 13.79 | 0.0645 | 0.6361 |
| | 20.68 | 0.1118 | 0.7356 |
| | 27.58 | 0.1588 | 0.7840 |
| | 34.47 | 0.2053 | 0.8121 |
| | 41.37 | 0.2511 | 0.8280 |
| | 48.26 | 0.2956 | 0.8361 |
| | 55.15 | 0.3387 | 0.8387 |
| | 62.05 | 0.3813 | 0.8389 |
| | 68.95 | 0.4239 | 0.8387 |
| | 75.84 | 0.4648 | 0.8352 |
| | 82.74 | 0.5043 | 0.8281 |
| | 89.63 | 0.5448 | 0.8139 |
| | 96.53 | 0.5942 | 0.7861 |
| | 102.11 | 0.698 | 0.698 |

**COMPONENTS:**

(1) Ethane; $C_2H_6$; [74-84-0]

(2) Dodecane; $C_{12}H_{26}$; [112-40-3]

**ORIGINAL MEASUREMENTS:**

Meskel-Lesavre, M.; Richon, D.; Renon, H.

*Ind. Eng. Chem. Fundam.* 1981, *20*, 284-289.

**VARIABLES:**

$T$/K: 308.15-373.15

$P$/MPa: 0.36-5.84

**PREPARED BY:**

C. L. Young

**EXPERIMENTAL VALUES:**

| $T$/K | $P$/MPa | Mole fraction of ethane $x_{C_2H_6}$ | Molar volume /dm$^3$ | $T$/K | $P$/MPa | Mole fraction of ethane $x_{C_2H_6}$ | Molar volume /dm$^3$ |
|---|---|---|---|---|---|---|---|
| 308.15 | 0.36 | 0.0875 | 0.2167 | 338.15 | 0.49 | 0.0874 | 0.2234 |
| | 0.94 | 0.2365 | 0.1930 | | 1.34 | 0.2365 | 0.1993 |
| | 1.41 | 0.3481 | 0.1748 | | 2.07 | 0.3481 | 0.1804 |
| | 1.60 | 0.3923 | 0.1677 | | 2.36 | 0.3923 | 0.1736 |
| | 2.42 | 0.5470 | 0.1428 | | 3.09 | 0.4752 | 0.1601 |
| | 3.00 | 0.6440 | 0.1274 | | 3.67 | 0.5470 | 0.1479 |
| | 3.67 | 0.7600 | 0.1110 | | 4.65 | 0.6440 | 0.1333 |
| | 4.22 | 0.8470 | 0.0990 | | 5.84 | 0.7600 | 0.1165 |
| 323.15 | 0.42 | 0.0875 | 0.2205 | 373.15 | 0.66 | 0.0874 | 0.2325 |
| | 1.13 | 0.2365 | 0.1962 | | 1.86 | 0.2365 | 0.2077 |
| | 1.73 | 0.3481 | 0.1779 | | 2.92 | 0.3481 | 0.1885 |
| | 1.97 | 0.3923 | 0.1704 | | 3.36 | 0.3923 | 0.1820 |
| | 2.56 | 0.4752 | 0.1569 | | 4.41 | 0.4752 | 0.1680 |
| | 3.03 | 0.5470 | 0.1455 | | 5.27 | 0.5470 | 0.1556 |
| | 3.80 | 0.6440 | 0.1305 | | | | |
| | 4.74 | 0.7600 | 0.1134 | | | | |
| | 5.50 | 0.8470 | 0.1025 | | | | |

**AUXILIARY INFORMATION**

**METHOD/APPARATUS/PROCEDURE:**

Variable volume equilibrium cell. The pressure-volume diagram determined for given temperature and composition. Discontinuity in slope of PV diagram gives bubble point. Temperature measured with thermocouple and pressure measured with transducer and null differential pressure indicator. Details in source.

**SOURCE AND PURITY OF MATERIALS:**

1. Airgaz sample, minimum purity 99.95 per cent (by volume).

2. Fluka sample, purity 99 mole per cent.

**ESTIMATED ERROR:**

$\delta T$/K = ±0.1; $\delta P$/MPa = ±0.01;

$\delta x_{C_2H_6}$ = ±0.1% or less.

**REFERENCES:**

| COMPONENTS: | ORIGINAL MEASUREMENTS: |
|---|---|
| (1) Ethane; $C_2H_6$; [74-84-0]<br>(2) Dodecane; $C_{12}H_{26}$; [112-40-3] | Legret, D.; Richon, D.; Renon, H., *Ind. Eng. Chem. Fundam.* 1980, *19*, 122-126. |
| **VARIABLES:** | **PREPARED BY:** |
| $T$/K: 298.15-373.15<br>$P$/MPa: 0.29-5.61 | C. L. Young |

EXPERIMENTAL VALUES:

| $T$/K | $P$/MPa | Mole fraction of ethane in liquid, $x_{C_2H_6}$ | $T$/K | $P$/MPa | Mole fraction of ethane in liquid, $x_{C_2H_6}$ |
|---|---|---|---|---|---|
| 298.15 | 0.29(5) | 0.097 | 323.15 | 0.78(0) | 0.172 |
| | 0.56(5) | 0.180 | | 1.25(5) | 0.267 |
| | 0.90(0) | 0.275 | | 1.76(5) | 0.357(8) |
| | 1.25(0) | 0.367 | | 2.48(5) | 0.462 |
| | 1.54(0) | 0.432 | | 3.12(5) | 0.552(3) |
| | 1.87(5) | 0.503 | | 3.49(0) | 0.601 |
| | 2.29(0) | 0.593 | | 4.30(5) | 0.704 |
| | 2.53(5) | 0.641 | | 5.52(0) | 0.857(8) |
| | 3.00(0) | 0.736 | 348.15 | 0.51(0) | 0.089 |
| | 3.31(5) | 0.808 | | 0.99(5) | 0.166 |
| | 3.69(0) | 0.888 | | 1.62(0) | 0.260 |
| | 3.81(5) | 0.919 | | 2.33(5) | 0.350 |
| 313.15 | 0.35(0) | 0.094 | | 3.10(5) | 0.431 |
| | 0.68(5) | 0.176 | | 3.93(0) | 0.519 |
| | 1.10(0) | 0.269 | | 4.40(5) | 0.568(9) |
| | 1.52(5) | 0.361 | | 5.61(5) | 0.673 |
| | 2.24(0) | 0.478 | 373.15 | 0.62(0) | 0.086 |
| | 2.77(5) | 0.568 | | 1.22(0) | 0.163 |
| | 3.09(5) | 0.6167 | | 1.99(0) | 0.255 |
| | 3.76(0) | 0.7178 | | 2.88(0) | 0.344 |
| | 4.74(5) | 0.872 | | 3.61(5) | 0.406 |
| | 5.00(0) | 0.912 | | 4.70(5) | 0.495(4) |
| 323.15 | 0.40(5) | 0.091 | | 5.29(5) | 0.543 |

AUXILIARY INFORMATION

METHOD/APPARATUS/PROCEDURE:

Static equilibrium still with magnetic stirrer. Pressure measurements made using a membrane gauge pressure transducer and temperature measured with thermocouple. Calculation of mole fraction involved an iterative technique which assumed the mole fraction of dodecane in vapor is very small. Details in source.

SOURCE AND PURITY OF MATERIALS:

1. AIR GAZ sample, minimum purity 99.95 volume per cent.
2. Fluka sample, minimum purity 99 mole per cent.

ESTIMATED ERROR:

$\delta T$/K = ±0.1; $\delta P$/MPa = ±0.1; $\delta x_{C_2H_6}$ = ±0.009 or less.

REFERENCES:

COMPONENTS:

(1) Ethane; $C_2H_6$; [74-84-0]

(2) Dodecane; $C_{12}H_{26}$; [112-40-3]

ORIGINAL MEASUREMENTS:

Lee, K.H.; Kohn, J.P.

*J. Chem. Eng. Data* 1969, *14*, 292-5

VARIABLES:

$T$/K: 273.15-373.15

$P$/MPa: 0.405-6.28

PREPARED BY:

C.L. Young

EXPERIMENTAL VALUES:

| $T$/K | $P$/atm | $P$/MPa | Mole fraction of ethane in liquid, $x_{C_2H_6}$ |
|---|---|---|---|
| 273.15 | 4.00 | 0.405 | 0.190 |
| | 8.00 | 0.811 | 0.365 |
| | 12.00 | 1.216 | 0.522 |
| | 16.00 | 1.621 | 0.677 |
| | 20.00 | 2.027 | 0.843 |
| | 22.00 | 2.229 | 0.935 |
| 298.15 | 4.00 | 0.405 | 0.125 |
| | 8.00 | 0.811 | 0.240 |
| | 12.00 | 1.216 | 0.345 |
| | 16.00 | 1.621 | 0.440 |
| | 20.00 | 2.027 | 0.530 |
| | 24.00 | 2.432 | 0.614 |
| | 28.00 | 2.837 | 0.700 |
| | 32.00 | 3.242 | 0.790 |
| | 36.00 | 3.648 | 0.882 |
| | 40.00 | 4.053 | 0.980 |
| 323.15 | 4.00 | 0.405 | 0.088 |
| | 8.00 | 0.811 | 0.174 |
| | 12.00 | 1.216 | 0.254 |
| | 16.00 | 1.621 | 0.327 |
| | 20.00 | 2.027 | 0.393 |
| | 24.00 | 2.432 | 0.454 |

AUXILIARY INFORMATION

METHOD/APPARATUS/PROCEDURE:

Borosilicate glass cell. Temperature measured with Platinum resistance thermometer. Pressure measured on Bourdon gauge. Samples of ethane added to dodecane, equilibrated. Liquid phase composition estimated from known overall composition and volume of both phases. Details in source.

SOURCE AND PURITY OF MATERIALS:

1. Matheson Co. "pure grade" fractionated, final purity about 99.7 mole %.

2. Humphrey Wilkinson Inc. sample purity 99.0 mole %.

ESTIMATED ERROR:

$\delta T$/K = ±0.07; $\delta P$/bar = ±0.07

$\delta x_{C_2H_6}$ = ±0.0025.

REFERENCES:

| COMPONENTS: | ORIGINAL MEASUREMENTS: |
|---|---|
| (1) Ethane; $C_2H_6$; [74-84-0] | Lee, K.H.; Kohn, J.P. |
| (2) Dodecane; $C_{12}H_{26}$; [112-40-3] | *J. Chem. Eng. Data* 1969, *14*, 292-5. |

EXPERIMENTAL VALUES:

| $T$/K | $P$/atm | $P$/MPa | Mole fraction of ethane in liquid, $x_{C_2H_6}$ |
|---|---|---|---|
| 323.15 | 28.00 | 2.837 | 0.512 |
| | 32.00 | 3.242 | 0.565 |
| | 36.00 | 3.648 | 0.616 |
| | 40.00 | 4.053 | 0.666 |
| | 44.00 | 4.458 | 0.715 |
| | 48.00 | 4.864 | 0.764 |
| | 52.00 | 5.269 | 0.813 |
| | 56.00 | 5.674 | 0.861 |
| | 60.00 | 6.080 | 0.910 |
| | 62.00 | 6.282 | 0.936 |
| 348.15 | 4.00 | 0.405 | 0.068 |
| | 8.00 | 0.811 | 0.129 |
| | 12.00 | 1.216 | 0.198 |
| | 16.00 | 1.621 | 0.259 |
| | 20.00 | 2.027 | 0.314 |
| | 24.00 | 2.432 | 0.365 |
| | 28.00 | 2.837 | 0.414 |
| | 32.00 | 3.242 | 0.458 |
| | 36.00 | 3.648 | 0.498 |
| | 40.00 | 4.053 | 0.536 |
| | 44.00 | 4.458 | 0.578 |
| | 48.00 | 4.864 | 0.608 |
| | 52.00 | 5.269 | 0.644 |
| | 56.00 | 5.674 | 0.676 |
| | 60.00 | 6.080 | 0.710 |
| | 62.00 | 6.282 | 0.727 |
| 373.15 | 4.00 | 0.405 | 0.050 |
| | 8.00 | 0.811 | 0.101 |
| | 12.00 | 1.216 | 0.152 |
| | 16.00 | 1.621 | 0.200 |
| | 20.00 | 2.027 | 0.247 |
| | 24.00 | 2.432 | 0.292 |
| | 28.00 | 2.837 | 0.334 |
| | 32.00 | 3.242 | 0.373 |
| | 36.00 | 3.648 | 0.409 |
| | 40.00 | 4.053 | 0.444 |
| | 44.00 | 4.458 | 0.475 |
| | 48.00 | 4.864 | 0.506 |
| | 52.00 | 5.269 | 0.536 |
| | 56.00 | 5.674 | 0.565 |
| | 60.00 | 6.080 | 0.594 |
| | 62.00 | 6.282 | 0.608 |

**COMPONENTS:**

(1) Ethane; $C_2H_6$; [74-84-0]

(2) Eicosane; $C_{20}H_{42}$; [112-95-8]

**ORIGINAL MEASUREMENTS:**

Puri, S.; Koln, J. P.

*J. Chem. Eng. Data*

1970, *15*, 372-374.

**VARIABLES:**

$T$/K: 333.15

$P$/MPa: 0.51-6.08

**PREPARED BY:**

C. L. Young

**EXPERIMENTAL VALUES:**

| $T$/K | $P$/MPa | Mole fraction of ethane, $x_{C_2H_6}$ |
|---|---|---|
| 333.15 | 0.51 | 0.1200 |
| | 1.01 | 0.2338 |
| | 1.52 | 0.3225 |
| | 2.03 | 0.3975 |
| | 2.53 | 0.4635 |
| | 3.04 | 0.5230 |
| | 3.55 | 0.5750 |
| | 4.05 | 0.6220 |
| | 4.56 | 0.6637 |
| | 5.07 | 0.7020 |
| | 5.57 | 0.7360 |
| | 6.08 | 0.7700 |

Additional solid-liquid-vapor equilibrium in source.

AUXILIARY INFORMATION

**METHOD/APPARATUS/PROCEDURE:**

Borosilicate glass cell. Temperature measured with Platinum resistance thermometer. Pressure measured with Bourdon gauge. Samples of ethane added to eicosane and equilibrated. Liquid phase composition estimated from known overall composition and volume of both phases. Details in ref. (1).

**SOURCE AND PURITY OF MATERIALS:**

1. Matheson pure grade, further purified by distillation and absorption. Final purity better than 99.5 mole per cent.
2. Humphrey Wilkinson sample, minimum purity 99 mole per cent.

**ESTIMATED ERROR:**

$\delta T$/K = ±0.25; $\delta P$/MPa = ±0.05; $\delta x_{C_2H_6}$ = ±0.002.

**REFERENCES:**

1. Lee, K. H.; Koln, J. P. *J. Chem. Eng. Data* 1969, *14*, 292.

COMPONENTS:

(1) Ethane; $C_2H_6$; [74-84-0]

(2) Butane; $C_4H_{10}$; [106-97-8]

(3) Pentane; $C_5H_{12}$; [109-66-0]

ORIGINAL MEASUREMENTS:

Herlihy, J.C.; Thodos, G.

*J. Chem. Eng. Data* 1962, *7*, 346-351.

VARIABLES:

$T$/K: 338.71
$P$/MPa: 3.489-5.914
$x_2$/mol fraction: 0.03-0.378

PREPARED BY:

C.L. Young

EXPERIMENTAL VALUES:

$T$/K = 338.71

| $P$/psia | $P$/MPa | Liquid Mole Fraction Ethane $x_{C_2H_6}$ | Liquid Mole Fraction Butane $x_{C_4H_{10}}$ | Liquid Mole Fraction Pentane $x_{C_5H_{12}}$ | Vapor Mole Fraction Ethane $y_{C_2H_6}$ | Vapor Mole Fraction Butane $y_{C_4H_{10}}$ | Vapor Mole Fraction Pentane $y_{C_5H_{12}}$ |
|---|---|---|---|---|---|---|---|
| 506.0 | 3.489 | 0.517 | 0.274 | 0.209 | 0.846 | 0.102 | 0.052 |
| 606.2 | 4.180 | 0.602 | 0.229 | 0.169 | 0.875 | 0.086 | 0.039 |
| 737.9 | 5.088 | 0.713 | 0.175 | 0.112 | 0.886 | 0.077 | 0.037 |
| 797.7 | 5.500 | 0.765 | 0.144 | 0.091 | 0.882 | 0.081 | 0.037 |
| 825.7 | 5.693 | 0.794 | 0.128 | 0.078 | 0.872 | 0.085 | 0.043 |
| 514.9 | 3.550 | 0.510 | 0.378 | 0.112 | 0.828 | 0.149 | 0.023 |
| 611.5 | 4.216 | 0.599 | 0.312 | 0.089 | 0.858 | 0.122 | 0.020 |
| 708.0 | 4.881 | 0.690 | 0.245 | 0.065 | 0.871 | 0.111 | 0.018 |
| 767.2 | 5.290 | 0.740 | 0.208 | 0.052 | 0.871 | 0.110 | 0.019 |
| 816.7 | 5.631 | 0.788 | 0.170 | 0.042 | 0.865 | 0.115 | 0.020 |
| 527.7 | 3.638 | 0.560 | 0.131 | 0.309 | 0.893 | 0.046 | 0.061 |
| 621.8 | 4.287 | 0.641 | 0.113 | 0.246 | 0.898 | 0.043 | 0.059 |
| 712.9 | 4.915 | 0.718 | 0.093 | 0.189 | 0.904 | 0.041 | 0.055 |
| 808.7 | 5.576 | 0.793 | 0.076 | 0.131 | 0.911 | 0.038 | 0.051 |
| 838.7 | 5.783 | 0.823 | 0.068 | 0.109 | 0.910 | 0.038 | 0.052 |
| 857.7 | 5.914 | 0.836 | 0.061 | 0.103 | 0.904 | 0.041 | 0.055 |
| 517.5 | 3.568 | 0.534 | 0.049 | 0.417 | 0.899 | 0.016 | 0.085 |
| 616.2 | 4.249 | 0.617 | 0.044 | 0.339 | 0.910 | 0.020 | 0.070 |
| 711.5 | 4.906 | 0.703 | 0.039 | 0.258 | 0.917 | 0.018 | 0.065 |
| 808.7 | 5.576 | 0.797 | 0.033 | 0.170 | 0.921 | 0.017 | 0.062 |
| 878.7 | 6.058 | 0.858 | 0.030 | 0.112 | 0.919 | 0.017 | 0.064 |

AUXILIARY INFORMATION

METHOD/APPARATUS/PROCEDURE:

Static cell with moveable piston which enabled the volume and pressure of the cell contents to be varied. Fitted with magnetic stirrer. Pressure measured with dead weight gauge and temperature measured with thermocouple. Samples analysed by gas chromatography. Details in ref. (1) and source.

SOURCE AND PURITY OF MATERIALS:

1. Phillips Petroleum Co., sample purity 99.91 mole per cent.
2. Phillips Petroleum Co. sample purity 99.91 mole per cent.
3. Phillips Petroleum Co. sample purity 99.80 mole per cent.

ESTIMATED ERROR:

$\delta T$/K = 0.1; $\delta P$/MPa = ±0.015; $\delta x$, $\delta y$, = ±0.005.

REFERENCES:

1. Rigas, T.J.; Mason, D.F.; Thodos, G.

*Ind. Eng. Chem.* 1958, *50*, 1297.

| COMPONENTS: | EVALUATOR: |
|---|---|
| (1) Ethane; $C_2H_6$; [74-84-0]<br><br>(2) Non-polar, non-paraffin solvents. | Walter Hayduk<br>Department of Chemical Engineering<br>University of Ottawa<br>Ottawa, Canada K1N 9B4 |

CRITICAL EVALUATION:

Ethane solubilities in non-polar solvents were tested for consistency utilizing solvent solubility parameters (1, 2). In no case was there a sufficient number of data for a single solvent to draw firm conclusions as to the accuracy of the data. Instead, in most cases, only one or two sources of data were available; hence, a consistency test was most helpful in revealing major deviations from expected solubility behavior, or the likely more accurate one of two differing results. The consistency test is shown for the solvents considered to form regular solutions with ethane, in the figure below which shows the mole fraction solubility at 298.15 K at a gas partial pressure of 101.325 kPa (1 atm) as a function of the solvent solubility parameter evaluated at 298.15 K. The solvents and sources of data considered in order of increasing solubility parameter are: (a) perfluorotributylamine (3,4), (b) perfluoroheptane (5,6), (c) Freon 113 (7,8), (d) cyclohexane (9,10), (e) bicyclohexyl (11), (f) decalin (12), (g) carbon tetrachloride (13,14), (h) toluene (15), (i) benzene (8,13,14), (j) chlorobenzene (13), (k) methyl naphthalene (16), and (l) carbon disulfide (3,6).

The critical evaluation for the solvents will be made in the order listed.

Of the two sources for the ethane solubility in perfluorotributylamine, the more recent value of Powell (3) is considered reliable whereas the earlier one by Kobatake and Hildebrand (4) differing by some 46% is rejected. The latter value is not

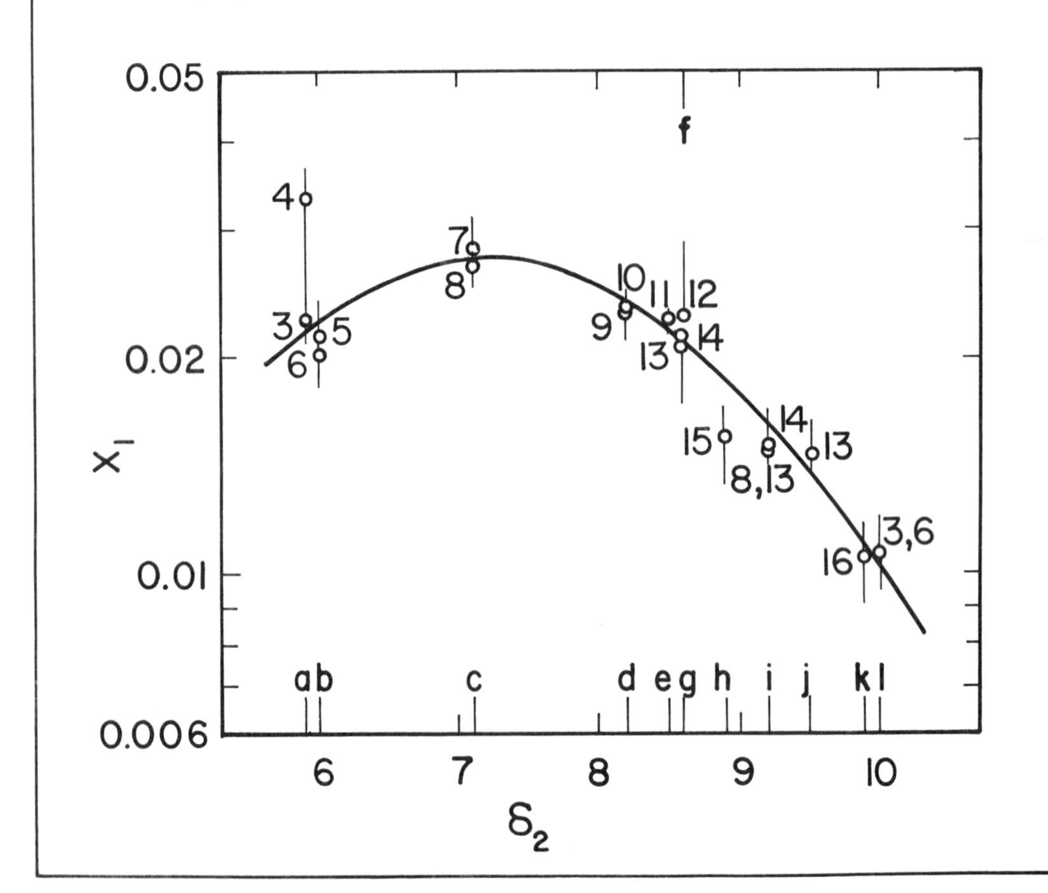

| COMPONENTS: | EVALUATOR: |
|---|---|
| (1) Ethane; $C_2H_6$; [74-84-0]<br>(2) Non-polar, non-paraffin solvents. | Walter Hayduk<br>Department of Chemical Engineering<br>University of Ottawa<br>Ottawa, Canada K1N 9B4 |

CRITICAL EVALUATION: continued

mentioned in the book by Hildebrand et al. (1). The most recent data for the solubility in perfluoroheptane (5) are considered the more accurate of the two (5,6), both having been determined in the same laboratory. The most recent (8) of the two values for the solubilities in Freon 113 (7,8) is also considered the more accurate, one of the authors having been involved in both measurements. The average of two solubilities in cyclohexane by completely independent workers (9,10) is considered to represent the data best. The single extrapolated value for the solubility in bicyclohexyl (11) appears to correspond well to a regular solution. The solubility in decalin (12) conforms approximately to that of a regular solution; however, the low pressure, gas chromatographic method for solubility derterminations of Lenoir et al. (12) has been found to give erratic results in certain instances, as discussed in the Critical Evaluation of solubilities in polar solvents. Hence, considerable doubt is cast on the accuracy reported for decalin.

The average of the two solubilities in carbon tetrachloride (13,14) which are within 2% of one another, is probably the most accurate. Although another source is indicated for ethane solubility in carbon tetrachloride in "Regular Solutions" (2), the paper referred to does not have the data in question. Solubilities in toluene (15) are reported over a temperature range permitting extrapolation to 298.15 K. The extrapolated solubility at 298.15 K is significantly less than expected for regular solutions and hence is questioned. The possible error could be in the order of 15%. Another value listed in "Regular Solutions" (2) is assumed to be a calculated value since the paper referred to does not contain the solubility data in toluene. The two results in benzene (13,14) check within 1% while the single value of Armitage et al. (8) at 301.51 K (28.36°C) checks the Horiuti (13) interpolated result within 0.5%. One may conclude that the solubility in benzene at 298.15 K is the average estimated from the three sources. Single sources are available for the solubilities in chlorobenzene (13) and methyl naphthalene (16). Two sources which check one another within 1% are available for the solubility in carbon disulfide (1,6). It is considered that the average of the two represents the best estimate of the true value.

All the solubility values discussed are classified as tentative values. It is estimated that when an average of two (or more) results has been recommended, its deviation from the true solubility is less than 2%. If only one solubility value has been considered, its deviation from the true solubility is difficult to assess and may be considerably in excess of 2%.

A table summarizing the ethane solubilities at 298.15 K and 101.325 kPa pressure in the non-polar solvents reviewed, follows:

| Solvent, Source(s) | Mole Fraction solubility at 298.15 K | Probable Error |
|---|---|---|
| 1. Perfluorotributylamine (3,4) | 0.0227 | - |
| 2. Perfluoroheptane (5,6) | 0.0216 | - |
| 3. Freon 113 (7,8) | 0.0269 | - |
| 4. Cyclohexane (9,10) | 0.0233 | ±2% |
| 5. Bicyclohexyl (11) | 0.0228 | - |
| 6. Decalin (12) | 0.0224 | - |

| COMPONENTS: | EVALUATOR: |
| --- | --- |
| (1) Ethane; $C_2H_6$; [74-84-0]<br>(2) Non-polar, non-paraffin solvents. | Walter Hayduk<br>Department of Chemical Engineering<br>University of Ottawa<br>Ottawa, Ontario K1N 9B4 |

CRITICAL EVALUATION:

...continued

| | Solvent, Source(s)* | Mole Fraction solubility at 298.15 K | Probable Error |
| --- | --- | --- | --- |
| 7. | Carbon tetrachloride (13,14) | 0.0211 | ±2% |
| 8. | Toluene (15) | 0.0155 | - |
| 9. | Benzene (8,13,14) | 0.0149 | ±2% |
| 10. | Chlorobenzene (13) | 0.0146 | ±2% |
| 11. | Methyl naphthalene (16) | 0.0106 | - |
| 12. | Carbon disulfide (3,6) | 0.0107 | ±2% |

Some ethane solubilities at 298 K are available in mixed solvent solutions composed of Freon 113 and benzene (7,8). The more recent work of Armitage et al. (8) appears to be more thorough, reporting the solubilities for a larger number of solvent compositions. Except for the solubility in Freon 113 itself, the data check one another closely. The solubility in the mixed solvent solution has been interpreted by Hildebrand et al. (1) in terms of a volume fraction average mixed solvent solubility parameter of the solvent components. The effect of temperature on the ethane solubility in a mixed solvent containing 0.756 mole fraction benzene and the remainder Freon 113, is also available (8). Although the solvent is predominantly benzene, the relatively small amount of Freon 113 that is present has a large influence on the solubility. The temperature coefficient of solubility appears to be as for a pure solvent; that is, the relation is approximately linear when ln $x_1$ is plotted versus $1/T$.

References


1. Hildebrand, J.H.; Prausnitz, J.M.; Scott, R.L. "*Regular and Related Solutions*", Van Nostrand Reinhold, New York, 1970, Appendix 5, 135, 207.

2. Hildebrand, J.H.; Scott, R.L. "*Regular Solutions*", Prentice-Hall, Englewood Cliffs, N.J. 1962, 25, 162.

3. Powell, R.J. *J. Chem. Eng. Data* 1972, *17*, 302-304.

4. Kobatake, Y.; Hildebrand, J.H. *J. Phys. Chem.* 1961, *65*, 331-335.

5. Thomsen, E.S.; Gjaldbaek, J.C. *Acta Chem. Scand.* 1963, *17*, 127-133.

6. Gjaldbaek, J.C.; Niemann, H. *Acta Chem. Scand.* 1958, *12*, 611-614.

7. Linford, R.G.; Hildebrand, J.H. *Trans. Faraday Soc.* 1970, *66*, 577-581.

8. Armitage, D.A.; Linford, R.G.; Thornhill, D.G.T. *Ind. Eng. Chem. Fundam.* 1978, *17*, 362-364.

9. Ben-Naim, A.; Yaacobi, M. *J. Phys. Chem.* 1974, *78*, 175-178.

10. Dymond, J.H. *J. Phys. Chem.* 1967, *71*, 1829-1831.

COMPONENTS:

(1) Ethane; $C_2H_6$; [74-84-0]

(2) Non-polar, non-paraffin solvents.

EVALUATOR:

Walter Hayduk
Department of Chemical Engineering
University of Ottawa
Ottawa, Canada K1N 9B4

CRITICAL EVALUATION:

...continued


11. Cukor, P.M.; Prausnitz, J.M. *J. Phys. Chem.* 1972, *76*,

12. Lenoir, J-Y.; Renault, P.; Renon, H.; *J. Chem. Eng. Data* 1971, *16*, 340-342.

13. Horiuti, J. *Sci. Pap. Inst. Phys. Chem. Res. (Japan)* 1931-32, *17*, 125-256.

14. Jadot, R. *J. Chim. Phys.* 1972, *69*, 1036-1040.

15. Waters, J.A.; Mortimer, G.A.; Clements, H.E. *J. Chem. Eng. Data* 1970, *15*, 174-176.

16. Chappelow, C.C.; Prausnitz, J.M. *Am. Inst. Chem. Engnrs. Jr.* 1974, *20*, 1097-1104.

**COMPONENTS:**

(1) Ethane; $C_2H_6$; [74-84-0]

(2) 1,1,2,2,3,3,4,4,4-Nonafluoro-N, N-bis(nonafluorobutyl)-1-butanamine or perfluorotributylamine; $(C_4F_9)_3N$; [311-89-7]

**ORIGINAL MEASUREMENTS:**

Powell, R. J.

*J. Chem. Eng. Data* 1972, *17*, 302 - 304.

**VARIABLES:**

$T$/K: 288.15 - 318.15

$p_1$/kPa: 101.325 (1 atm)

**PREPARED BY:**

P. L. Long
H. L. Clever

**EXPERIMENTAL VALUES:**

| $T$/K | Mol Fraction $10^4 x_1$ | Bunsen Coefficient $\alpha/cm^3$ (STP) $cm^{-3}$ $atm^{-1}$ | Ostwald Coefficient $L/cm^3$ $cm^{-3}$ | $N = R \frac{\Delta \log x_1}{\Delta \log T}$ |
|---|---|---|---|---|
| 298.15 | 227.4 | 1.46 | 1.59 | -6.36 |

The Bunsen and Ostwald coefficients were calculated by the compiler.

The author states that the solubility measurements were made over the temperature interval of about 288.15 to 303.15 K, but only the solubility value at 298.15 K was given in the paper. The slope, $N=R(\Delta \log x_1/\Delta \log T)$, was given.

Smoothed Data: For use between 288.15 and 318.15 K

The smoothed data were calculated by the compiler from the slope, N, in the form

$$\log_{10} x_1 = \log_{10}(227.4 \times 10^{-4}) - (6.36/R) \log (T/298.15)$$

with $R$ = 1.9872 cal $K^{-1}$ $mol^{-1}$.

| $T$/K | Mol Fraction $x_1$ |
|---|---|
| 288.15 | 0.02536 |
| 293.15 | 0.02400 |
| 298.15 | 0.02274 |
| 303.15 | 0.02156 |
| 308.15 | 0.02046 |
| 313.15 | 0.01943 |
| 318.15 | 0.01847 |

**AUXILIARY INFORMATION**

**METHOD/APPARATUS/PROCEDURE:**

The apparatus is the Dymond and Hildebrand (1) apparatus which uses an all glass pumping system to spray slugs of degassed solvent into the gas. The amount of gas dissolved is calculated from the initial and final pressures. The solvent is degassed by freezing, pumping, and followed by boiling under reduced pressure.

**SOURCE AND PURITY OF MATERIALS:**

(1) Ethane. Source not given. Research grade, dried over $CaCl_2$ before use.

(2) 1,1,2,2,3,3,4,4,4-Nonafluoro-N, N-bis(nonafluorobutyl)-butanamine. Minnesota Mining & Manufacturing Co. Distilled, used portion boiling between 447.85 - 448.64 K which gave a single GLC peak. $\rho_{298.15}$ = 1.880 g $cm^{-3}$.

**ESTIMATED ERROR:**

$\delta x_1/x_1 = \pm 0.002$

$\delta N$/cal $K^{-1}$ $mol^{-1}$ = ± 0.1

**REFERENCES:**

1. Dymond, J. H.; Hildebrand, J. H. *Ind. Eng. Chem. Fundam.* 1967, *6*, 130.

**COMPONENTS:**

(1) Ethane; $C_2H_6$; [74-84-0]

(2) 1,1,2,2,3,3,4,4,4-Nonafluoro-N, N-bis(nonafluorobutyl)-1-butanamine or perfluorotributylamine; $(C_4F_9)_3N$; [311-89-7]

**ORIGINAL MEASUREMENTS:**

Kobatake, Y.; Hildebrand, J. H.

*J. Phys. Chem.* 1961, *65*, 331 - 335.

**VARIABLES:**

$T$/K: 286.70 - 304.04

$P$/kPa: 101.325 (1 atm)

**PREPARED BY:**

M. E. Derrick

H. L. Clever

**EXPERIMENTAL VALUES:**

| Temperature $t$/°C | Temperature $T$/K | Mol Fraction $10^2 x_1$ | Bunsen Coefficient $\alpha$/cm$^3$(STP) cm$^{-3}$ atm$^{-1}$ | Ostwald Coefficient $L$/cm$^3$ cm$^{-3}$ |
|---|---|---|---|---|
| 13.55 | 286.70 | 3.940 | 2.64 | 2.77 |
| 20.60 | 293.75 | 3.541 | 2.33 | 2.51 |
| 25.00 | 298.15 | 3.327 | 2.16 | 2.36 |
| 25.61 | 298.76 | 3.300 | 2.14 | 2.34 |
| 30.89 | 304.04 | 3.063 | 1.96 | 2.18 |

The Bunsen and Ostwald coefficients were calculated by the compiler.

Smoothed Data: For use between 286.70 and 309.09 K

$$\ln x_1 = -7.6395 + 12.6301/(T/100\text{K})$$

The standard error about the regression line is $2.44 \times 10^{-5}$.

| $T$/K | Mol Fraction $10^2 x_1$ |
|---|---|
| 288.15 | 3.853 |
| 293.15 | 3.575 |
| 298.15 | 3.326 |
| 303.15 | 3.102 |

AUXILIARY INFORMATION

**METHOD/APPARATUS/PROCEDURE:**

The apparatus consists of a gas measuring buret, an absorption pipet, and a reservoir for the solvent. The buret is thermostated at 25°C, the pipet at any temperature from 5 to 30 °C. The pipet contains an iron bar in glass for magnetic stirring. The pure solvent is degassed by freezing with liquid nitrogen, evacuating, then boiling with a heat lamp. The degassing process is repeated three times. The solvent is flowed into the pipet where it is again boiled for final degassing. Manipulation of the apparatus is such that the solvent never comes in contact with stopcock grease. The liquid in the pipet is sealed off by mercury. Its volume is the difference between the capacity of the pipet and the volume of mercury that confines it. Gas is admitted into the pipet. Its exact amount is determined by P-V measurements in the buret before and after introduction of the gas into the pipet. The stirrer is set in motion. Equilibrium is attained within 24 hours.

**SOURCE AND PURITY OF MATERIALS:**

(1) Ethane. Matheson Co., Inc. Research grade. Dried by passage over $P_2O_5$ followed by multiple trap vaporization and evacuation at liquid $N_2$ temperature.

(2) Perfluorotributylamine. Minnesota Mining and Manufacturing Co. Dried, fractionated, boiling point 178.5 - 179.0°C. Density, $\rho$/g cm$^{-3}$ = 1.872.

**ESTIMATED ERROR:**

$\delta T$/K = 0.02

$\delta x_1/x_1$ = 0.003

**REFERENCES:**

**COMPONENTS:**

(1) Ethane; $C_2H_6$; [74-84-0]

(2) Hexadecafluoroheptane or perfluoroheptane; $C_7F_{16}$; [335-57-9]

**ORIGINAL MEASUREMENTS:**

Thomsen, E. S.; Gjaldbaek, J. C.

*Acta Chem. Scand.* 1963, *17*, 127 - 133.

**VARIABLES:**

$T$/K: 298.15

$p_1$/kPa: 101.325 (1 atm)

**PREPARED BY:**

E. S. Thomsen

**EXPERIMENTAL VALUES:**

| $T$/K | Mol Fraction $10^2x_1$ | Bunsen Coefficient $\alpha/cm^3 (STP) cm^{-3} atm^{-1}$ | Ostwald Coefficient $L/cm^3 cm^{-3}$ |
|---|---|---|---|
| 298.15 | 2.12 | 2.12 | 2.31 |
| 298.15 | 2.17 | 2.17 | 2.37 |
| 298.15 | 2.18 | 2.18 | 2.38 |

The mole fraction and Ostwald solubility values were calculated by the compiler.

**AUXILIARY INFORMATION**

**METHOD/APPARATUS/PROCEDURE:**

A calibrated all-glass combined manometer and bulb was enclosed in an air thermostat and shaken until equilibrium. Mercury was used for calibration and as the confining liquid. The solvents were degassed in the apparatus. Details are in references 1 and 2.

The absorped volume of gas was calculated from the initial and final amounts, both saturated with solvent vapor. The amount of solvent was determined by the weight of displaced mercury.

The saturation of the liquid with the gas was carried out close to atmospheric pressure. The solubility values were reported for one atmosphere gas pressure assuming Henry's law is obeyed.

**SOURCE AND PURITY OF MATERIALS:**

(1) Ethane. Phillips Petroleum Co. Research grade. Contained 0.2 per cent air and 0.1 per cent unidentified impurity.

(2) Hexadecafluoroheptane. Source not given. Passed through silica gel column; fractionated; boiling point 82.55 - 82.56°C at 760 mmHg. Extinction coefficient 0.02 at 216 nm.

**ESTIMATED ERROR:**

$\delta T/K = 0.05$

$\delta x_1/x_1 = 0.015$

**REFERENCES:**

1. Lannung, A. *J. Am. Chem. Soc.* 1930, *52*, 68.

2. Gjaldbaek, J. C. *Acta Chem. Scand.* 1952, *6*, 623.

**COMPONENTS:**

(1) Ethane; $C_2H_6$; [74-84-0]

(2) Hexadecafluoroheptane; $C_7F_{16}$; [335-57-9]

**ORIGINAL MEASUREMENTS:**

Gjaldbaek, J. C.; Niemann, H.

*Acta Chem. Scand.* 1958, *12*, 611 - 614.

**VARIABLES:**

$T$/K: 298.15
$P$/kPa: 101.325 (1 atm)

**PREPARED BY:**

J. Chr. Gjaldbaek

**EXPERIMENTAL VALUES:**

| $T$/K | Mol Fraction $10^2 x_1$ | Bunsen Coefficient $\alpha$/cm$^3$ (STP) cm$^{-3}$ atm$^{-1}$ | Ostwald Coefficient $L$/cm$^3$ cm$^{-3}$ |
|---|---|---|---|
| 298.15 | 2.032 | 2.032 | 2.218 |
| 298.15 | 2.027 | 2.027 | 2.212 |

The Ostwald and mole fraction solubility values were calculated by the compiler. The solubilities are reported for an ethane partial pressure of 101.325 kPa (1 atm) assuming Henry's law is obeyed.

**AUXILIARY INFORMATION**

**METHOD/APPARATUS/PROCEDURE:**

A calibrated all-glass combined manometer and bulb was enclosed in an air thermostat and shaken until equilibrium. Mercury was used for calibration and as the confining liquid. The solvents were degassed in the apparatus. Details are in references 1 and 2.

The absorbed volume of gas is calculated from the initial and final amounts, both saturated with solvent vapor. The amount of solvent is determined by the weight of displaced mercury.

**SOURCE AND PURITY OF MATERIALS:**

(1) Ethane. Phillips Petroleum Co. Research grade. According to combination analysis 99.6-100.4 per cent. Butane and higher hydrocarbons were absent, and ethene was less than 0.5 per cent.

(2) Hexadecafluoroheptane. Source not given. Fractionated; purified according to Glew and Reeves. Boiling point $t$/°C 82.55-82.56. Extinction coefficient 0.02 and 216 nm.

**ESTIMATED ERROR:**

$\delta T$/K = 0.05
$\delta x_1/x_1$ = 0.015

**REFERENCES:**

1. Lannung, A. *J. Am. Chem. Soc.* 1930, *52*, 68.

2. Gjaldbaek, J. C. *Acta Chem. Scand.* 1952, *6*, 623.

**COMPONENTS:**

(1) Ethane; $C_2H_6$; [74-84-0]

(2) 1,1,2-Trichloro-1,2,2-trifluoroethane or Freon 113; $C_2Cl_3F_3$; [76-13-1]

**ORIGINAL MEASUREMENTS:**

Linford, R. G.; Hildebrand, J. H.

*Trans. Faraday Soc.* 1970, *66*, 577 - 581.

**VARIABLES:**

$T$/K: 286.25 - 301.35
$p$/kPa: 101.325 (1 atm)

**PREPARED BY:**

P. L. Long
H. L. Clever

**EXPERIMENTAL VALUES:**

| Temperature $t/^0C$ | Temperature $T/K$ | Mol Fraction $10^2 x_1$ | Bunsen Coefficient $\alpha/cm^3(STP)cm^{-3}atm^{-1}$ | Ostwald Coefficient $L/cm^3cm^{-3}$ |
|---|---|---|---|---|
| 13.10 | 286.25 | 3.396 | 6.69 | 7.01 |
| 15.00 | 288.15 | 3.322 | 6.52 | 6.88 |
| 19.35 | 292.50 | 3.118 | 6.10 | 6.53 |
| 25.00 | 298.15 | 2.852 | 5.49 | 5.99 |
| 28.20 | 301.35 | 2.737 | 5.23 | 5.78 |

The Bunsen and Ostwald coefficients were calculated by the compiler assuming ideal gas behavior.

Smoothed Data: For use between 286.25 and 301.35 K.

$$\ln x_1 = -7.7758 + 12.5876/(T/100\ K)$$

The standard error about the regression line is $1.37 \times 10^{-4}$.

| $T$/K | Mol Fraction $10^2 x_1$ |
|---|---|
| 288.15 | 3.313 |
| 293.15 | 3.075 |
| 298.15 | 2.861 |

AUXILIARY INFORMATION

**METHOD/APPARATUS/PROCEDURE:**

The authors used the Dymond-Hildebrand (1) apparatus which uses an all-glass pumping system to spray slugs of degassed solvent into the gas. The liquid is saturated with gas at a partial pressure of gas equal to one atm. The amount of gas dissolved is calculated from initial and final pressures.

**SOURCE AND PURITY OF MATERIALS:**

(1) Ethane. Source not given. Stated to be purest commercially available sample, dried.

(2) 1,1,2-Trichloro-1,2,2-trifluoroethane. Matheson, Coleman and Bell Co. Spectroquality.

**ESTIMATED ERROR:**

$\delta x_1/x_1 = \pm 0.005$

**REFERENCES:**

1. Dymond, J. H.; Hildebrand, J. H. *Ind. Eng. Chem., Fundam.* 1967, *6*, 130.

**COMPONENTS:**

(1) Ethane; $C_2H_6$; [74-84-0]

(2) 1,1,2-Trichloro-1,2,2- trifluoroethane (Freon 113); $C_2Cl_3F_3$; [76-13-1]

**ORIGINAL MEASUREMENTS:**

Armitage, D.A.; Linford, R.G.; Thornhill, D.G.T.

*Ind. Eng. Chem. Fundam.* 1978, *17*, 362-364.

**VARIABLES:**

$T$/K: 284.01-298
$P$/kPa: 101.325 (1 atm)

**PREPARED BY:**

W. Hayduk

**EXPERIMENTAL VALUES:**

| $T$/K | Mole fraction[1] $10^4 x_1$ | Bunsen Coefficient[3] $\alpha/cm^3(STP)cm^{-3}atm^{-1}$ | Ostwald Coefficient[3] $L/cm^3cm^{-3}$ |
|---|---|---|---|
| 284.01 | 348.4 (349.1)[4] | 6.89 | 7.18 |
| 287.71 | 323.9 (325.0) | 6.35 | 6.69 |
| 291.01 | 306.1 (305.5) | 5.97 | 6.36 |
| 291.06 | 307.9 (305.2) | 5.99 | 6.38 |
| 294.01 | 288.1 (289.1) | 5.58 | 6.01 |
| 298.09 | 267.1 (268.6) | 5.14 | 5.60 |
| 298.08 | 269.7 (268.7) | 5.19 | 5.66 |
| 298.11 | 268.4[2] (268.5) | 5.16 | 5.63 |

[1] Original data

[2] Value extrapolated by authors

[3] Bunsen and Ostwald coefficients calculated by compiler assuming ideal gas behavior.

[4] From equation of smoothed data:

$$\ln x_1 = 1575.0/T - 8.9008$$

Correlation coefficient = 0.9988

AUXILIARY INFORMATION

**METHOD/APPARATUS/PROCEDURE:**

The method is volumetric utilizing a glass apparatus containing a known volume of gas in a contact chamber through which degassed solvent is circulated using a magnetically operated glass pump. The solubility is determined from observed drop in in pressure. Solvent is degassed by evacuating boiling solvent and frozen solvent in sequence.

**SOURCE AND PURITY OF MATERIALS:**

1. Cambrian Chemicals; minimum specified purity 99.0 mole per cent.

2. British Drug Houses; minimum specified purity 99.8 mole per cent.

**ESTIMATED ERROR:**

$\delta x_1/x_1 = 0.01$ (authors)

**REFERENCES:**

**COMPONENTS:**

(1) Ethane; $C_2H_6$; [74-84-0]

(2) Cyclohexane; $C_6H_{12}$; [110-82-7]

**ORIGINAL MEASUREMENTS:**

Ben-Naim, A.; Yaacobi, M.

*J. Phys. Chem.* 1974, *78*, 175-8

**VARIABLES:**

$T$/K: 283.15-303.15

**PREPARED BY:**

C.L. Young

**EXPERIMENTAL VALUES:**

| $T$/K | Ostwald coefficient,* $L$ | Mole fraction[+] at partial pressure of 101.3kPa, $x_{C_2H_6}$ |
|---|---|---|
| 283.15 | 6.470 | 0.0289 (0.0290)** |
| 288.15 | 6.067 | 0.0268 (0.0268) |
| 293.15 | 5.673 | 0.0249 (0.0247) |
| 298.15 | 5.291 | 0.0230 (0.0229) |
| 303.15 | 4.921 | 0.0212 (0.0213) |

* Smoothed values obtained from the equation.

$kT \ln L = -2{,}712.0 + 29.932\,(T/K) - 0.05878\,(T/K)^2$ cal mol$^{-1}$
where k is in units of cal mol$^{-1}$ K$^{-1}$

\+ calculated by compiler assuming ideal gas law for ethane.

** From alternate equation of smoothed data:

$$\ln x_1 = 1325.5/T - 8.2206$$

Correlation coefficient = 0.9992

AUXILIARY INFORMATION

**METHOD/APPARATUS/PROCEDURE:**

The apparatus was similar to that described by Ben-Naim and Baer (1) and Wen and Hung (2). It consists of three main parts, a dissolution cell of 300 to 600 cm$^3$ capacity, a gas volume measuring column, and a manometer. The solvent is degassed in the dissolution cell, the gas is introduced and dissolved while the liquid is kept stirred by a magnetic stirrer immersed in the water bath. Dissolution of the gas results in the change in the height of a column of mercury which is measured by a cathetometer.

**SOURCE AND PURITY OF MATERIALS:**

1. Matheson sample, purity 99.9 mol per cent.

2. AR grade.

**ESTIMATED ERROR:**

$\delta T$/K = ±0.1; $\delta x_{C_2H_6}$ = ±2%
(estimated by compiler)

**REFERENCES:**

1. Ben-Naim, A.; Baer, S. *Trans. Faraday Soc.* 1963, *59*, 2735.

2. Wen, W.-Y.; Hung, J.H. *J. Phys. Chem.* 1970, *74*, 170.

**COMPONENTS:**

(1) Ethane; $C_2H_6$; [74-84-0]

(2) Cyclohexane; $C_6H_{12}$; [110-82-7]

**ORIGINAL MEASUREMENTS:**

Dymond, J. H.

*J. Phys. Chem.* 1967, *71*, 1829-1831.

**VARIABLES:**

$T$/K: 292.35 - 307.95
$p$/kPa: 101.325 (1 atm)

**PREPARED BY:**

M. E. Derrick
H. L. Clever

**EXPERIMENTAL VALUES:**

| Temperature t/°C | Temperature $T$/K | Mol Fraction $10^2 x_1$ | Bunsen Coefficient $\alpha$/cm$^3$ (STP) cm$^{-3}$ atm$^{-1}$ | Ostwald Coefficient $L$/cm$^3$ cm$^{-3}$ |
|---|---|---|---|---|
| 19.20 | 292.35 | 2.580 | 5.50 | 5.89 |
| 25.45 | 298.60 | 2.340 | 4.93 | 5.39 |
| 31.20 | 304.35 | 2.150 | 4.49 | 5.00 |
| 34.80 | 307.95 | 2.055 | 4.27 | 4.81 |

The Bunsen and Ostwald coefficients were calculated by the compiler.

Smoothed Data: For use between 292.35 and 308.15 K.

$$\ln x_1 = -8.1748 + 13.2020/(T/100\text{K})$$

The standard error about the regression line is $6.95 \times 10^{-5}$.

| $T$/K | Mol Fraction $10^2 x_1$ |
|---|---|
| 293.15 | 2.544 |
| 298.15 | 2.359 |
| 303.15 | 2.193 |
| 308.15 | 2.043 |

**AUXILIARY INFORMATION**

**METHOD/APPARATUS/PROCEDURE:**

The liquid is saturated with the gas at a partial pressure of one atm.

The apparatus is that described by Dymond and Hildebrand (1). It uses an all-glass pumping system to spray slugs of degassed solvent into the gas. The amount of gas dissolved is calculated from the initial and final gas pressures.

**SOURCE AND PURITY OF MATERIALS:**

(1) Ethane. Phillips Petroleum Co. Research grade, dried.

(2) Cyclohexane. Matheson, Coleman and Bell chromatoquality reagent. Dried and fractionally frozen. m.p. 6.45°C.

**ESTIMATED ERROR:**

$$\delta x_1/x_1 = 0.01$$

**REFERENCES:**

1. Dymond, J.; Hildebrand, J. H. *Ind. Eng. Chem. Fundam.* 1967, *6*, 130.

**COMPONENTS:**

(1) Ethane; $C_2H_6$; [74-84-0]

(2) 1,1'-Bicyclohexyl; $C_{12}H_{22}$; [92-51-3]

**ORIGINAL MEASUREMENTS:**

Cukor, P.M.; Prausnitz, J.M.

*J. Phys. Chem.* 1972, *76*, 598-601

**VARIABLES:**

$T$/K: 300-475

**PREPARED BY:**

C.L. Young

**EXPERIMENTAL VALUES:**

| $T$/K | Henry's Constant[a] /atm | Mole fraction of ethane[b] in liquid, $x_{C_2H_6}$ |
|---|---|---|
| 300 | 42.6 | 0.0235 (0.0223)[c] |
| 325 | 59.2 | 0.0169 (0.0167) |
| 350 | 78.7 | 0.0127 (0.0130) |
| 375 | 99.8 | 0.0100 (0.0105) |
| 400 | 121 | 0.00826(0.00873) |
| 425 | 140 | 0.00714(0.00740) |
| 450 | 154 | 0.00649(0.00639) |
| 475 | 163 | 0.00613(0.00561) |

[a] Quoted in supplementary material for original paper

[b] Calculated by compiler for a partial pressure of 1 atmosphere

[c] From equation of smoothed data:

$$\ln x_1 = 1122.9/T - 7.5477$$

AUXILIARY INFORMATION

**METHOD/APPARATUS/PROCEDURE:**

Volumetric apparatus similar to that described by Dymond and Hildebrand (1). Pressure measured with a null detector and precision gauge. Details in ref. (2).

**SOURCE AND PURITY OF MATERIALS:**

No details given

**ESTIMATED ERROR:**

$\delta T$/K = ±0.05; $\delta x_{H_2}$ = ±2%

**REFERENCES:**


1. Dymond, J.; Hildebrand, J.H. *Ind. Eng. Chem. Fundam.* 1967, *6*, 130.

2. Cukor, P.M.; Prausnitz, J.M. *Ind. Eng. Chem. Fundam.* 1971, *10* 638.

**COMPONENTS:**

(1) Ethane; $C_2H_6$; [74-84-0]

(2) Decahydronaphthalene, (Decalin); $C_{10}H_{18}$; [91-17-8]

**ORIGINAL MEASUREMENTS:**

Lenoir, J-Y.; Renault, P.; Renon, H.

*J. Chem. Eng. Data* 1971, *16*, 340-2

**VARIABLES:**

$T$/K: 298.15-323.15

**PREPARED BY:**

C.L. Young

**EXPERIMENTAL VALUES:**

| $T$/K | Henry's constant $H_{C_2H_6}$/atm | Mole fraction at 1 atm* $/x_{C_2H_6}$ |
|---|---|---|
| 298.15 | 44.7 | 0.0224 (0.0224)† |
| 323.15 | 54.4 | 0.0184 (0.0184) |

* Calculated by compiler assuming a linear function of $p_{C_2H_6}$ vs $x_{C_2H_6}$, i.e. $x_{C_2H_6}$(1 atm) = $1/H_{C_2H_6}$

† Calculated from equation through data points:

$$\ln x_1 = 758.10/T - 6.3414$$

AUXILIARY INFORMATION

**METHOD/APPARATUS/PROCEDURE:**

A conventional gas-liquid chromatographic unit fitted with a thermal conductivity detector was used. The carrier gas was helium. The value of Henry's law constant was calculated from the retention time. The value applies to very low partial pressures of gas and there may be a substantial difference from that measured at 1 atm. pressure. There is also considerable uncertainty in the value of Henry's constant since surface adsorption was not allowed for although its possible existence was noted.

**SOURCE AND PURITY OF MATERIALS:**

(1) L'Air Liquide sample, minimum purity 99.9 mole per cent.

(2) Touzart and Matignon or Serlabo sample, purity 99 mole per cent.

**ESTIMATED ERROR:**

$\delta T$/K = $\pm 0.1$; $\delta H$/atm = $\pm 6\%$ (estimated by compiler).

**REFERENCES:**

**COMPONENTS:**

(1) Ethane; $C_2H_6$; [74-84-0]

(2) Tetrachloromethane or carbon tetrachloride; $CCl_4$; [56-23-5]

**ORIGINAL MEASUREMENTS:**

Horiuti, J.

*Sci. Pap. Inst. Phys. Chem. Res. (Jpn)* 1931/32, *17*, 125 - 256.

**VARIABLES:**

$T$/K: 273.15 - 313.15
$p_1$/kPa: 101.325 (1 atm)

**PREPARED BY:**

M. E. Derrick
H. L. Clever

**EXPERIMENTAL VALUES:**

| $T$/K | Mol Fraction $10^2x_1$ | Bunsen Coefficient $\alpha$/cm$^3$(STP)cm$^{-3}$atm$^{-1}$ | Ostwald Coefficient $L$/cm$^3$cm$^{-3}$ |
|---|---|---|---|
| 273.15 | 3.115 | 7.648 | 7.648 |
| 278.15 | 2.866 | 6.978 | 7.106 |
| 283.15 | 2.639 | 6.371 | 6.604 |
| 288.15 | 2.438 | 5.839 | 6.160 |
| 293.15 | 2.242 | 5.326 | 5.716 |
| 298.15 | 2.085 | 4.916 | 5.366 |
| 303.15 | 1.932 | 4.520 | 5.016 |
| 308.15 | 1.804 | 4.188 | 4.725 |
| 313.15 | 1.683 | 3.878 | 4.446 |

The mole fraction and Bunsen coefficient values were calculated by the compiler with the assumption the gas is ideal and that Henry's law is obeyed.

Smoothed Data: For use between 273.15 and 313.15 K.

$$\ln x_1 = -8.3052 + 13.2190/(T/100\text{K})$$

The standard error about the regression line is 5.92 x $10^{-5}$.

| $T$/K | Mol Fraction $10^2x_1$ |
|---|---|
| 273.15 | 3.125 |
| 283.15 | 2.634 |
| 288.15 | 2.429 |
| 293.15 | 2.246 |
| 298.15 | 2.083 |
| 303.15 | 1.936 |
| 313.15 | 1.684 |

**AUXILIARY INFORMATION**

**METHOD/APPARATUS/PROCEDURE:**

The apparatus consists of a gas buret, a solvent reservoir, and an absorption pipet. The volume of the pipet is determined at various meniscus heights by weighing a quantity of water. The meniscus height is read with a cathetometer.

The dry gas is introduced into the degassed solvent. The gas and solvent are mixed with a magnetic stirrer until saturation. Care is taken to prevent solvent vapor from mixing with the solute gas in the gas buret. The volume of gas is determined from the gas buret readings, the volume of solvent is determined from the meniscus height in the absorption pipet.

**SOURCE AND PURITY OF MATERIALS:**

(1) Ethane. A 60% solution of potassium acetate was electrolyzed between Pt and Cu electrodes. The gas was passed through several wash solutions, dried, and fractionated from liquid air. Boiling point (760 mmHg) -88.3°C.

(2) Tetrachloromethane. Kahlbaum. Dried over $P_2O_5$ and distilled. Boiling point (760 mmHg) 76.74°C.

**ESTIMATED ERROR:**

$\delta T$/K = 0.05
$\delta x_1/x_1$ = 0.01

**REFERENCES:**

COMPONENTS:

(1) Ethane; $C_2H_6$; [74-84-0]

(2) Tetrachloromethane; $CCl_4$; [56-23-5]

ORIGINAL MEASUREMENTS:

Jadot, R.

*J. Chim. Phys.* 1972, *69*, 1036-40

VARIABLES:

$T$/K: 298.15

PREPARED BY:

C.L. Young

EXPERIMENTAL VALUES:

| $T$/K | Henry's Law Constant, $H$/atm | Mole fraction[+] at partial pressure of 101.3kPa $x_{C_2H_6}$ | # $\Delta H^\infty$ /cal mol$^{-1}$ (/J mol$^{-1}$) |
|---|---|---|---|
| 298.15 | 46.93 | 0.02131 | 378 (1582) |

\+ Calculated by compiler assuming $x_{C_2H_6} = 1/H$

\# Excess partial molar enthalpy of solution at infinite dilution.

AUXILIARY INFORMATION

METHOD/APPARATUS/PROCEDURE:

The conventional gas chromatographic technique was used. The carrier gas was helium. The value of Henry's law constant was calculated from the retention time. The value applies to very low partial pressures of gas and there may be a substantial difference from that measured at 1 atm. pressure. There is also considerable uncertainty in the value of Henry's constant since no allowance was made for surface adsorption.

SOURCE AND PURITY OF MATERIALS:

No details given

ESTIMATED ERROR:

$\delta T$/K = ±0.05; $\delta H$ = ±2%

REFERENCES:

**COMPONENTS:**

(1) Ethane; $C_2H_6$; [74-84-0]

(2) Methylbenzene or toluene; $C_7H_8$; [108-88-3]

**ORIGINAL MEASUREMENTS:**

Waters, J. A.; Mortimer, G. A. Clements, H. E.

*J. Chem. Eng. Data* 1970, *15*, 174 - 176.

**VARIABLES:**

$T$/K: 253.15 - 297.95

**PREPARED BY:**

P. L. Long
H. L. Clever

**EXPERIMENTAL VALUES:**

| Temperature $t/^{0}C$ | $T$/K | Pressure $p_1$/mmHg | Ethane $c_1$/ mol dm$^{-3}$atm$^{-1}$ | Mol Fraction $10^2 x_1$ | Bunsen Coefficient $\alpha$[1] | Ostwald Coefficient $L$/cm$^3$cm$^{-3}$ |
|---|---|---|---|---|---|---|
| -20 | 253.15 | 759.0 | 0.290 | 2.87 | 6.499 | 6.023 |
| 0 | 273.15 | 754.0 | 0.212 | 2.16 | 4.760 | 4.760 |
| 0 | 273.15 | 760.0 | 0.214 | 2.18 | 4.805 | 4.805 |
| 24.8 | 297.95 | 760.0 | 0.145 | 1.53 | 3.254 | 3.549 |

[1] $\alpha$/cm$^3$(STP)cm$^{-3}$atm$^{-1}$

The mole fraction and Ostwald coefficient values were calculated by the compiler assuming ideal gas behavior.

Smoothed Data: For use between 253.15 and 298.15 K

$$\ln x_1 = -7.7254 + 10.6024/(T/100\ \mathrm{K})$$

The standard error about the regression line is 4.38 x $10^{-4}$.

| $T$/K | Mol Fraction $10^2 x_1$ |
|---|---|
| 253.15 | 2.91 |
| 263.15 | 2.48 |
| 273.15 | 2.14 |
| 283.15 | 1.87 |
| 293.15 | 1.64 |

AUXILIARY INFORMATION

**METHOD/APPARATUS/PROCEDURE:**

The authors describe two methods of gas solubility measurement. The ethane solubilities were measured by their method B.

The gas absorbed by a known volume of solvent is determined by measuring the pressure change in a gas reservoir of known volume. A correction for the non-ideality of the gas is applied.

The apparatus consists of a steel bomb (the gas reservoir) connected to a pressure gage, two regulators, and a 500 cm$^3$ absorption vessel. The solvent is degassed in the absorption vessel.

**SOURCE AND PURITY OF MATERIALS:**

(1) Ethane. Matheson Co., Inc. Research grade. Stated to be 99.90 mole per cent pure.

(2) Methylbenzene. Fisher Co. Spectrophotometric grade.

**ESTIMATED ERROR:**

$\delta T$/K = ± 0.2
$\delta p$/atm = ± 0.01
$\delta\alpha/\alpha$ = ± 0.02

**REFERENCES:**

**COMPONENTS:**

(1) Ethane; $C_2H_6$; [74-84-0]

(2) Benzene; $C_6H_6$; [71-43-2]

**ORIGINAL MEASUREMENTS:**

Armitage, D.A.; Linford, R.G; Thornhill, D.G.T.

*Ind. Eng. Chem. Fundam.* 1978, *17*, 362-364.

**VARIABLES:**

$T$/K: 301.51
$P$/kPa: 101.325 (1 atm)

**PREPARED BY:**

W. Hayduk

**EXPERIMENTAL VALUES:**

| $t^1$/°C | $T$/K | Mole Fraction[1] $10^4 x_1$ | Bunsen Coefficient[2] $\alpha$/cm$^3$(STP)cm$^{-3}$atm$^{-1}$ | Ostwald Coefficient[2] $L$/cm$^3$cm$^{-3}$ |
|---|---|---|---|---|
| 28.36 | 301.51 | 141.5 | 3.58 | 3.95 |

[1] Original data

[2] Bunsen and Ostwald coefficients calculated by compiler assuming ideal gas behavior.

**AUXILIARY INFORMATION**

**METHOD/APPARATUS/PROCEDURE:**

The method is volumetric utilizing a glass apparatus containing a known volume of gas in a contact chamber through which degassed solvent is circulated using a magnetically operated glass pump. The solubility is determined from observed drop in pressure. Solvent is degassed by evacuating boiling solvent and frozen solvent in sequence.

**SOURCE AND PURITY OF MATERIALS:**

1. Cambrian Chemicals; minimum specified purity 99.0 mole per cent.

2. British Drug Houses; minimum specified purity 99.8 mole per cent.

**ESTIMATED ERROR:**

$\delta x_1/x_1 = 0.01$ (authors)

**REFERENCES:**

**COMPONENTS:**

(1) Ethane; $C_2H_6$; [74-84-0]

(2) Benzene; $C_6H_6$; [71-43-2]

**ORIGINAL MEASUREMENTS:**

Horiuti, J.

*Sci. Pap. Inst. Phys. Chem. Res. (Jpn)* 1931/32, *17*, 125 - 256.

**VARIABLES:**

$T$/K: 278.15 - 323.15

$p_1$/kPa: 101.325 (1 atm)

**PREPARED BY:**

M. E. Derrick

H. L. Clever

**EXPERIMENTAL VALUES:**

| $T$/K | Mol Fraction $10^2x_1$ | Bunsen Coefficient $\alpha/cm^3(STP)cm^{-3}atm^{-1}$ | Ostwald Coefficient $L/cm^3cm^{-3}$ |
|---|---|---|---|
| 278.15 | 1.947 | 5.097 | 5.190 |
| 283.15 | 1.813 | 4.712 | 4.885 |
| 288.15 | 1.691 | 4.364 | 4.604 |
| 293.15 | 1.585 | 4.063 | 4.360 |
| 298.15 | 1.484 | 3.775 | 4.120 |
| 303.15 | 1.398 | 3.533 | 3.921 |
| 308.15 | 1.317 | 3.305 | 3.728 |
| 313.15 | 1.243 | 3.098 | 3.552 |
| 318.15 | 1.178 | 2.915 | 3.395 |
| 323.15 | 1.119 | 2.751 | 3.255 |

The mole fraction and Bunsen coefficient values were calculated by the compiler with the assumption the gas is ideal and that Henry's law is obeyed.

Smoothed Data: For use between 278.15 and 323.15 K.

$$\ln x_1 = -10.7165 + 15.0700/(T/100\text{K}) + 1.3295 \ln (T/100\text{K})$$

The standard error about the regression line is $1.26 \times 10^{-5}$.

| $T$/K | Mol Fraction $10^2x_1$ |
|---|---|
| 278.15 | 1.948 |
| 288.15 | 1.691 |
| 298.15 | 1.485 |
| 308.15 | 1.317 |
| 318.15 | 1.178 |
| 328.15 | 1.063 |

**AUXILIARY INFORMATION**

**METHOD/APPARATUS/PROCEDURE:**

The apparatus consists of a gas buret, a solvent reservoir, and an absorption pipet. The volume of the pipet is determined at various meniscus heights by weighing a quantity of water. The meniscus height is read with a cathetometer.

The dry gas is introduced into the degassed solvent. The gas and solvent are mixed with a magnetic stirrer until saturation. Care is taken to prevent solvent vapor from mixing with the solute gas in the gas buret. The volume of gas is determined from the gas buret readings, the volume of solvent is determined from the meniscus height in the absorption pipet.

**SOURCE AND PURITY OF MATERIALS:**

(1) Ethane. A 60% solution of potassium acetate was electrolyzed between Pt and Cu electrodes. The gas was passed through several wash solutions, dried, and fractionated from liquid air. Boiling point (766 mmHg) -88.3°C.

(2) Benzene. Merck. Extra pure and free of sulfur. Refluxed with sodium amalgam, distilled. Boiling point (760 mmHg) 80.18°C.

**ESTIMATED ERROR:**

$\delta T$/K = 0.05

$\delta x_1/x_1$ = 0.01

**REFERENCES:**

COMPONENTS:

(1) Ethane; $C_2H_6$; [74-84-0]

(2) Benzene; $C_6H_6$; [71-43-2]

ORIGINAL MEASUREMENTS:

Jadot, R.

*J. Chim. Phys.* 1972, *69*, 1036-40

VARIABLES:

$T$/K: 298.15

PREPARED BY:

C.L. Young

EXPERIMENTAL VALUES:

| $T$/K | Henry's Law Constant, $H$/atm | Mole fraction[+] at partial pressure of 101.3kPa, $x_{C_2H_6}$ | #$\Delta H^\infty$ /cal $mol^{-1}$ (/J $mol^{-1}$) |
|---|---|---|---|
| 298.15 | 66.66 | 0.01500 | 555(2322) |

\+ Calculated by compiler assuming $x_{C_2H_6} = 1/H$

\# Excess partial molar enthalpy of solution at infinite dilution.

AUXILIARY INFORMATION

METHOD/APPARATUS/PROCEDURE:

The conventional gas chromatographic technique was used. The carrier gas was helium. The value of Henry's law constant was calculated from the retention time. The value applies to very low partial pressures of gas and there may be a substantial difference from that measured at 1 atm. pressure. There is also considerable uncertainty in the value of Henry's constant since no allowance was made for surface adsorption.

SOURCE AND PURITY OF MATERIALS:

No details given

ESTIMATED ERROR:

$\delta T$/K = ±0.05; $\delta H$ = ±2%

REFERENCES:

COMPONENTS:

(1) Ethane; $C_2H_6$; [74-84-0]

(2) Chlorobenzene; $C_6H_5Cl$; [108-90-7]

ORIGINAL MEASUREMENTS:

Horiuti, J.

*Sci. Pap. Inst. Phys. Chem. Res. (Jpn)* 1931/32, *17*, 125 - 256.

EXPERIMENTAL VALUES:

| $T$/K | Mol Fraction $10^3 x_1$ | Bunsen Coefficient $\alpha$/cm$^3$ (STP) cm$^{-3}$ atm$^{-1}$ | Ostwald Coefficient $L$/cm$^3$ cm$^{-3}$ |
|---|---|---|---|
| 273.15 | 21.35 | 4.900 | 4.900 |
| 278.15 | 19.67 | 4.486 | 4.568 |
| 283.15 | 18.18 | 4.119 | 4.270 |
| 288.15 | 16.79 | 3.781 | 3.989 |
| 293.15 | 15.61 | 3.494 | 3.750 |
| 298.15 | 14.56 | 3.238 | 3.534 |
| 303.15 | 13.61 | 3.009 | 3.340 |
| 308.15 | 12.79 | 2.811 | 3.171 |
| 313.15 | 12.02 | 2.628 | 3.013 |
| 318.15 | 11.34 | 2.464 | 2.870 |
| 323.15 | 10.73 | 2.320 | 2.745 |
| 328.15 | 10.15 | 2.183 | 2.622 |
| 333.15 | 9.625 | 2.057 | 2.509 |
| 338.15 | 9.165 | 1.948 | 2.412 |
| 343.15 | 8.706 | 1.840 | 2.312 |
| 348.15 | 8.260 | 1.736 | 2.213 |
| 353.15 | 7.942 | 1.660 | 2.146 |

The mole fraction and Bunsen coefficient values were calculated by the compiler with the assumption the gas is ideal and that Henry's law is obeyed.

Smoothed Data: For use between 273.15 and 353.15 K.

$$\ln x_1 = -12.3790 + 17.9594/(T/100\text{K}) + 1.9475 \ln (T/100\text{K})$$

The standard error about the regression line is $9.24 \times 10^{-5}$.

| $T$/K | Mol Fraction $10^2 x_1$ |
|---|---|
| 273.15 | 2.134 |
| 288.15 | 1.682 |
| 298.15 | 1.458 |
| 308.15 | 1.279 |
| 318.15 | 1.133 |
| 328.15 | 1.013 |
| 338.15 | 0.914 |
| 353.15 | 0.794 |

Continued on next page.

**COMPONENTS:**

(1) Ethane; $C_2H_6$; [74-84-0]

(2) Chlorobenzene; $C_6H_5Cl$; [108-90-7]

**ORIGINAL MEASUREMENTS:**

Horiuti, J.

*Sci. Pap. Inst. Phys. Chem. Res. (Jpn)* 1931/32, *17*, 125 - 256.

**VARIABLES:**

$T$/K: 273.15 - 353.15

$p_1$/kPa: 101.325 (1 atm)

**PREPARED BY:**

M. E. Derrick
H. L. Clever

**EXPERIMENTAL VALUES:**

See preceeding page.

**AUXILIARY INFORMATION**

**METHOD/APPARATUS/PROCEDURE:**

The apparatus consists of a gas buret, a solvent reservoir, and an absorption pipet. The volume of the pipet is determined at various meniscus heights by weighing a quantity of water. The meniscus height is read with a cathetometer.

The dry gas is introduced into the degassed solvent. The gas and solvent are mixed with a magnetic stirrer until saturation. Care is taken to prevent solvent vapor from mixing with the solute gas in the gas buret. The volume of gas is determined from the gas buret readings, the volume of solvent is determined from the meniscus height in the absorption pipet.

**SOURCE AND PURITY OF MATERIALS:**

(1) Ethane. A 60% solution of potassium acetate was electrolyzed between Pt and Cu electrodes. The gas was passed through several wash solutions, dried, and fractionated from liquid air.

(2) Chlorobenzene. Kahlbaum. Dried and distilled. Boiling point (760 mmHg) 131.96°C.

**ESTIMATED ERROR:**

$\delta T$/K = 0.05

$\delta x_1/x_1$ = 0.01

**REFERENCES:**

**COMPONENTS:**

(1) Ethane; $C_2H_6$; [74-84-0]

(2) Napthalene, 1-methyl-; $C_{11}H_{10}$; [1321-94-4]

**ORIGINAL MEASUREMENTS:**

Chappelow, C.C.; Prausnitz, J.M. *Am. Inst. Chem. Engnrs. Jr.* 1974, *20*, 1097-1104.

**VARIABLES:**

$T$/K: 300-475

**PREPARED BY:**

C.L. Young

**EXPERIMENTAL VALUES:**

| $T$/K | Henry's Constant[a] /atm | Mole fraction[b] at 1 atm partial pressure $/10^4 x_{C_2H_6}$ |
|---|---|---|
| 300 | 90.9 | 110 (104)[c] |
| 325 | 120 | 83.3 (82.9) |
| 350 | 150 | 66.7 (68.2) |
| 375 | 182 | 54.9 (57.6) |
| 400 | 212 | 47.2 (49.7) |
| 425 | 238 | 42.0 (43.6) |
| 450 | 257 | 38.9 (38.8) |
| 475 | 260 | 38.5 (35.0) |

[a] Authors stated measurements were made at several pressures and values of solubility used were all within the Henry's Law region.

[b] Calculated by compiler assuming linear relationship between mole fraction and pressure.

[c] From equation of smoothed data:

$$\ln x_1 = 887.82/T - 7.5245$$

**AUXILIARY INFORMATION**

**METHOD/APPARATUS/PROCEDURE:**

Volumetric apparatus similar to that described by Dymond and Hildebrand (1). Pressure measured with a null detector and precision gauge. Details in ref. (2).

**SOURCE AND PURITY OF MATERIALS:**

Solvents degassed, no other details given.

**ESTIMATED ERROR:**

$\delta T/K = \pm 0.1$; $\delta x_{C_2H_6} = \pm 1\%$

**REFERENCES:**


1. Dymond, J.; Hildebrand, J.H. *Ind. Eng. Chem. Fundam.* 1967, *6*, 130.
2. Cukor, P.M.; Prausnitz, J.M. *Ind. Eng. Chem. Fundam.* 1971, *10*, 638.

**COMPONENTS:**

(1) Ethane; $C_2H_6$; [74-84-0]

(2) Carbon disulfide; $CS_2$; [75-15-0]

**ORIGINAL MEASUREMENTS:**

Powell, R. J.

*J. Chem. Eng. Data* 1972, *17*, 302 - 304.

**VARIABLES:**

$T$/K: 273.15 - 303.15
$p_1$/kPa: 101.325 (1 atm)

**PREPARED BY:**

P. L. Long
H. L. Clever

**EXPERIMENTAL VALUES:**

| $T$/K | Mol Fraction $10^4 x_1$ | Bunsen Coefficient $\alpha$/cm$^3$ (STP) cm$^{-3}$ atm$^{-1}$ | Ostwald Coefficient $L$/cm$^3$ cm$^{-3}$ | $N = R \frac{\Delta \log x_1}{\Delta \log T}$ |
|---|---|---|---|---|
| 298.15 | 107.9 | 4.03 | 4.40 | -7.55 |

The Bunsen and Ostwald coefficients were calculated by the compiler.

The author states that the solubility measurements were made over the temperature interval of about 273.15 to 303.15 K, but only the solubility value at 298.15 K was given in the paper. The slope, N = R($\Delta$log $x_1$/$\Delta$log $T$), was given.

Smoothed Data: For use between 273.15 and 303.15 K

The smoothed data were calculated by the compiler from the slope, N, in the form

$$\log x_1 = \log (107.9 \times 10^{-4}) - (7.55/R) \log (T/298.15)$$

with $R$ = 1.9872 cal K$^{-1}$ mol$^{-1}$.

| $T$/K | Mol Fraction $x_1$ |
|---|---|
| 273.15 | 0.01505 |
| 278.15 | 0.01405 |
| 283.15 | 0.01313 |
| 293.15 | 0.01228 |
| 293.15 | 0.01151 |
| 298.15 | 0.01079 |
| 303.15 | 0.01013 |

AUXILIARY INFORMATION

**METHOD/APPARATUS/PROCEDURE:**

The apparatus is the Dymond and Hildebrand (1) apparatus which uses an all glass pumping system to spray slugs of degassed solvent into the gas. The amount of gas dissolved is calculated from the initial and final pressures. The solvent is degassed by freezing, pumping, and followed by boiling under reduced pressure.

**SOURCE AND PURITY OF MATERIALS:**

(1) Ethane. Source not given. Stated to be manufacturer's research grade, dried over $CaCl_2$ before use.

(2) Carbon disulfide. Source not given. Stated to be manufacturer's spectrochemical grade.

**ESTIMATED ERROR:**

$\delta x_1/x_1 = \pm 0.002$
$\delta N$/cal K$^{-1}$mol$^{-1}$ = ± 0.1

**REFERENCES:**

1. Dymond, J. H.; Hildebrand, J. H. *Ind. Eng. Chem. Fundam.* 1967, *6*, 130.

**COMPONENTS:**

(1) Ethane; $C_2H_6$; [74-84-0]

(2) Carbon disulfide; $CS_2$; [75-15-0]

**ORIGINAL MEASUREMENTS:**

Gjaldbaek, J. C.; Niemann, H.

*Acta Chem. Scand.* 1958, *12*, 611 - 614.

**VARIABLES:**

$T$/K: 298.15

$P$/kPa: 101.325 (1 atm)

**PREPARED BY:**

J. Chr. Gjaldbaek

**EXPERIMENTAL VALUES:**

| $T$/K | Mol Fraction $10^2 x_1$ | Bunsen Coefficient $\alpha$/cm$^3$ (STP) cm$^{-3}$ atm$^{-1}$ | Ostwald Coefficient $L$/cm$^3$ cm$^{-3}$ |
|---|---|---|---|
| 298.15 | 1.07 | 3.927 | 4.286 |
| 298.15 | 1.07 | 3.944 | 4.305 |

The Ostwald and mole fraction solubility values were calculated by the compiler. The solubilities are reported for an ethane partial pressure of 101.325 kPa (1 atm) assuming Henry's law is obeyed.

**AUXILIARY INFORMATION**

**METHOD/APPARATUS/PROCEDURE:**

A calibrated all-glass combined manometer and bulb was enclosed in an air thermostat and shaken until equilibrium. Mercury was used for calibration and as the confining liquid. The solvents were degassed in the apparatus. Details are in references 1 and 2.

The absorped volume of gas is calculated from the initial and final amounts, both saturated with solvent vapor. The amount of solvent is determined by the weight of displaced mercury.

**SOURCE AND PURITY OF MATERIALS:**

(1) Ethane. Phillips Petroleum Co. Research grade. According to combination analysis 99.6-100.4 per cent. Butane and higher hydrocarbons were absent, and ethene was less than 0.5 per cent.

(2) Carbon disulfide. Analytical reagent grade. Distilled, normal boiling point 46.03°C.

**ESTIMATED ERROR:**

$\delta T$/K = 0.05

$\delta x_1/x_1$ = 0.015

**REFERENCES:**

1. Lannung, A. *J. Am. Chem. Soc.* 1930, *52*, 68.

2. Gjaldbaek, J. C. *Acta Chem. Scand.* 1952, *6*, 623.

**COMPONENTS:**

(1) Ethane; $C_2H_6$; [74-84-0]

(2) Benzene; $C_6H_6$; [71-43-2]

(3) 1,1,2-Trichloro-1,2,2-trifluoroethane or Freon 113; $C_2Cl_3F_3$; [76-13-1]

**ORIGINAL MEASUREMENTS:**

Linford, R. G.; Hildebrand, J. H.

*J. Phys. Chem.* 1969, *73*, 4410-4411.

**VARIABLES:**

$T$/K: 298.15

$p$/kPa: 101.325 (1 atm)

$C_6H_6/x_2$: 0.260 - 0.776

**PREPARED BY:**

H. L. Clever

**EXPERIMENTAL VALUES:**

| $T$/K | Mol Fraction Benzene $x_2$ | Mol Fraction $10^2x_1$ | Bunsen Coefficient $\alpha/cm^3(STP)cm^{-3}atm^{-1}$ | Ostwald Coefficient $L/cm^3cm^{-3}$ |
|---|---|---|---|---|
| 298.15 | 0.0 | 2.858 | 5.50 | 6.00 |
| | 0.260 | 2.654 | | |
| | 0.510 | 2.380 | | |
| | 0.776 | 1.932 | | |
| | 1.000 | 1.510[1] | 3.84 | 4.19 |

[1] Value from Horiuti, J. *Sci. Papers Inst. Phys. Chem. Res. Tokyo* 1931, *17*, 125.

The Bunsen and Ostwald coefficients were calculated by the compiler.

**AUXILIARY INFORMATION**

**METHOD/APPARATUS/PROCEDURE:**

The liquid is saturated with the gas at a partial pressure of one atm.

The apparatus is that described by Dymond and Hildebrand (1). It uses an all-glass pumping system to spray slugs of degassed solvent into the gas. The amount of gas dissolved is calculated from the initial and final gas pressures.

**SOURCE AND PURITY OF MATERIALS:**

(1) Ethane. Source not given. Highest purity commercially available. Dried.

(2) Benzene

(3) 1,1,2-Trichloro-1,2,2-trifluoroethane. Both from Matheson, Coleman and Bell Co. Spectroquality.

**ESTIMATED ERROR:**

$\delta x_1/x_1 = 0.005$

**REFERENCES:**

1. Dymond, J. H.; Hildebrand, J. H. *I. & E. C. Fundam.* 1967, *6*, 130.

**COMPONENTS:**

(1) Ethane; $C_2H_6$; [74-84-0]

(2) Benzene; $C_6H_6$; [71-43-2]

(3) 1,1,2-Trichloro-1,2,2-trifluoroethane (Freon 113); $C_2Cl_3F_3$; (76-13-1]

**ORIGINAL MEASUREMENTS:**

Armitage, D.A.; Linford, R.G.; Thornhill, D.G.T.

*Ind. Eng. Chem. Fundam.* 1978, *17*, 362-364.

**VARIABLES:**

$T$/K: 298.11
$P$/kPa: 101.325 (1 atm)
$C_6H_6/x_2$: 0.104-0.776

**PREPARED BY:**

W. Hayduk

**EXPERIMENTAL VALUES:**

| $T$/K | Mole Fraction Benzene[1] $x_2$ | Mole Fraction Ethane in mixed solvent[1] $10^4x_1$ |
|---|---|---|
| 298.11 | 0.104 | 276.8 |
| | 0.203 | 271.0 |
| | 0.260 | 267.8[2] |
| | 0.308 | 261.4 |
| | 0.404 | 248.4 |
| | 0.502 | 239.2 |
| | 0.510 | 238.1[2] |
| | 0.602 | 223.9 |
| | 0.701 | 207.5 |
| | 0.756 | 199.6 |
| | 0.776 | 195.5[2] |

[1] From original data

[2] Values interpolated by authors

AUXILIARY INFORMATION

**METHOD/APPARATUS/PROCEDURE:**

The method is volumetric utilizing a glass apparatus containing a known volume of gas in a contact chamber through which degassed solvent is circulated using a magnetically operated glass pump. The solubility is determined from observed drop in pressure. Solvent is degassed by evacuating boiling solvent and frozen solvent in sequence.

**SOURCE AND PURITY OF MATERIALS:**

1. Cambrian Chemicals; minimum specified purity 99.0 mole per cent.
2. British Drug Houses; minimum specified purity 99.8 mole per cent.
3. British Drug Houses; minimum specified purity 99.8 mole per cent.

**ESTIMATED ERROR:**

$\delta T/\mathrm{K} = 0.02$ (compiler)
$\delta x_2/x_2 = 0.002$ (compiler)
$\delta x_1/x_1 = 0.01$ (authors)

**REFERENCES:**

**COMPONENTS:**

(1) Ethane; $C_2H_6$; [74-84-0]

(2) Benzene; $C_6H_6$; [71-43-2]

(3) 1,1,2-Trichloro-1,2,2-trifluro-ethane (Freon 113); $C_2Cl_3F_3$; [76-13-1]

**ORIGINAL MEASUREMENTS:**

Armitage, D.A.; Linford, R.G.; Thornhill, D.G.T.

*Ind. Eng. Chem. Fundam.* 1978, *17*, 362-364.

**VARIABLES:**

$T$/K: 287.71-308.04
$P$/kPa: 101.325 (1 atm)
$C_6H_6/x_2$: 0.756

**PREPARED BY:**

W. Hayduk

**EXPERIMENTAL VALUES:**

| $T$/K | Mole Fraction Benzene[1] $/x_2$ | Mole Fraction Ethane[1] $10^4x_1$ |
|---|---|---|
| 287.71 | 0.756 | 231.8 (231.5)[2] |
| 293.09 | 0.756 | 214.8 (214.4) |
| 298.11 | 0.756 | 199.6 (200.1) |
| 303.02 | 0.756 | 187.7 (187.4) |
| 308.04 | 0.756 | 175.9 (175.7) |

[1] Original data

[2] From equation of smoothed data between 287.71 and 308.04 K:

$$\ln x_1 = 1203.9/T - 7.950$$

**AUXILIARY INFORMATION**

**METHOD/APPARATUS/PROCEDURE:**

The method is volumetric utilizing a glass apparatus containing a known volume of gas in a contact chamber through which degassed solvent is circulated using a magnetically operated glass pump. The solubility is determined from observed drop in pressure. Solvent is degassed by evacuating boiling solvent and frozen solvent in sequence.

**SOURCE AND PURITY OF MATERIALS:**

1. Cambrian chemicals; minimum specified purity 99.0 mole per cent.
2. British Drug Houses; minimum specified purity 99.8 mole per cent.
3. British Drug Houses; minimum specified purity 99.8 mole per cent.

**ESTIMATED ERROR:**

$\delta T/K = 0.02$ (compiler)
$\delta x_2/x_2 = 0.002$ (compiler)
$\delta x_1/x_1 = 0.01$ (authors)

**REFERENCES:**

| COMPONENTS: | EVALUATOR: |
|---|---|
| (1) Ethane; $C_2H_6$; [74-84-0]<br>(2) Alcohols | Walter Hayduk<br>Department of Chemical Engineering<br>University of Ottawa<br>Ottawa, Canada K1N 9B4 |

CRITICAL EVALUATION:

The solubilities at atmospheric pressure in normal alcohols have been measured by four groups of researchers (1,2,3,4) with relatively consistent results with the exception of those of McDaniel (4) as shown in the figure below. The relation between the mole fraction solubility at 298.15 K and the number of carbon atoms per alcohol molecule is shown on logarithmic scales. McDaniel's (4) early results are known to be significantly lower than those of other workers and are rejected.

Boyer and Bircher (2) developed an equation describing their data at 298.15 K (and also at 308.15 K) in normal alcohols of carbon numbers up to eight:

$$\log x_1 = 0.668 \log C_n - 2.365$$

Although their equation describes their data well, it does not represent the combined data of Ben-Naim and Yaacobi (1) and Gjaldbaek and Niemann (3) as well as that of Boyer and Bircher (2). The combined data were correlated with the following equation which yielded a correlation coefficient of 0.9982:

$$\ln x_1 = 0.70796 \ln C_n - 5.5244$$

The above equation shown as a dotted line in the figure, represents the combined data with a maximum deviation of 5%, usually less, and may be considered tentative for solubilities in normal alcohols from methanol to octanol. Only Boyer and Bircher's data are available for heptanol and octanol; but these seem entirely consistent with all the other data.

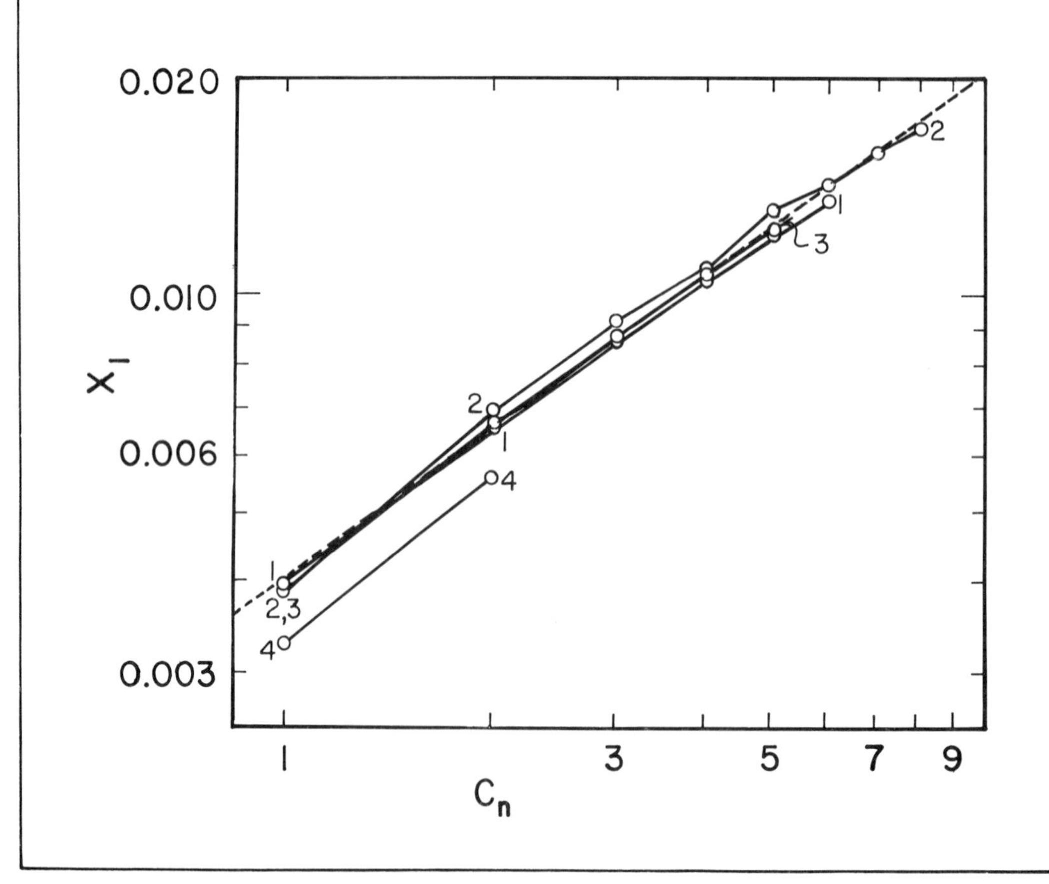

COMPONENTS:

(1) Ethane; $C_2H_6$; [74-84-0]

(2) Alcohols

EVALUATOR:

Walter Hayduk
Department of Chemical Engineering
University of Ottawa
Ottawa, Canada K1N 9B4

CRITICAL EVALUATION:

...continued

Attention is drawn to the fact that only for some of the data in normal alcohols (2,4) was the ethane molar volume corresponding to that of an ideal gas used in the conversion of Ostwald coefficients to mole fractions. In the other cases a true ethane molar volume was used. The difference is approximately 0.7% at 298.15 K depending on the actual molar volume data used.

The two sets of data that are available for the solubilities at 298.15 K in ethylene glycol (3,5) differ substantially-by a factor of three. In view of the nature of the low pressure chromatographic technique with its possible sources of error, it is considered that the data of Lenoir et al. (5) in ethylene glycol are unreliable. For dilute solutions and at low pressures, Henry's law usually applies to a good approximation. Because the effective solubilities measured were extremely low, surface effects may have become greater than simple solubility effects in the Lenoir et al. experiments, and hence made interpretation of the results difficult. Similarly, the data of these authors in benzyl alcohol, phenol and dipropylene glycol are also questioned and considered doubtful.

On the other hand, the solubility in ethylene glycol at atmospheric pressure of Gjaldbaek and Niemann (3) is considered tentative. It appears consistent with the solubilities of other gases in ethylene glycol (6). The solubility of the same authors in cyclohexanol is very roughly two thirds that in normal hexanol, a fraction which is similar for solubilities in cyclohexane compared with those in hexane.

The solubilities of Ezhelva and Zorin (7) in ethylcellosolve are considered tentative.

## References

1. Ben-Naim, A.; Yaacobi, M. *J. Phys. Chem.* 1974, *78*, 175-178.

2. Boyer, F.L.; Bircher, L.J. *J. Phys. Chem.* 1960, *64*, 1330-1331.

3. Gjaldbaek, J.C.; Niemann, H. *Acta Chem. Scand.* 1958, *12*, 1015-1023.

4. McDaniel, A.S. *J. Phys. Chem.* 1911, *15*, 587-610.

5. Lenoir, J-Y.; Renault, P.; Renon, H. *J. Chem. Eng. Data* 1971, *16*, 340-342.

6. Hayduk, W.; Laudie, H. *Am. Inst. Chem. Eng.* J. 1973, *19*, 1233-1238.

7. Ezheleva, A.E.; Zorin, A.D. *Tr. Khim. Khim. Tech. (Gorkii)* 1961, *37*, 37-40.

**COMPONENTS:**

(1) Ethane; $C_2H_6$; [74-84-0]

(2) Methanol; $CH_4O$; [67-56-1]

**ORIGINAL MEASUREMENTS:**

Boyer, F. L.; Bircher, L. J.

*J. Phys. Chem.* 1960, *64*, 1330 - 1331.

**VARIABLES:**

$T$/K: 298.15
$P$/kPa: 101.325 (1 atm)

**PREPARED BY:**

M. E. Derrick
H. L. Clever

**EXPERIMENTAL VALUES:**

| $T$/K | Mol Fraction $10^4 x_1$ | Bunsen Coefficient[1] $\alpha$ | Ostwald Coefficient $L$/cm$^3$ cm$^{-3}$ |
|---|---|---|---|
| 298.15 | 40.5 (38.8) | 2.14 | 2.34 ± 0.09 |

[1] $\alpha$/cm$^3$(STP) cm$^{-3}$ atm$^{-1}$

The Bunsen coefficient was calculated by the compiler.

The mole fraction solubility was taken from Boyer's thesis (1).

The authors observed a linear relationship between the logarithm of the mole fraction solubility and the number of linear alcohol carbon atoms. Boyer's thesis gives the equations:

$$\log x_1 = -2.365 + 0.668 \log C \text{ for } 298.15 \text{ K}$$
$$\log x_1 = -2.441 + 0.700 \log C \text{ for } 308.15 \text{ K}$$

where C is the number of alcohol carbon atoms. Most of the mole fraction solubility values given in Table II of the paper were calculated from the equation for 298.15 K.

The mole fraction solubility calculated by the compilers differs from that of the authors and is shown in brackets.

**AUXILIARY INFORMATION**

**METHOD/APPARATUS/PROCEDURE:**

A commercial Van Slyke blood gas apparatus (E. H. Sargent Co.) was modified by the authors.

The total pressure of the gas and the solvent vapor in the solution chamber was adjusted to a pressure of one atm. The pressure was maintained at one atm during the solution process. The saturated solution was transferred to a bulb below the lower stopcock of the extraction vessel and sealed off. The gas and solvent vapor were then brought to volume over mercury. See (2) for details of the extraction procedure.

**SOURCE AND PURITY OF MATERIALS:**

(1) Ethane. Phillips Petroleum Co. Stated to be 99.9 mole per cent.

(2) Methanol. Source not given. Treated by standard methods to remove aldehydes and ketones, then dried and distilled.

**ESTIMATED ERROR:**

$\delta T$/K = ± 0.01
$\delta L$/cm$^3$ = ± 0.09

**REFERENCES:**

1. Boyer, F. L., Ph.D. thesis, 1959 Vanderbilt Univ., Nashville, TN

2. Peters, J. P.; Van Slyke, D. D. *Quantitative Clinical Chemistry* Baltimore, MD, 1932, Volume II.

COMPONENTS:

(1) Ethane; $C_2H_6$; [74-84-0]

(2) Methanol; $CH_4O$; [67-56-1]

ORIGINAL MEASUREMENTS:

Ben-Naim, A.; Yaacobi, M.

*J. Phys. Chem.*, 1974, *78*, 175-8

VARIABLES:

$T$/K: 283.15-303.15

PREPARED BY:

C.L. Young

EXPERIMENTAL VALUES:

| $T$/K | Ostwald coefficient,* $L$ | Mole fraction$^+$ at partial pressure of 101.3 kPa, $x_{C_2H_6}$ |
|---|---|---|
| 283.15 | 2.864 | 0.00491 (0.00490)** |
| 288.15 | 2.678 | 0.00454 (0.00455) |
| 293.15 | 2.518 | 0.00422 (0.00423) |
| 298.15 | 2.379 | 0.00395 (0.00395) |
| 303.15 | 2.258 | 0.00370 (0.00369) |

* Smoothed values obtained from the equation

$$kT \ln L = 4,979.8 - 25.237 (T/K) + 0.03440 (T/K)^2 \text{ cal mol}^{-1}$$

where k is in units of cal $mol^{-1}$ $K^{-1}$

\+ calculated by compiler assuming the ideal gas law for ethane.

** From alternate equation:

$$\ln x_1 = 1211.15/T - 9.5964$$

Correlation coefficient = 0.9998

AUXILIARY INFORMATION

METHOD/APPARATUS/PROCEDURE:

The apparatus was similar to that described by Ben-Naim and Baer (1) and Wen and Hung (2). It consists of three main parts, a dissolution cell of 300 to 600 $cm^3$ capacity, a gas volume measuring column, and a manometer. The solvent is degassed in the dissolution cell, the gas is introduced and dissolved while the liquid is kept stirred by a magnetic stirrer immersed in the water bath. Dissolution of the gas results in the change in the height of a column of mercury which is measured by a cathetometer.

SOURCE AND PURITY OF MATERIALS:

1. Matheson sample, purity 99.9 mol per cent.

2. AR grade.

ESTIMATED ERROR:

$\delta T/K = \pm 0.1$; $\delta x_{C_2H_6} = \pm 2\%$

(estimated by compiler)

REFERENCES:

1. Ben-Naim, A.; Baer, S. *Trans. Faraday Soc.* 1963, *59*, 2735

2. Wen, W.-Y.; Hung, J.H. *J. Phys. Chem.* 1974, *74*, 170

**COMPONENTS:**

(1) Ethane; $C_2H_6$; [74-84-0]

(2) Methanol; $CH_4O$; [67-56-1]

**ORIGINAL MEASUREMENTS:**

Gjaldbaek, J. C.; Niemann, H.

*Acta Chem. Scand.* 1958, *12*, 1015 - 1023.

**VARIABLES:**

$T$/K: 298.15
$P$/kPa: 101.325 (1 atm)

**PREPARED BY:**

J. Chr. Gjaldbaek

**EXPERIMENTAL VALUES:**

| $T$/K | Mol Fraction $10^3 x_1$ | Bunsen Coefficient $\alpha/cm^3(STP)cm^{-3}atm^{-1}$ | Ostwald Coefficient $L/cm^3 cm^{-3}$ |
|---|---|---|---|
| 298.15 | 3.88 | 2.124 | 2.318 |
| 298.16 | 3.96 | 2.161 | 2.359 |
| 298.15 | 3.80 | 2.075 | 2.265 |

The Ostwald and mole fraction solubility values were calculated by the compiler. The solubilities are reported for an ethane partial pressure of 101.325 kPa (1 atm) assuming Henry's law is obeyed.

**AUXILIARY INFORMATION**

**METHOD/APPARATUS/PROCEDURE:**

A calibrated all-glass combined manometer and bulb was enclosed in an air thermostat and shaken for equilibration. Mercury was used for calibration and as the confining liquid. The solvents were degassed in the apparatus. Details are in references 1 and 2.

The absorbed volume of gas is calculated from the initial and final amounts, both saturated with solvent vapor. The amount of solvent is determined by the weight of displaced mercury.

**SOURCE AND PURITY OF MATERIALS:**

(1) Ethane. Phillips Petroleum Co. Research grade. According to combination analysis 99.6-100.4 per cent. Butane and higher hydrocarbons were absent, and ethene was less than 0.5 per cent.

(2) Methanol. Source not given. Dried with magnesium and fractionated. Boiling point, $t$/°C 64.60-64.63. Refractive index (NaD, 20°C) 1.3285.

**ESTIMATED ERROR:**

$\delta T/K = 0.05$
$\delta x_1/x_1 = 0.015$

**REFERENCES:**

1. Lannung, A. *J. Am. Chem. Soc.* 1930, *52*, 68.

2. Gjaldbaek, J. C. *Acta Chem. Scand.* 1952, *6*, 623.

**COMPONENTS:**

(1) Ethane; $C_2H_6$; [74-84-0]

(2) Methanol; $CH_4O$; [67-56-11]

**ORIGINAL MEASUREMENTS:**

McDaniel, A.S.

*J. Phys. Chem.*, 1911, *15*, 587-610.

**VARIABLES:**

$T$/K: 295.65 - 318.35
$P$/kPa: 101.325 (1 atm)

**PREPARED BY:**

W. Hayduk

**EXPERIMENTAL VALUES:**

| $t$/°C | $T$/K | Ostwald coefficient[1] $L$/cm$^3$cm$^{-3}$ | Bunsen coefficient[3] $\alpha$/cm$^3$(STP)cm$^{-3}$atm$^{-1}$ | Mole fraction[3] $10^4x_1$ |
|---|---|---|---|---|
| 22.5 | 295.65 | 2.02 | 1.87 | 33.7 |
| 25.0 | 298.15 | 1.98[2] | 1.81 | 32.7 |
| 30.1 | 303.25 | 1.88 | 1.69 | 30.8 |
| 45.2 | 318.35 | 1.73 | 1.48 | 27.4 |

[1] Original data listed as Absorption coefficient interpreted by compiler to be equivalent to Ostwald coefficient as listed here.

[2] Ostwald coefficient (Absorption coefficient) as estimated at 298.15 K by author.

[3] Bunsen coefficient and mole fraction solubility calculated by compiler assuming ideal gas behavior.

[4] McDaniel's results are consistently from 20 to 80 per cent too low when compared with more reliable data.

**AUXILIARY INFORMATION**

**METHOD/APPARATUS/PROCEDURE:**

Glass apparatus consisting of a gas burette connected to a solvent contacting chamber. Gas pressure or volume adjusted using mercury displacement. Equilibration achieved at atmospheric pressure by hand shaking apparatus, and incrementally adding gas to contacting chamber. Solubility measured by obtaining total uptake of gas by known volume of solvent.

**SOURCE AND PURITY OF MATERIALS:**

1. Prepared by reaction of ethyl iodide with zinc-copper Purity not measured.

2. Source not given; purity specified as 99 per cent.

**ESTIMATED ERROR:**

$\delta L/L = -0.20$
(estimated by compiler; see note [4] above)

**REFERENCES:**

**COMPONENTS:**

(1) Ethane; $C_2H_6$; [74-84-0]

(2) Ethanol; $C_2H_6O$; [64-17-5]

**ORIGINAL MEASUREMENTS:**

Boyer, F. L.; Bircher, L. J.

*J. Phys. Chem.* 1960, *64*, 1330 - 1331.

**VARIABLES:**

$T$/K: 298.15
$P$/kPa: 101.325 (1 atm)

**PREPARED BY:**

M. E. Derrick
H. L. Clever

**EXPERIMENTAL VALUES:**

| $T$/K | Mol Fraction $10^4x_1$ | Bunsen Coefficient[1] $\alpha$ | Ostwald Coefficient $L$/cm$^3$ cm$^{-3}$ |
|---|---|---|---|
| 298.15 | 68.2 | 2.63 | 2.87 ± 0.04 |

[1] $\alpha$/cm$^3$(STP) cm$^{-3}$ atm$^{-1}$

The Bunsen coefficient was calculated by the compiler.

The mole fraction solubility was taken from Boyer's thesis (1).

See the methanol data sheet for the equations relating the mole fraction solubility and the number of normal alcohol carbon numbers.

AUXILIARY INFORMATION

**METHOD/APPARATUS/PROCEDURE:**

A commercial Van Slyke blood gas apparatus (E. H. Sargent Co.) was modified by the authors.

The total pressure of the gas and the solvent vapor in the solution chamber was adjusted to a pressure of one atm. The pressure was maintained at one atm during the solution process. The saturated solution was transferred to a bulb below the lower stopcock of the extraction vessel and sealed off. The gas and solvent vapor were then brought to volume over mercury. See (2) for details of the extraction procedure.

**SOURCE AND PURITY OF MATERIALS:**

(1) Ethane. Phillips Petroleum Co. Stated to be 99.9 mol per cent.

(2) Ethanol. Source not given. Treated by standard methods to remove aldehydes and ketones, then dried and distilled.

**ESTIMATED ERROR:**

$\delta T$/K = ± 0.01
$\delta L$/cm$^3$ = ± 0.04

**REFERENCES:**

1. Boyer, F. L., Ph.D. thesis, 1959 Vanderbilt Univ., Nashville, TN

2. Peters, J. P.; Van Slyke, D. D. *Quantitative Clinical Chemistry* Baltimore, MD, 1932, Volume II.

**COMPONENTS:**

(1) Ethane; $C_2H_6$; [74-84-0]

(2) Ethanol; $C_2H_6O$; [74-17-5]

**ORIGINAL MEASUREMENTS:**

Yaacobi, M.; Ben-Naim, A.[4]

*J. Solution Chem.* 1973, *2*, 425-443.

**VARIABLES:**

$T$/K: 283.15-303.15
$P$/kPa: 101.325 (1 atm)

**PREPARED BY:**

W. Hayduk

**EXPERIMENTAL VALUES:**

| $t$/°C | $T$/K | Ostwald Coefficient[1] $L$/cm$^3$cm$^{-3}$ | Bunsen Coefficient[2] $\alpha$/cm$^3$(STP)cm$^{-3}$atm$^{-1}$ | Mole Fraction[2] /$10^4 x_1$ |
|---|---|---|---|---|
| 10 | 283.15 | 3.361 | 3.24 | 82.8 (82.5)[3] |
| 15 | 288.15 | 3.117 | 2.95 | 75.9 (76.1) |
| 20 | 293.15 | 2.910 | 2.71 | 70.1 (70.3) |
| 25 | 298.15 | 2.730 | 2.50 | 65.1 (65.2) |
| 30 | 303.15 | 2.580 | 2.33 | 60.8 (60.6) |

[1] From original data

[2] Bunsen coefficient and mole fraction calculated by compiler assuming ideal gas behavior.

[3] From equation of smoothed data between 283.15 and 303.15 K:

$$\ln x_1 = 1326.8/T - 9.4833$$

Correlation coefficient = 0.9997

[4] Same data also subsequently reported by Ben-Naim and Yaacobi in *J. Phys. Chem.* 1974, *73*, 175-178.

**AUXILIARY INFORMATION**

**METHOD/APPARATUS/PROCEDURE:**

The method is volumetric utilizing an all-glass apparatus consisting of a dissolution cell of 300 to 600 cm$^3$ capacity, a gas volume measuring column, and a manometer. The solvent is degassed in the dissolution cell. Gas dissolves while the liquid is stirred using a magnetic stirrer. The volume of gas confined over mercury is read initially and after equilibration, by means of a cathetometer.

The apparatus is described by Ben-Naim and Baer (1) but it includes the modification introduced by Wen and Hung (2) of replacing the stopcocks with Teflon needle valves.

**SOURCE AND PURITY OF MATERIALS:**

1. Matheson; minimum specified purity 99.9 mole per cent.

2. Absolute ethanol; supplier not specified.

**ESTIMATED ERROR:**

$\delta T$/K = 0.05
$\delta L/L$ = 0.01
Estimated by compiler

**REFERENCES:**

1. Ben-Naim, A.; Baer, S. *Trans. Faraday Soc.* 1963, *59*, 2735.

2. Wen, W.-Y.; Hung, J.H. *J. Phys. Chem.* 1970, *74*, 170.

**COMPONENTS:**

(1) Ethane; $C_2H_6$; [74-84-0]

(2) Ethanol; $C_2H_6O$; [64-17-5]

**ORIGINAL MEASUREMENTS:**

Gjaldbaek, J. C.; Niemann, H.

*Acta Chem. Scand.* 1958, *12*, 1015 - 1023.

**VARIABLES:**

$T$/K: 298.15
$P$/kPa: 101.325 (1 atm)

**PREPARED BY:**

J. Chr. Gjaldbaek

**EXPERIMENTAL VALUES:**

| $T$/K | Mol Fraction $10^3 x_1$ | Bunsen Coefficient $\alpha$/cm$^3$ (STP) cm$^{-3}$ atm$^{-1}$ | Ostwald Coefficient $L$/cm$^3$ cm$^{-3}$ |
|---|---|---|---|
| 298.15 | 6.64 | 2.523 | 2.754 |
| 298.15 | 6.63 | 2.519 | 2.749 |

The Ostwald and mole fraction solubility values were calculated by the compiler. The solubilities are reported for an ethane partial pressure of 101.325 kPa (1 atm) assuming Henry's law is obeyed.

**AUXILIARY INFORMATION**

**METHOD/APPARATUS/PROCEDURE:**

A calibrated all-glass combined manometer and bulb was enclosed in an air thermostat and shaken for equilibration. Mercury was used for calibration and as the confining liquid. The solvents were degassed in the apparatus. Details are in references 1 and 2.

The absorbed volume of gas is calculated from the initial and final amounts, both saturated with solvent vapor. The amount of solvent is determined by the weight of displaced mercury.

**SOURCE AND PURITY OF MATERIALS:**

(1) Ethane. Phillips Petroleum Co. Research grade. According to combination analysis 99.6-100.4 per cent. Butane and higher hydrocarbons were absent, and ethene was less than 0.5 per cent.

(2) Ethanol. Source not given. Dried with magnesium and fractionated. Boiling point, $t$/°C 78.49. Refractive index (NaD, 20°C) 1.3614.

**ESTIMATED ERROR:**

$\delta T$/K = 0.05
$\delta x_1/x_1$ = 0.015

**REFERENCES:**

1. Lannung, A. *J. Am. Chem. Soc.* 1930, *52*, 68.

2. Gjaldbaek, J. C. *Acta Chem. Scand.* 1952, *6*, 623.

COMPONENTS:

(1) Ethane; $C_2H_6$; [74-84-0]
(2) Ethanol; $C_2H_6O$; [64-17-5]

ORIGINAL MEASUREMENTS:

McDaniel, A.S.

*J. Phys. Chem.*, 1911, *15*, 587-610.

VARIABLES:

$T$/K: 295.15 - 323.15
$P$/kPa: 101.325 (1 atm)

PREPARED BY:

W. Hayduk

EXPERIMENTAL VALUES:

| $t$/°C | $T$/K | Ostwald coefficient[1] $L$/cm$^3$cm$^{-3}$ | Bunsen coefficient [3] $\alpha$/cm$^3$(STP)cm$^{-3}$atm$^{-1}$ | Mole fraction[3] $10^4x_1$ |
|---|---|---|---|---|
| 22.0 | 295.15 | 2.33 | 2.16 | 56.7 |
| 25.0 | 298.15 | 2.29[2] | 2.10 | 55.2 |
| 30.0 | 303.15 | 2.22 | 2.00 | 52.9 |
| 40.0 | 313.15 | 2.07 | 1.80 | 48.2 |
| 50.0 | 323.15 | 1.85 | 1.57 | 42.5 |

[1] Original data listed as Absorption coefficient interpreted by compiler to be equivalent to Ostwald coefficient as listed here.

[2] Ostwald coefficient (Absorption coefficient) as estimated at 298.15 K by author.

[3] Bunsen coefficient and mole fraction solubility calculated by compiler assuming ideal gas behavior.

[4] McDaniel's results are consistently from 20 to 80 per cent too low when compared with more reliable data.

AUXILIARY INFORMATION

METHOD/APPARATUS/PROCEDURE:

Glass apparatus consisting of a gas burette connected to a solvent contacting chamber. Gas pressure or volume adjusted using mercury displacement. Equilibration achieved at atmospheric pressure by hand shaking apparatus, and incrementally adding gas to contacting chamber. Solubility measured by obtaining total uptake of gas by known volume of solvent.

SOURCE AND PURITY OF MATERIALS:

1. Prepared by reaction of ethyl iodide with zinc-copper. Purity not measured.

2. Source not given; purity specified as 99 per cent.

ESTIMATED ERROR:

$\delta L/L$ = -0.20
(estimated by compiler; see note [4] above)

REFERENCES:

**COMPONENTS:**

(1) Ethane; $C_2H_6$; [74-84-0]

(2) 1-Propanol; $C_3H_8O$; [71-23-8]

**ORIGINAL MEASUREMENTS:**

Boyer, F. L.; Bircher, L. J.

*J. Phys. Chem.* 1960, *64*, 1330 - 1331.

**VARIABLES:**

$T$/K: 298.15, 308.15

$P$/kPa: 101.325 (1 atm)

**PREPARED BY:**

M. E. Derrick
H. L. Clever

**EXPERIMENTAL VALUES:**

| $T$/K | Mol Fraction $10^4 x_1$ | Bunsen Coefficient[1] $\alpha$ | Ostwald Coefficient $L/cm^3\ cm^{-3}$ |
|---|---|---|---|
| 298.15 | 91.7 | 2.73 | 2.98 ± 0.01 |
| 308.15 | 78.2 | 2.36 | 2.66 ± 0.05 |

[1] $\alpha/cm^3(STP)\ cm^{-3}\ atm^{-1}$

The Bunsen coefficients were calculated by the compiler.

The mole fraction solubilities were taken from Boyer's thesis (1).

See the methanol data sheet for the equations relating the mole fraction solubility and the number of normal alcohol carbon numbers.

**AUXILIARY INFORMATION**

**METHOD/APPARATUS/PROCEDURE:**

A commercial Van Slyke blood gas apparatus (E. H. Sargent Co.) was modified by the authors.

The total pressure of the gas and the solvent vapor in the solution chamber was adjusted to a pressure of one atm. The pressure was maintained at one atm during the solution process. The saturated solution was transferred to a bulb below the lower stopcock of the extraction vessel and sealed off. The gas and solvent vapor were then brought to volume over mercury. See (2) for details of the extraction procedure.

**SOURCE AND PURITY OF MATERIALS:**

(1) Ethane. Phillips Petroleum Co. Stated to be 99.9 mol per cent.

(2) 1-Propanol. Source not given. Treated by standard methods to remove aldehydes and ketones, then dried and distilled.

**ESTIMATED ERROR:**

$\delta T/K = \pm 0.01$

$\delta L/cm^3 = \pm 0.01$ (at 298.15)
$\pm 0.05$ (at 308.15)

**REFERENCES:**

1. Boyer, F. L., Ph.D. thesis, 1959 Vanderbilt Univ., Nashville, TN

2. Peters, J. P.; Van Slyke, D. D. *Quantitative Clinical Chemistry* Baltimore, MD, 1932, Volume II.

COMPONENTS:

(1) Ethane; $C_2H_6$; [74-84-0]

(2) 1-Propanol; $C_3H_8O$; [71-23-8]

ORIGINAL MEASUREMENTS:

Ben-Naim, A.; Yaacobi, M.

*J. Phys. Chem.* 1974, *78*, 175-8

VARIABLES:

$T$/K: 283.15-303.15

$P$/KPa: 101.325 (1atm)

PREPARED BY:

C.L. Young

EXPERIMENTAL VALUES:

| $T$/K | Ostwald coefficient,* $L$ | Mole fraction+ at partial pressure of 101.3 kPa, $x_{C_2H_6}$ |
|---|---|---|
| 283.15 | 3.492 | 0.0110 (0.0110)** |
| 288.15 | 3.241 | 0.0101 (0.0101) |
| 293.15 | 3.014 | 0.00928 (0.00927) |
| 298.15 | 2.808 | 0.00855 (0.00855) |
| 303.15 | 2.621 | 0.00789 (0.00790) |

* Smoothed values obtained from the equation.

$kT \ln L = 1{,}928.0 - 2.606\ (T/K) - 0.00607\ (T/K)^2$ cal mol$^{-1}$
where k is in units of cal mol$^{-1}$ K$^{-1}$

\+ calculated by compiler assuming the ideal gas law for ethane.

** From alternate equation:

$$\ln x_1 = 1427.0/T - 9.5484$$

AUXILIARY INFORMATION

METHOD/APPARATUS/PROCEDURE:

The apparatus was similar to that described by Ben-Naim and Baer (1) and Wen and Hung (2). It consists of three main parts, a dissolution cell of 300 to 600 $cm^3$ capacity, a gas volume measuring column, and a manometer. The solvent is degassed in the dissolution cell, the gas is introuduced and dissolved while the liquid is kept stirred by a magnetic stirrer immersed in the water bath. Dissolution of the gas results in the change in the height of a column of mercury which is measured by a cathetometer.

SOURCE AND PURITY OF MATERIALS:

1. Matheson sample, purity 99.9 mol per cent.
2. AR grade.

ESTIMATED ERROR:

$\delta T/K = \pm 0.1$; $\delta x_{C_2H_6} = \pm 2\%$
(estimated by compiler)

REFERENCES:

1. Ben-Naim, A.; Baer, S. *Trans.Faraday Soc.* 1963, *59*, 2735.
2. Wen, W.-Y.; Hung, J.H. *J. Phys. Chem.* 1970, *74*, 170.

**COMPONENTS:**

(1) Ethane; $C_2H_6$; [74-84-0]

(2) 1-Propanol; $C_3H_8O$; [71-23-8]

**ORIGINAL MEASUREMENTS:**

Gjaldbaek, J. C.; Niemann, H.

*Acta Chem. Scand.* 1958, *12*, 1015 - 1023.

**VARIABLES:**

$T$/K: 298.15, 308.15

$p_1$/kPa: 101.325 (1 atm)

**PREPARED BY:**

J. Chr. Gjaldbaek

**EXPERIMENTAL VALUES:**

| $T$/K | Mol Fraction $10^3 x_1$ | Bunsen Coefficient $\alpha$/cm$^3$ (STP) cm$^{-3}$ atm$^{-1}$ | Ostwald Coefficient $L$/cm$^3$ cm$^{-3}$ |
|---|---|---|---|
| 298.15 | 8.75 | 2.599 | 2.837 |
| 298.16 | 8.72 | 2.593 | 2.830 |
| 308.15 | 7.57 | 2.235 | 2.521 |
| 308.15 | 7.58 | 2.239 | 2.526 |

The mole fraction and Ostwald coefficients were calculated by the compiler.

Smoothed Data: For use between 298.15 and 308.15 K.

$$\ln x_1 = -9.1333 + 13.0976/(T/100\text{K})$$

| $T$/K | Mol Fraction $10^3 x_1$ |
|---|---|
| 298.15 | 8.745 |
| 303.15 | 8.125 |
| 308.15 | 7.575 |

**AUXILIARY INFORMATION**

**METHOD/APPARATUS/PROCEDURE:**

A calibrated all-glass combined manometer and bulb was enclosed in an air thermostat and shaken for equilibration. Mercury was used for calibration and as the confining liquid. The solvents were degassed in the apparatus. Detials are in references 1 and 2.

The absorbed volume of gas is calculated from the initial and final amounts, both saturated with solvent vapor. The amount of solvent is determined by the weight of displaced mercury.

**SOURCE AND PURITY OF MATERIALS:**

(1) Ethane. Phillips Petroleum Co. Research grade. According to combination analysis 99.6-100.4 per cent. Butane and higher hydrocarbons were absent, and ethene was less than 0.5 per cent.

(2) 1-Propanol. Source not given. Boiling point, $t$/°C 97.1 - 97.4. Refractive index (NaD 20°C) 1.3856.

**ESTIMATED ERROR:**

$\delta T/\text{K} = 0.05$

$\delta x_1/x_1 = 0.015$

**REFERENCES:**

1. Lannung, A. *J. Am. Chem. Soc.* 1930, *52*, 68.

2. Gjaldbaek, J. C. *Acta Chem. Scand.* 1952, *6*, 623.

**COMPONENTS:**

(1) Ethane; $C_2H_6$; [74-84-0]

(2) 1-Butanol; $C_4H_{10}O$; [71-36-3]

**ORIGINAL MEASUREMENTS:**

Boyer, F. L.; Bircher, L. J.

*J. Phys. Chem.* 1960, *64*, 1330 - 1331.

**VARIABLES:**

$T$/K: 298.15, 308.15
$P$/kPa: 101.325 (1 atm)

**PREPARED BY:**

M. E. Derrick
H. L. Clever

**EXPERIMENTAL VALUES:**

| $T$/K | Mol Fraction $10^4 x_1$ | Bunsen Coefficient[1] $\alpha$ | Ostwald Coefficient $L/cm^3\ cm^{-3}$ |
|---|---|---|---|
| 298.15 | 109.0 | 2.68 | 2.93 ± 0.01 |
| 308.15 | 93.4 | 2.27 | 2.56 ± 0.03 |

[1] $\alpha/cm^3$(STP) $cm^{-3}$ $atm^{-1}$

The Bunsen coefficients were calculated by the compiler.

The mole fraction solubilities were taken from Boyer's thesis (1).

See the methanol data sheet for the equations relating the mole fraction solubility and the number of normal alcohol carbon numbers.

AUXILIARY INFORMATION

**METHOD/APPARATUS/PROCEDURE:**

A commercial Van Slyke blood gas apparatus (E. H. Sargent Co.) was modified by the authors.

The total pressure of the gas and the solvent vapor in the solution chamber was adjusted to a pressure of one atm. The pressure was maintained at one atm during the solution process. The saturated solution was transferred to a bulb below the lower stopcock of the extraction vessel and sealed off. The gas and solvent vapor were then brought to volume over mercury. See (2) for details of the extraction procedure.

**SOURCE AND PURITY OF MATERIALS:**

(1) Ethane. Phillips Petroleum Co. Stated to be 99.9 mol per cent.

(2) 1-Butanol. Source not given. Treated by standard methods to remove aldehydes and ketones, then dried and distilled.

**ESTIMATED ERROR:**

$\delta T/K = \pm\ 0.01$

$\delta L/cm^3 = \pm\ 0.01$ (at 298.15)
$\pm\ 0.03$ (at 308.15)

**REFERENCES:**

1. Boyer, F. L., Ph.D. thesis, 1959 Vanderbilt Univ., Nashville, TN

2. Peters, J. P.; Van Slyke, D. D. *Quantitative Clinical Chemistry* Baltimore, MD, 1932, Volume II.

**COMPONENTS:**

(1) Ethane; $C_2H_6$; [74-84-0]

(2) 1-Butanol; $C_4H_{10}O$; [71-36-3]

**ORIGINAL MEASUREMENTS:**

Ben-Naim, A.; Yaacobi, M.

*J. Phys. Chem.* 1974, *78*, 175-178

**VARIABLES:**

$T$/K: 283.15-303.15

**PREPARED BY:**

C.L. Young

**EXPERIMENTAL VALUES:**

| $T$/K | Ostwald coefficient* $L$ | Mole fraction+ at partial pressure of 101.3 kPa, $x_{C_2H_6}$ |
|---|---|---|
| 283.15 | 3.491 | 0.0135 (0.0143)** |
| 288.15 | 3.225 | 0.0123 (0.0123) |
| 293.15 | 3.008 | 0.0113 (0.0114) |
| 298.15 | 2.830 | 0.0105 (0.0106) |
| 303.15 | 2.686 | 0.00988(0.00979) |

* Smoothed values obtained from the equation:

$$kT \ln L = 9{,}001.0 - 51{,}627\ (T/K) + 0.07884\ (T/K)^2 \text{cal mol}^{-1}$$

where k = 1.987 cal $mol^{-1}$ $K^{-1}$

+ calculated by compiler assuming the ideal gas law for ethane.

** From alternate equation of smoothed data:

$$\ln x_1 = 1345.4/T - 9.0640$$

Correlation coefficient = 0.9981

**AUXILIARY INFORMATION**

**METHOD/APPARATUS/PROCEDURE:**

The apparatus was similar to that described by Ben-Naim and Baer (1) and Wen and Hung (2). It consists of three main parts, a dissolution cell of 300 to 600 $cm^3$ capacity, a gas volume measuring column, and a manometer. The solvent is degassed in the dissolution cell, the gas is introduced and dissolved while the liquid is kept stirred by a magnetic stirrer immersed in the water bath. Dissolution of the gas results in the change in the height of a column of mercury which is measured by a cathetometer.

**SOURCE AND PURITY OF MATERIALS:**

1. Matheson sample, purity 99.9 mole per cent.

2. AR grade.

**ESTIMATED ERROR:**

$\delta T$/K = ±0.1; $\delta x_{C_2H_6}$ = ±2%

(estimated by compiler)

**REFERENCES:**


1. Ben-Naim, A.; Baer, S. *Trans. Faraday Soc.* 1963, *59*, 2735.

2. Wen, W.-Y.; Hung, J.H. *J. Phys. Chem.* 1970, *74*, 170.

**COMPONENTS:**

(1) Ethane; $C_2H_6$; [74-84-0]

(2) 1-Butanol; $C_4H_{10}O$; [71-36-3]

**ORIGINAL MEASUREMENTS:**

Gjaldbaek, J. C.; Niemann, H.

*Acta Chem. Scand.* 1958, *12*, 1015 - 1023.

**VARIABLES:**

$T$/K: 298.15

$p_1$/kPa: 101.325 (1 atm)

**PREPARED BY:**

J. Chr. Gjaldbaek

**EXPERIMENTAL VALUES:**

| $T$/K | Mol Fraction $10^2x_1$ | Bunsen Coefficient $\alpha$/cm$^3$ (STP) cm$^{-3}$ atm$^{-1}$ | Ostwald Coefficient $L$/cm$^3$ cm$^{-3}$ |
|---|---|---|---|
| 298.15 | 1.08 | 2.623 | 2.863 |
| 298.16 | 1.07 | 2.597 | 2.835 |

The Ostwald and mole fraction solubility values were calculated by the compiler. The solubilities are reported for an ethane partial pressure of 101.325 kPa (1 atm) assuming Henry's law is obeyed.

**AUXILIARY INFORMATION**

**METHOD/APPARATUS/PROCEDURE:**

A calibrated all-glass combined manometer and bulb was enclosed in an air thermostat and shaken for equilibration. Mercury was used for calibration and as the confining liquid. The solvents were degassed in the apparatus. Details are in references 1 and 2.

The absorbed volume of gas is calculated from the initial and final amounts, both saturated with solvent vapor. The amount of solvent is determined by the weight of displaced mercury.

**SOURCE AND PURITY OF MATERIALS:**

(1) Ethane. Phillips Petroleum Co. Research grade. According to combination analysis 99.6-100.4 per cent. Butane and higher hydrocarbons were absent, and ethene was less than 0.5 per cent.

(2) 1-Butanol. Source not given. Boiling point, $t$/°C 117.75 - 117.83. Refractive index (NaD, 20°C) 1.3995.

**ESTIMATED ERROR:**

$\delta T/K = 0.05$

$\delta x_1/x_1 = 0.015$

**REFERENCES:**

1. Lannung, A. *J. Am. Chem. Soc.* 1930, *52*, 68.

2. Gjaldbaek, J. C. *Acta Chem. Scand.* 1952, *6*, 623.

**COMPONENTS:**

(1) Ethane; $C_2H_6$; [74-84-0]

(2) 1-Pentanol; $C_5H_{12}O$; [71-41-0]

**ORIGINAL MEASUREMENTS:**

Boyer, F. L.; Bircher, L. J.

*J. Phys. Chem.* 1960, *64*, 1330 - 1331.

**VARIABLES:**

$T$/K: 298.15, 308.15
$P$/kPa: 101.325 (1 atm)

**PREPARED BY:**

M. E. Derrick
H. L. Clever

**EXPERIMENTAL VALUES:**

| $T$/K | Mol Fraction $10^4 x_1$ | Bunsen Coefficient[1] $\alpha$ | Ostwald Coefficient $L/cm^3\ cm^{-3}$ |
|---|---|---|---|
| 298.15 | 130.0 | 2.70 | 2.95 ± 0.01 |
| 308.15 | 114.0 | 2.34 | 2.64 ± 0.03 |

[1] $\alpha/cm^3(STP)\ cm^{-3}\ atm^{-1}$

The Bunsen coefficients were calculated by the compiler.

The mole fraction solubilities were taken from Boyer's thesis (1).

See the methanol data sheet for the equations relating the mole fraction solubility and the number of normal alcohol carbon numbers.

**AUXILIARY INFORMATION**

**METHOD/APPARATUS/PROCEDURE:**

A commercial Van Slyke blood gas apparatus (E. H. Sargent Co.) was modified by the authors.

The total pressure of the gas and the solvent vapor in the solution chamber was adjusted to a pressure of one atm. The pressure was maintained at one atm during the solution process. The saturated solution was transferred to a bulb below the lower stopcock of the extraction vessel and sealed off. The gas and solvent vapor were then brought to volume over mercury. See (2) for details of the extraction procedure.

**SOURCE AND PURITY OF MATERIALS:**

(1) Ethane. Phillips Petroleum Co. Stated to be 99.9 mol per cent.

(2) 1-Pentanol. Source not given. Treated by standard methods to remove aldehydes and ketones, then dried and distilled.

**ESTIMATED ERROR:**

$\delta T$/K = ± 0.01

$\delta L/cm^3$ = ± 0.01 (at 298.15)
± 0.03 (at 308.15)

**REFERENCES:**

1. Boyer, F. L., Ph.D. thesis, 1959 Vanderbilt Univ., Nashville, TN

2. Peters, J. P.; Van Slyke, D. D. *Quantitative Clinical Chemistry* Baltimore, MD, 1932, Volume II.

COMPONENTS:

(1) Ethane; $C_2H_6$; [76-84-0]

(2) 1-Pentanol; $C_5H_{12}O$; [71-41-0]

ORIGINAL MEASUREMENTS:

Ben-Naim, A.; Yaacobi, M.

*J. Phys. Chem.* 1974, *78*, 175-8

VARIABLES:

$T$/K = 283.15-303.15

PREPARED BY:

C.L. Young

EXPERIMENTAL VALUES:

| $T$/K | Ostwald coefficient,* $L$ | Mole fraction$^+$ at partial pressure of 101.3 kPa, $x_{C_2H_6}$ |
|---|---|---|
| 283.15 | 3.403 | 0.0154 |
| 288.15 | 3.155 | 0.0142 |
| 293.15 | 2.935 | 0.0130 |
| 298.15 | 2.740 | 0.0120 |
| 303.15 | 2.566 | 0.0112 |

* Smoothed values obtained from the equation :

$kT \ln L = 3{,}395.6 - 12.817\ (T/K) + 0.01155\ (T/K)^2$ cal mol$^{-1}$
where k is in units of cal mol$^{-1}$ K$^{-1}$

\+ calculated by compiler assuming the ideal gas law for ethane ; alternate equation:

$$\ln x_1 = 1383.0/T - 9.0576$$

AUXILIARY INFORMATION

METHOD/APPARATUS/PROCEDURE:

The apparatus was similar to that described by Ben-Naim and Baer (1) and Wen and Hung (2). It consists of three main parts, a dissolution cell of 300 to 600 cm$^3$ capacity, a gas volume measuring column, and a manometer. The solvent is degassed in the dissolution cell, the gas is introduced and dissolved while the liquid is kept stirred by a magnetic stirrer immersed in the water bath. Dissolution of the gas results in the change in the height of a column of mercury which is measured by a cathetometer.

SOURCE AND PURITY OF MATERIALS:

1. Matheson sample, purity 99.9 mol per cent.

2. AR grade.

ESTIMATED ERROR:

$\delta T$/K = ±0.1; $\delta x_{C_2H_6}$ = ±2% (estimated by compiler).

REFERENCES:


1. Ben-Naim, A.; Baer, S. *Trans. Faraday Soc.* 1963, *59*, 2735.

2. Wen, W.-Y.; Hung, J.H. *J. Phys. Chem.* 1970, *74*, 170.

**COMPONENTS:**

(1) Ethane; $C_2H_6$; [74-84-0]

(2) 1-Pentanol; $C_5H_{12}O$; [71-41-0]

**ORIGINAL MEASUREMENTS:**

Gjaldbaek, J. C.; Niemann, H.

*Acta Chem. Scand.* 1958, *12*, 1015 - 1023.

**VARIABLES:**

$T$/K: 298.15, 308.15

$p_1$/kPa: 101.325 (1 atm)

**PREPARED BY:**

J. Chr. Gjaldbaek

**EXPERIMENTAL VALUES:**

| $T$/K | Mol Fraction $10^2 x_1$ | Bunsen Coefficient $\alpha$/cm$^3$ (STP) cm$^{-3}$ atm$^{-1}$ | Ostwald Coefficient $L$/cm$^3$ cm$^{-3}$ |
|---|---|---|---|
| 298.16 | 1.23 | 2.520 | 2.75 |
| 298.15 | 1.25 | 2.565 | 2.80 |
| 298.15 | 1.24 | 2.498 | 2.73 |
| 308.15 | 1.07 | 2.190 | 2.47 |
| 308.14 | 1.06 | 2.176 | 2.45 |

The mole fraction and Ostwald coefficient values were calculated by the compiler.

Smoothed Data: For use between 298.15 and 308.15 K, by compiler.

$$\ln x_1 = -9.0818 + 13.9884/(T/100\text{K})$$

| $T$/K | Mol Fraction $10^2 x_1$ |
|---|---|
| 298.15 | 1.24 |
| 303.15 | 1.15 |
| 308.15 | 1.065 |

**AUXILIARY INFORMATION**

**METHOD/APPARATUS/PROCEDURE:**

A calibrated all-glass combined manometer and bulb was enclosed in an air thermostat and shaken for equilibration. Mercury was used for calibration and as the confining liquid. The solvents were degassed in the apparatus. Details are in references 1 and 2.

The absorbed volume of gas is calculated from the initial and final amounts, both saturated with solvent vapor. The amount of solvent is determined by the weight of displaced mercury.

**SOURCE AND PURITY OF MATERIALS:**

(1) Ethane. Phillips Petroleum Co. Research grade. According to combination analysis 99.6-100.4 per cent. Butane and higher hydrocarbons were absent, and ethene was less than 0.5 per cent.

(2) 1-Pentanol. Source not given. Fractionated at about 115 mmHg. Boiling point, $t$/°C 137.83 - 137.90 at 760 mmHg. Refractive index (NaD, 20°C) 1.412.

**ESTIMATED ERROR:**

$\delta T$/K = 0.05

$\delta x_1/x_1$ = 0.015

**REFERENCES:**

1. Lannung, A.
*J. Am. Chem. Soc.* 1930, *52*, 68.

2. Gjaldbaek, J. C.
*Acta Chem. Scand.* 1952, *6*, 623.

**COMPONENTS:**

(1) Ethane; $C_2H_6$; [74-84-0]

(2) 1-Hexanol; $C_6H_{14}O$; [111-27-3]

**ORIGINAL MEASUREMENTS:**

Boyer, F. L.; Bircher, L. J.

*J. Phys. Chem.* 1960, *64*, 1330 - 1331.

**VARIABLES:**

$T$/K: 298.15
$P$/kPa: 101.325 (1 atm)

**PREPARED BY:**

M. E. Derrick
H. L. Clever

**EXPERIMENTAL VALUES:**

| $T$/K | Mol Fraction $10^4 x_1$ | Bunsen Coefficient[1] $\alpha$ | Ostwald Coefficient $L/cm^3\ cm^{-3}$ |
|---|---|---|---|
| 298.15 | 143.0 | 2.60 | 2.84 ± 0.03 |

[1] $\alpha/cm^3(STP)\ cm^{-3}\ atm^{-1}$

The Bunsen coefficient was calculated by the compiler.

The mole fraction solubility was taken from Boyer's thesis (1).

See the methanol data sheet for the equations relating the mole fraction solubility and the number of normal alcohol carbon numbers.

**AUXILIARY INFORMATION**

**METHOD/APPARATUS/PROCEDURE:**

A commercial Van Slyke blood gas apparatus (E. H. Sargent Co.) was modified by the authors.

The total pressure of the gas and the solvent vapor in the solution chamber was adjusted to a pressure of one atm. The pressure was maintained at one atm during the solution process. The saturated solution was transferred to a bulb below the lower stopcock of the extraction vessel and sealed off. The gas and solvent vapor were then brought to volume over mercury. See (2) for details of the extraction procedure.

**SOURCE AND PURITY OF MATERIALS:**

(1) Ethane. Phillips Petroleum Co. Stated to be 99.9 mol per cent.

(2) 1-Hexanol. Source not given. Treated by standard methods to remove aldehydes and ketones, then dried and distilled.

**ESTIMATED ERROR:**

$\delta T$/K = ± 0.01

$\delta L$ = ± 0.03

**REFERENCES:**

1. Boyer, F. L., Ph.D. thesis, 1959 Vanderbilt Univ., Nashville, TN

2. Peters, J. P.; Van Slyke, D. D. *Quantitative Clinical Chemistry* Baltimore, MD, 1932, Volume II.

**COMPONENTS:**

(1) Ethane; $C_2H_6$; [74-84-0]

(2) 1-Hexanol; $C_6H_{14}O$; [111-27-3]

**ORIGINAL MEASUREMENTS:**

Ben-Naim, A.; Yaacobi, M.

*J. Phys. Chem.* 1974, *78*, 175-8

**VARIABLES:**

$T$/K = 283.15-303.15

**PREPARED BY:**

C.L. Young

**EXPERIMENTAL VALUES:**

| $T$/K | Ostwald coefficient,* $L$ | Mole fraction$^+$ at partial pressure of 101.3 kPa, $x_{C_2H_6}$ |
|---|---|---|
| 283.15 | 3.252 | 0.0170 |
| 288.15 | 3.034 | 0.0157 |
| 293.15 | 2.842 | 0.0145 |
| 298.15 | 2.674 | 0.0135 |
| 303.15 | 2.526 | 0.0126 |

* Smoothed values obtained from the equation

$kT \ln L = 4{,}183.0 - 19.124\ (T/K) + 0.02364\ (T/K)^2$ cal mol$^{-1}$ where $k = 1.987$ cal mol$^{-1}$ K$^{-1}$

\+ calculated by compiler assuming ideal gas law for ethane ; alternate equation: $\ln x_1 = 1288.1/T - 8.6249$

**AUXILIARY INFORMATION**

**METHOD/APPARATUS/PROCEDURE:**

The apparatus was similar to that described by Ben-Naim and Baer (1) and Wen and Hung (2). It consists of three main parts, a dissolution cell of 300 to 600 $cm^3$ capacity, a gas volume measuring column, and a manometer. The solvent is degassed in the dissolution cell, the gas is introduced and dissolved while the liquid is kept stirred by a magnetic stirrer immersed in the water bath. Dissolution of the gas results in the change in the height of a column of mercury which is measured by a cathetometer.

**SOURCE AND PURITY OF MATERIALS:**

1. Matheson sample, purity 99.9 mol per cent.

2. AR grade.

**ESTIMATED ERROR:**

$\delta T$/K = ±0.1; $\delta x_{C_2H_6}$ = ±2% (estimated by compiler)

**REFERENCES:**

1. Ben-Naim, A.; Baer, S. *Trans. Faraday Soc.* 1963, *59*, 2735.

2. Wen, W.-Y.; Hung, J.H. *J. Phys. Chem.* 1970, *74*, 170.

**COMPONENTS:**

(1) Ethane; $C_2H_6$; [74-84-0]

(2) 1-Heptanol; $C_7H_{16}O$; [111-70-6]

**ORIGINAL MEASUREMENTS:**

Boyer, F. L.; Bircher, L. J.

*J. Phys. Chem.* 1960, *64*, 1330 - 1331.

**VARIABLES:**

$T$/K: 298.15
$P$/kPa: 101.325 (1 atm)

**PREPARED BY:**

M. E. Derrick
H. L. Clever

**EXPERIMENTAL VALUES:**

| $T$/K | Mol Fraction $10^4 x_1$ | Bunsen Coefficient[1] $\alpha$ | Ostwald Coefficient $L/\text{cm}^3\ \text{cm}^{-3}$ |
|---|---|---|---|
| 298.15 | 158.0 | 2.53 | 2.76 ± 0.03 |

[1] $\alpha/\text{cm}^3(\text{STP})\ \text{cm}^{-3}\ \text{atm}^{-1}$

The Bunsen coefficient was calculated by the compiler.

The mole fraction solubility was taken from Boyer's thesis (1).

See the methanol data sheet for the equations relating the mole fraction solubility and the number of normal alcohol carbon numbers.

**AUXILIARY INFORMATION**

**METHOD/APPARATUS/PROCEDURE:**

A commercial Van Slyke blood gas apparatus (E. H. Sargent Co.) was modified by the authors.

The total pressure of the gas and the solvent vapor in the solution chamber was adjusted to a pressure of one atm. The pressure was maintained at one atm during the solution process. The saturated solution was transferred to a bulb below the lower stopcock of the extraction vessel and sealed off. The gas and solvent vapor were then brought to volume over mercury. See (2) for details of the extraction procedure.

**SOURCE AND PURITY OF MATERIALS:**

(1) Ethane. Phillips Petroleum Co. Stated to be 99.9 mol per cent.

(2) 1-Heptanol. Source not given. Treated by standard methods to remove aldehydes and ketones, then dried and distilled.

**ESTIMATED ERROR:**

$\delta T/\text{K} = \pm 0.01$

$\delta L = \pm 0.03$

**REFERENCES:**

1. Boyer, F. L., Ph.D. thesis, 1959 Vanderbilt Univ., Nashville, TN

2. Peters, J. P.; Van Slyke, D. D. *Quantitative Clinical Chemistry* Baltimore, MD, 1932, Volume II.

**COMPONENTS:**

(1) Ethane; $C_2H_6$; [74-84-0]

(2) 1-Octanol; $C_8H_{18}O$; [111-87-5]

**ORIGINAL MEASUREMENTS:**

Boyer, F. L.; Bircher, L. J.

*J. Phys. Chem.* 1960, *64*, 1330 - 1331.

**VARIABLES:**

$T$/K: 298.15, 308.15
$P$/kPa: 101.325 (1 atm)

**PREPARED BY:**

M. E. Derrick
H. L. Clever

**EXPERIMENTAL VALUES:**

| $T$/K | Mol Fraction $10^4 x_1$ | Bunsen Coefficient[1] $\alpha$ | Ostwald Coefficient $L$/cm$^3$ cm$^{-3}$ |
|---|---|---|---|
| 298.15 | 170.0 | 2.44 | 2.66 ± 0.03 |
| 308.15 | 153.0 | 2.16 | 2.44 ± 0.02 |

[1] $\alpha$/cm$^3$(STP) cm$^{-3}$ atm$^{-1}$

The Bunsen coefficients were calculated by the compiler.

The mole fraction solubilities were taken from Boyer's thesis (1).

See the methanol data sheet for the equations relating the mole fraction solubility and the number of normal alcohol carbon numbers.

AUXILIARY INFORMATION

**METHOD/APPARATUS/PROCEDURE:**

A commercial Van Slyke blood gas apparatus (E. H. Sargent Co.) was modified by the authors.

The total pressure of the gas and the solvent vapor in the solution chamber was adjusted to a pressure of one atm. The pressure was maintained at one atm during the solution process. The saturated solution was transferred to a bulb below the lower stopcock of the extraction vessel and sealed off. The gas and solvent vapor were then brought to volume over mercury. See (2) for details of the extraction procedure.

**SOURCE AND PURITY OF MATERIALS:**

(1) Ethane. Phillips Petroleum Co. Stated to be 99.9 mol per cent.

(2) 1-Octanol. Source not given. Treated by standard methods to remove aldehydes and ketones, then dried and distilled.

**ESTIMATED ERROR:**

$\delta T$/K = ± 0.01

$\delta L$ = ± 0.03 (at 298.15)
± 0.02 (at 308.15)

**REFERENCES:**

1. Boyer, F. L., Ph.D. thesis, 1959 Vanderbilt Univ., Nashville, TN

2. Peters, J. P.; Van Slyke, D. D. *Quantitative Clinical Chemistry* Baltimore, MD, 1932, Volume II.

**COMPONENTS:**

(1) Ethane; $C_2H_6$; [74-84-0]

(2) Cyclohexanol; $C_6H_{12}O$; [108-93-0]

**ORIGINAL MEASUREMENTS:**

Gjaldbaek, J. C.; Niemann, H.

*Acta Chem. Scand.* 1958, *12*, 1015 - 1023.

**VARIABLES:**

$T$/K: 298.15

$p_1$/kPa: 101.325 (1 atm)

**PREPARED BY:**

J. Chr. Gjaldbaek

**EXPERIMENTAL VALUES:**

| $T$/K | Mol Fraction $10^3 x_1$ | Bunsen Coefficient $\alpha/cm^3 (STP) cm^{-3} atm^{-1}$ | Ostwald Coefficient $L/cm^3 cm^{-3}$ |
|---|---|---|---|
| 298.15 | 8.30 | 1.750 | 1.910 |
| 298.16 | 8.28 | 1.747 | 1.901 |

The Ostwald and mole fraction solubility values were calculated by the compiler. The solubilities are reported for an ethane partial pressure of 101.325 kPa (1 atm) assuming Henry's law is obeyed.

**AUXILIARY INFORMATION**

**METHOD/APPARATUS/PROCEDURE:**

A calibrated all-glass combined manometer and bulb was enclosed in an air thermostat and shaken for equilibration. Mercury was used for calibration and as the confining liquid. The solvents were degassed in the apparatus. Details are in references 1 and 2.

The absorbed volume of gas is calculated from the initial and final amounts, both saturated with solvent vapor. The amount of solvent is determined by the weight of displaced mercury.

**SOURCE AND PURITY OF MATERIALS:**

(1) Ethane. Phillips Petroleum Co. Research grade. According to combination analysis 99.6-100.4 per cent. Butane and higher hydrocarbons were absent, and ethene was less than 0.5 per cent.

(2) Cyclohexanol. Source not given. Fractionated by distillation at low pressure and fractionated by freezing. Completely solid at 24°C. Estimated water content was 0.05 per cent.

**ESTIMATED ERROR:**

$\delta T/K = 0.05$

$\delta x_1/x_1 = 0.015$

**REFERENCES:**

1. Lannung, A. *J. Am. Chem. Soc.* 1930, *52*, 68.

2. Gjaldbaek, J. C. *Acta Chem. Scand.* 1952, *6*, 623.

**COMPONENTS:**

(1) Ethane; $C_2H_6$; [74-84-0]

(2) 1,2-Ethanediol or ethylene glycol; $C_2H_6O_2$; [107-21-1]

**ORIGINAL MEASUREMENTS:**

Gjaldbaek, J. C.; Niemann, H.

*Acta Chem. Scand.* 1958, *12*, 1015 - 1023.

**VARIABLES:**

$T$/K: 298.16, 308.14
$p_1$/kPa: 101.325 (1 atm)

**PREPARED BY:**

J. Chr. Gjaldbaek

**EXPERIMENTAL VALUES:**

| $T$/K | Mol Fraction $10^4 x_1$ | Bunsen Coefficient $\alpha$/cm$^3$(STP)cm$^{-3}$atm$^{-1}$ | Ostwald Coefficient $L$/cm$^3$cm$^{-3}$ |
|---|---|---|---|
| 298.16 | 5.44 | 0.2159 | 0.2357 |
| 298.16 | 5.42 | 0.2150 | 0.2347 |
| 308.14 | 4.90 | 0.1949 | 0.2199 |
| 308.14 | 4.89 | 0.1945 | 0.2194 |

The mole fraction and Ostwald coefficient values were calculated by the compiler.

Smoothed Data: For use between 298.15 and 308.15 K.

$$\ln x_1 = -10.7210 + 9.5487/(T/100\text{K})$$

| $T$/K | Mol Fraction $10^4 x_1$ |
|---|---|
| 298.15 | 5.43 |
| 303.15 | 5.15 |
| 308.15 | 4.895 |

**AUXILIARY INFORMATION**

**METHOD/APPARATUS/PROCEDURE:**

A calibrated all-glass combined manometer and bulb was enclosed in an air thermostat and shaken for equilibration. Mercury was used for calibration and as the confining liquid. The solvents were degassed in the apparatus. Details are in references 1 and 2.

The absorbed volume of gas is calculated from the initial and final amounts, both saturated with solvent vapor. The amount of solvent is determined by the weight of displaced mercury.

**SOURCE AND PURITY OF MATERIALS:**

(1) Ethane. Phillips Petroleum Co. Research grade. According to combination analysis 99.6-100.4 per cent. Butane and higher hydrocarbons were absent, and ethene was less than 0.5 per cent.

(2) 1,2-Ethanediol. Fractionated in a column at low pressure. Boiling point $t$/°C 197.30 - 197.42. Refractive index (NaD, 20°C) 1.4320.

**ESTIMATED ERROR:**

$\delta T/\text{K} = 0.05$
$\delta x_1/x_1 = 0.015$

**REFERENCES:**

1. Lannung, A. *J. Am. Chem. Soc.* 1930, *52*, 68.

2. Gjaldbaek, J. C. *Acta Chem. Scand.* 1952, *6*, 623.

| COMPONENTS: | ORIGINAL MEASUREMENTS: |
|---|---|
| (1) Ethane; $C_2H_6$; [74-84-0]<br>(2) Phenol; $C_6H_6O$; [108-95-2] | Lenoir, J-Y.; Renault, P.; Renon, H.<br>*J. Chem. Eng. Data*, 1971, *16*, 340-3 |
| VARIABLES:<br>$T$/K: 323.2 | PREPARED BY:<br>C. L. Young |

EXPERIMENTAL VALUES:

| $T$/K | Henry's constant $H_{C_2H_6}$/atm | Mole fraction at 1 atm* $x_{C_2H_6}$ |
|---|---|---|
| 323.2 | 270 | 0.00370 |

* Calculated by compiler assuming a linear function of $p_{C_2H_6}$ vs $x_{C_2H_6}$, i.e. $x_{C_2H_6}$(1 atm) = $1/H_{C_2H_6}$

AUXILIARY INFORMATION

METHOD/APPARATUS/PROCEDURE:

A conventional gas-liquid chromatographic unit fitted with a thermal conductivity detector was used. The carrier gas was helium. The value of Henry's law constant was calculated from the retention time. The value applies to very low partial pressures of gas and there may be a substantial difference from that measured at 1 atm. pressure. There is also considerable uncertainty in the value of Henry's constant since surface adsorption was not allowed for although its possible existence was noted.

SOURCE AND PURITY OF MATERIALS:

(1) L'Air Liquide sample, minimum purity 99.9 mole per cent.

(2) Touzart and Matignon or Serlabo sample, purity 99 mole per cent.

ESTIMATED ERROR:

$\delta T$/K = $\pm 0.1$; $\delta H$/atm = $\pm 6$%
(estimated by compiler).

REFERENCES:

COMPONENTS:

(1) Ethane; $C_2H_6$; [74-84-0]

(2) 1,2-Ethanediol,(Ethylene glycol); $C_2H_6O_2$; [107-21-1]

ORIGINAL MEASUREMENTS:

Lenoir, J-Y.; Renault, P.; Renon, H.

*J. Chem. Eng. Data*, 1971, *16*, 340-2.

VARIABLES:

$T$/K: 298.2

PREPARED BY:

C. L. Young

EXPERIMENTAL VALUES:

| $T$/K | Henry's constant $H_{C_2H_6}$/atm | Mole fraction at 1 atm* $x_{C_2H_6}$ |
|---|---|---|
| 298.2 | 620 | 0.00161 |

* Calculated by compiler assuming a linear function of $p_{C_2H_6}$ vs $x_{C_2H_6}$, i.e., $x_{C_2H_6}$(1 atm) = $1/H_{C_2H_6}$.

AUXILIARY INFORMATION

METHOD/APPARATUS/PROCEDURE:

A conventional gas-liquid chromatographic unit fitted with a thermal conductivity detector was used. The carrier gas was helium. The value of Henry's law constant was calculated from the retention time. The value applies to very low partial pressures of gas and there may be a substantial difference from that measured at 1 atm. pressure. There is also considerable uncertainty in the value of Henry's constant since surface adsorption was not allowed for although its possible existence was noted.

SOURCE AND PURITY OF MATERIALS:

(1) L'Air Liquide sample, minimum purity 99.9 mole per cent.

(2) Touzart and Matignon or Serlabo sample, purity 99 mole per cent.

ESTIMATED ERROR:

$\delta T$/K = $\pm 0.1$; $\delta H$/atm = $\pm 6\%$ (estimated by compiler).

REFERENCES:

**COMPONENTS:**

(1) Ethane; $C_2H_6$; [74-84-0]

(2) Oxybispropanol, (Dipropylene glycol); $C_6H_{14}O_3$; [110-98-5]

**ORIGINAL MEASUREMENTS:**

Lenoir, J-Y; Renault, P.; Renon, H.

*J. Chem. Eng. Data*, 1971, *16*, 340-342.

**VARIABLES:**

$T$/K: 298.2-343.2

**PREPARED BY:**

C.L. Young

**EXPERIMENTAL VALUES:**

| $T$/K | Henry's Constant $H_{C_2H_6}$/atm | Ostwald Coefficient* $L/cm^3cm^{-3}$ | Mole fraction* $/x_{C_2H_6}$ |
|---|---|---|---|
| 298.2 | 197 | 0.948 | 0.00508 (0.00508)† |
| 323.2 | 246 | 0.803 | 0.00407 (0.00407) |
| 343.2 | 287 | 0.715 | 0.00348 (0.00348) |

* Calculated by compiler assuming a linear function of $p_{C_2H_6}$ vs $x_{C_2H_6}$, i.e., $x_{C_2H_6}$(1 atm) = $1/H_{C_2H_6}$

† From equation of smoothed data:

$$\ln x_1 = 859.91/T - 8.1657$$

Correlation coefficient = 0.9999

**AUXILIARY INFORMATION**

**METHOD/APPARATUS/PROCEDURE:**

A conventional gas-liquid chromatographic unit fitted with a thermal conductivity detector was used. The carrier gas was helium. The value of Henry's law constant was calculated from the retention time. The value applies to very low partial pressures of gas and there may be a substantial difference from that measured at 1 atm. pressure. There is also considerably uncertainty in the value of Henry's constant since surface adsorption was not allowed for although its possible existence was noted.

**SOURCE AND PURITY OF MATERIALS:**

(1) L'Air Liquide sample, minimum purity 99.9 mole per cent.

(2) Touzart and Matignon or Serlabo sample, purity 99 mole per cent.

**ESTIMATED ERROR:**

$\delta T$/K = $\pm$ 0.1; $\delta H$/atm = $\pm$ 6%

Estimated by compiler.

**REFERENCES:**

| COMPONENTS: | ORIGINAL MEASUREMENTS: |
|---|---|
| (1) Ethane; $C_2H_6$; [74-84-0]<br>(2) Benzenemethanol, (Benzyl alcohol); $C_7H_8O$; [100-51-6] | Lenoir, J-Y.; Renault, P.; Renon, H.<br>*J. Chem. Eng. Data*, 1971, *16*, 340-2 |
| **VARIABLES:** | **PREPARED BY:** |
| $T$/K: 298.15 | C.L. Young |

EXPERIMENTAL VALUES:

| $T$/K | Henry's Constant $H_{C_2H_6}$/atm | Mole fraction at 1 atm* $x_{C_2H_6}$ |
|---|---|---|
| 298.15 | 146 | 0.00685 |

* Calculated by compiler assuming a linear function of $p_{C_2H_6}$ vs $x_{C_2H_6}$, i.e., $x_{C_2H_6}$ (1 atm) = $1/H_{C_2H_6}$

AUXILIARY INFORMATION

METHOD/APPARATUS/PROCEDURE:

A conventional gas-liquid chromatographic unit fitted with a thermal conductivity detector was used. The carrier gas was helium. The value of Henry's law constant was calculated from the retention time. The value applies to very low partial pressures of gas and there may be a substantial difference from that measured at 1 atm. pressure. There is also considerable uncertainty in the value of Henry's constant since surface adsorption was not allowed for although its possible existence was noted.

SOURCE AND PURITY OF MATERIALS:

(1) L'Air Liquide sample, minimum purity 99.9 mole per cent.

(2) Touzart and Matignon or Serlabo sample, purity 99 mole per cent.

ESTIMATED ERROR:

$\delta T$/K = $\pm 0.1$; $\delta H$/atm = $\pm 6\%$ (estimated by compiler).

REFERENCES:

| COMPONENTS: | EVALUATOR: |
|---|---|
| (1) Ethane; $C_2H_6$; [74-84-0]<br>(2) Polar solvents excluding alcohols, water and aqueous solutions. | Walter Hayduk<br>Department of Chemical Engineering<br>University of Ottawa<br>Ottawa, Canada K1N 9B4 |

CRITICAL EVALUATION:

In most cases only single measurements of ethane solubility are available in the polar and/or hydrogen-bonding solvents excluding the alcohols, water and aqueous solutions. Hence, it was usually not possible to asses the accuracy of the data by comparing the results of different workers. Nor is there a consistency check equivalent to that for solvents which form regular solutions with ethane (please see Critical Evaluation for non-polar, non-paraffin solvents). However, useful solubility parameters can be calculated for polar or hydrogen-bonding solvents as discussed by Hansen and Beerbower (1) in their review. The total cohesive energy density or polar solubility parameter is considered to be made up of three components resulting in turn from non-polar, dipole and hydrogen-bonding interactions. While the relation between gas solubility and solubility parameter for polar solvents appears to be chiefly of a qualitative nature, it will be used in this discussion of ethane solubilities. It is emphasized that there is no expectation that the solubilities will form a single relation for the various polar and/or hydrogen-bonding solvents. Certain homologous solvents, however, are likely to show a consistent decrease in solubility with increasing solubility parameter. In the figure shown, the solubilities in some twenty solvents are shown as the mole fraction solubility at 298.15 K, interpolated in some instances, and at an ethane partial pressure of 101.325 kPa (1 atm) plotted against the solvent solubility parameter. Lines are also shown for the solubilities in solvents forming regular solutions as well as those in normal alcohols.

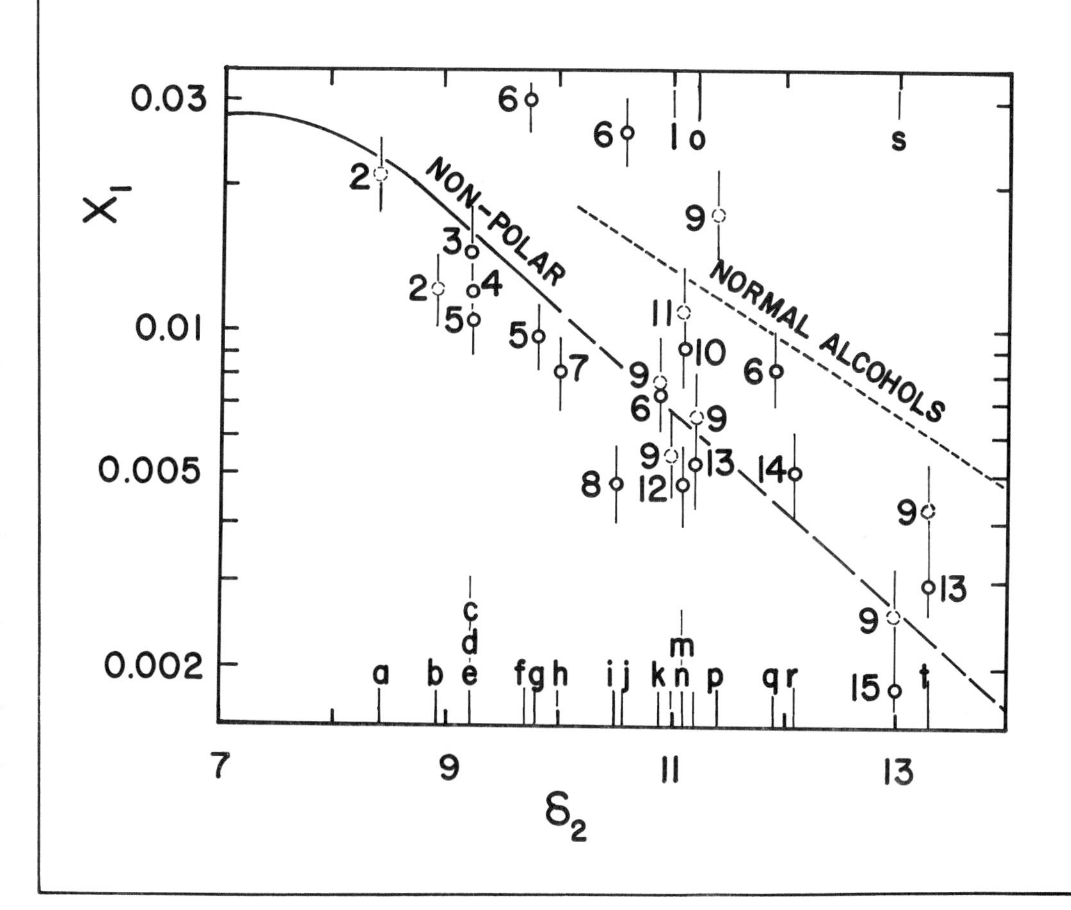

| COMPONENTS: | EVALUATOR: |
|---|---|
| (1) Ethane; $C_2H_6$; [74-84-0]<br>(2) Polar solvents excluding alcohols, water and aqueous solutions. | Walter Hayduk<br>Department of Chemical Engineering<br>University of Ottawa<br>Ottawa, Canada K1N 9B4 |

CRITICAL EVALUATION: continued

The solvents and sources of data considered in order of increasing solubility parameter are: (a) amyl acetate (2), (b) ethyl acetate (2), (c) cyclohexylamine (3), (d) diphenylmethane (4), (e) methyl acetate (5), (f) ethylcellosolve (6), (g) acetone (5), (h) dioxane (7), (i) acetic acid (8), (j) chlorex (6), (k) nitrobenzene (6,9), (l) aniline (9), (m) ethylene oxide (10,11), (n) N-methylacetamide (12), (o) methyl pyrrolidinone (9,13), (p) hexamethylphosphoric triamide (9), (q) furfural (6), (r) dimethyl formamide (14, (s) dimethyl sulfoxide (9,15) and (t) propylene carbonate (9,13).

In addition, solubilities in a number of other solvents are available. For some of these, polar solubility parameters, or properties to calculate them simply are not available or in three instances the solubility parameter or solubility was not in the range of the variables as plotted in the figure. These additional solvents and references are as follows: diglycolamine (13), monoethanolamine (13), octamethylcyclotetrasiloxane (16), sulfolane (13), heavy water (17), five esters of phosphoric acid (9) and finally the pseudo-liquid solvents, dog's blood and lung tissue (18).

The early data of McDaniel (2) have been previously shown to be inaccurate, being usually lower than those of more recent careful workers (see Critical Evaluation of solubilities in paraffin solvents) and are rejected. The copious data of Lenoir et al. (9) measured at low ethane partial pressures by means of a chromatographic technique also appear unreliable. They are from 6% to 46% higher than those of other workers for the four solvents for which comparable values are available as indicated in the table which follows. These comparisons place all the Lenoir et al. (9) data in doubt as being of qualitative value only, hence these data also are rejected. It is possible to assess the accuracy of data when only single values are available by applying some type of consistency test. Such a test was performed with some of the data of Ezheleva and Zorin (6). First it was considered unusual that the solubilities of those workers were reported to be as high as or higher than the ideal solubility of ethane of 0.0250 mole fraction at 298.15 K (19) in the polar solvents ethyl cellosolve and chlorex. This may be expected only if there is a chemical reaction between the solute gas and solvent, which is most unlikely for ethane. Hence it appears possible that the solubilities at least in those two solvents are erroneously high. Some further doubt is cast on the values of Ezheleva and Zorin on studying an example of their method of treatment of data as given in their paper (6) showing a linear plot of the mole fraction solubility versus gas partial pressure. When the resulting graph is extended, and the lines corresponding to different temperatures extrapolated, they do not pass through the origin even approximately, but were still apparently used to evaluate Henry's constants. If the same method was followed for the other gases and solvents, large errors would have resulted. The Ezheleva and Zorin data (6) are still considered to be tentative, since there is an insufficient basis for rejecting them in spite of some doubt cast on their accuracy.

The ethane solubility in octamethylcyclotetrasiloxane as reported by Chappelow and Prauznitz (16) when extrapolated to 298.15 K is considered unlikely because it is approximately twice the ideal solubility. Hildebrand, Prausnitz and Scott (20) discussed that the solvent power for iodine by this solvent is nearly equivalent to that of cyclohexane and that there is no complexing

COMPONENTS:

(1) Ethane; $C_2H_6$; [74-84-0]

(2) Polar solvents excluding alcohols, water and aqueous solutions.

EVALUATOR:

Walter Hayduk
Department of Chemical Engineering
University of Ottawa
Ottawa, Canada K1N 9B4

CRITICAL EVALUATION: continued

with either solvent. It may be considered then, that the solvent power of cyclohexane and the siloxane for ethane should also be similar; however, ethane solubility in the former solvent is 0.0233 mole fraction at 298.15 K (see Critical Evaluation in non-polar, non-paraffin solvents) and in the latter solvent is extrapolated to 0.0509. The solubility value in octamethylcyclotetrasiloxane appears erroneously high although that cannot be conclusively proven. The more recent measurements for the solubility in ethylene oxide of Olson (10) appear to be more accurate than those of Hess and Tilton (11), the latter data having been reported to fewer significant figures and apparently performed with a relatively high partial pressure of solvent vapor.

All the results not otherwise discussed are considered to be tentative. A table summarizes the data for ethane solubilities in polar solvents excluding those in alcohols, water and aqueous solutions, which are discussed elsewhere in this volume. The table shows the mole fraction solubility at 298.15 K, in some cases extrapolated to that temperature, for a gas partial pressure of 101.325 kPa. It is apparent that the available measurements for a number of these systems can only be called sketchy, and that many more measurements are required to produce definitive solubility data.

| Solvent, Source(s) | Mole fraction solubility at 298.15 K | Remarks |
|---|---|---|
| a. Amyl acetate (2) | 0.0211 | Rejected |
| b. Ethyl acetate (2) | 0.0122 | " |
| c. Cyclohexylamine (3) | 0.0145 | Tentative |
| d. Diphenyl methane (4) | 0.0120 | " |
| e. Methyl acetate (5) | 0.0105 | " |
| f. Ethyl cellosolve (6) | 0.0298 | " |
| g. Acetone (5) | 0.00972 | " |
| h. Dioxane (7) | 0.00816 | " |
| i. Acetic acid (9) | 0.00485 | " |
| j. Chlorex (6) | 0.0258 | " |
| k. Nitrobenzene (6) | 0.00735 | " |
| (9) | 0.00781 | Rejected, $\Delta$ = + 6% |
| l. Aniline (9) | 0.00546 | Rejected |
| m. Ethylene oxide (10) | 0.00917 | Tentative |
| (11) | 0.0110 | Rejected, $\Delta$ = + 20% |
| n. Methyl acetamide (12) | 0.00484 | Tentative |

| COMPONENTS: | EVALUATOR: |
| --- | --- |
| (1) Ethane; $C_2H_6$; [74-84-0]<br>(2) Polar solvents excluding alcohols, water and aqueous solutions. | Walter Hayduk<br>Department of Chemical Engineering<br>University of Ottawa<br>Ottawa, Ontario K1N 9B4 |

CRITICAL EVALUATION:

continued

| Solvent, Source(s) | Mole fraction solubility at 298.15 K | Remarks |
| --- | --- | --- |
| o. Methyl pyrrolidinone (13) | 0.00533 | Tentative |
| (9) | 0.00670 | Rejected, Δ=+26% |
| p. Hexamethyl phosphoramide (9) | 0.0174 | Rejected |
| q. Furfural (6) | 0.00861 | Tentative |
| r. Dimethyl formamide (14) | 0.00514 | " |
| s. Dimethyl sulfoxide (15) | 0.00178 | Tentative |
| (9) | 0.00259 | Rejected, Δ=+46% |
| t. Propylene carbonate (13) | 0.00300 | Tentative |
| (9) | 0.00431 | Rejected, Δ=+44% |
| 1. Diglycolamine (13) | 0.00215 | Tentative |
| 2. Monoethanolamine (13) | 0.000804 | " |
| 3. Octamethylcyclotetra-siloxane (16) | 0.0509 | " |
| 4. Sulfolane (16) | 0.00208 | " |
| 5. Heavy water (17) | 3.55 $(10^{-5})$ | " |
| 6. Ester of phosphoric acid (9) | - | Rejected |
| 7. Dog blood and lung tissue (18) | - | Tentative |

References


1. Hansen, C.M.; Beerbower, A., "*Solubility Parameters*", in Mark, H.F.; McKetta, J.J.; Othmer, D.F. (Eds.) "*Encyclopedia of Chemical Technology*" 2nd ed. suppl. vol., Interscience, New York, 1971, 889-910.

2. McDaniel, A.S. *J. Phys. Chem.* 1911, *15*, 587-610.

3. Keevil, T.A.; Taylor, D.R.; Streitwieser, A. Jr. *J. Chem. Eng. Data* 1978, *23*, 237-239.

4. Cukor, P.M.; Prausnitz, J.M. *J. Phys. Chem.* 1972, *76*, 598-601.

5. Horiuti, J. *Sci. Pap. Inst. Phys. Chem. Res. (Jpn)* 1931/1932, *17*, 125-256.

6. Ezhelveva, A.E.; Zorin, A.D. *Tr. Khim. Khim. Tech. (Gorkii)* 1961, *1*, 37-40.

| COMPONENTS: | EVALUATOR: |
|---|---|
| (1) Ethane; $C_2H_6$; [74-84-0]<br>(2) Polar solvents excluding alcohols, water and aqueous solutions. | Walter Hayduk<br>Department of Chemical Engineering<br>University of Ottawa<br>Ottawa, Canada K1N 9B4 |

CRITICAL EVALUATION:

continued


7. Ben-Naim, A.; Yaacobi, M. *J. Phys. Chem.* 1974, *78*, 175-178.
8. Barton, J.R. *Ph.D. Dissert. Chem. Eng.* 1970, Queen's University, Kingston, Ont. Canada.
9. Lenoir, J-Y.; Renault, P.; Renon, H. *J. Chem. Eng. Data* 1971, *16*, 340-342.
10. Olson, J.D. *J. Chem. Eng. Data* 1977, *22*, 326-329.
11. Hess, L.G.; Tilton, V.V. *Ind. Eng. Chem.* 1950, *42*, 1251-1258.
12. Wood, R.H.; DeLaney, D.E. *J. Phys. Chem.* 1968, *72*, 4651-4654.
13. Rivas, O.R.; Prausnitz, J.M. *Am. Inst. Chem. Eng.* 1979, *25*, 975-984.
14. Howard, W.B.; Schoch, E.P.; Mayforth, F.R. *Petrol. Refiner* 1954, *33*, 143-146.
15. Dymond, J.H. *J. Phys. Chem.* 1967, *71*, 1829-1831.
16. Chappelow, C.C.; Prausnitz, J.M. *Am. Inst. Chem. Eng. J.* 1974, *20*, 1097-1104.
17. Ben-Naim, A.; Wilf, J.; Yaacobi, M. *J. Phys. Chem.* 1973, *77*, 95-102.
18. Young, I.H.; Wagner, P.D. *J. Appl. Physiol.* 1979, *46*, 1207-1210.
19. Hayduk, W.; Laudie, H. *Am. Inst. Chem. Eng. J.* 1973, *19*, 1233-1238.
20. Hildebrand, J.H.; Prausnitz, J.M.; Scott, R.L. "*Regular and Related Solutions*" Van Nostrand Reinhold, New York, 1970, p. 61, 143.

**COMPONENTS:**

(1) Ethane; $C_2H_6$; [74-84-0]

(2) Acetic acid, pentyl ester (amyl acetate); $C_7H_{14}O_2$; [628-63-7]

**ORIGINAL MEASUREMENTS:**

McDaniel, A.S.

*J. Phys. Chem.* 1911, *15*, 587-610.

**VARIABLES:**

$T$/K: 295.15 - 323.15
$P$/kPa: 101.325 (1 atm)

**PREPARED BY:**

W. Hayduk

**EXPERIMENTAL VALUES:**

| $t$/°C | $T$/K | Ostwald coefficient[1] $L$/cm$^3$cm$^{-3}$ | Bunsen coefficient[3] $\alpha$/cm$^3$(STP)cm$^{-3}$atm$^{-1}$ | Mole fraction[3] $10^4x_1$ |
|---|---|---|---|---|
| 22.0 | 295.15 | 3.58 | 3.31 | 215 |
| 25.0 | 298.15 | 3.51[2] | 3.22 | 211 |
| 30.0 | 303.15 | 3.39 | 3.05 | 201 |
| 50.0 | 323.15 | 2.89 | 2.44 | 166 |

[1] Original data listed as Absorption coefficient interpreted by compiler to be equivalent to Ostwald coefficient as listed here.

[2] Ostwald coefficient (Absorption coefficient) as estimated at 298.15 K by author.

[3] Bunsen coefficient and mole fraction solubility calculated by compiler assuming ideal gas behavior.

[4] McDaniel's results are consistently from 20 to 80 per cent too low when compared with more reliable data.

**AUXILIARY INFORMATION**

**METHOD/APPARATUS/PROCEDURE:**

Glass apparatus consisting of a gas burette connected to a solvent contacting chamber. Gas pressure or volume adjusted using mercury displacement. Equilibration achieved at atmospheric pressure by hand shaking apparatus, and incrementally adding gas to contacting chamber. Solubility measured by obtaining total uptake of gas by known volume of solvent.

**SOURCE AND PURITY OF MATERIALS:**

1. Prepared by reaction of ethyl iodide with zinc-copper. Purity not measured.

2. Source not given; purity specified as 99 per cent.

**ESTIMATED ERROR:**

$\delta L/L = -0.20$
(estimated by compiler; see note [4] above)

**REFERENCES:**

**COMPONENTS:**

(1) Ethane; $C_2H_6$; [74-84-0]

(2) Acetic acid, ethyl ester (ethyl acetate); $C_4H_8O_2$; [141-78-6]

**ORIGINAL MEASUREMENTS:**

McDaniel, A.S.

*J. Phys. Chem.* 1911, *15*, 587-610.

**VARIABLES:**

$T$/K: 295.15 - 313.15
$P$/kPa: 101.325 (1 atm)

**PREPARED BY:**

W. Hayduk

**EXPERIMENTAL VALUES:**

| $t$/°C | $T$/K | Ostwald coefficient[1] $L$/cm$^3$cm$^{-3}$ | Bunsen coefficient [3] $\alpha$/cm$^3$(STP)cm$^{-3}$atm$^{-1}$ | Mole fraction[3] $10^4x_1$ |
|---|---|---|---|---|
| 22.0 | 295.15 | 3.08 | 2.85 | 123 |
| 25.0 | 298.15 | 3.07[2] | 2.82 | 122 |
| 30.0 | 303.15 | 3.06 | 2.76 | 121 |
| 40.0 | 313.15 | 3.00 | 2.62 | 116 |

[1] Original data listed as Absorption coefficient interpreted by compiler to be equivalent to Ostwald coefficient as listed here.

[2] Ostwald coefficient (Absorption coefficient) as estimated at 298.15 K by author.

[3] Bunsen coefficient and mole fraction solubility calculated by compiler assuming ideal gas behavior.

[4] McDaniel's results are consistently from 20 to 80 per cent too low when compared with more reliable data.

**AUXILIARY INFORMATION**

**METHOD/APPARATUS/PROCEDURE:**

Glass apparatus consisting of a gas burette connected to a solvent contacting chamber. Gas pressure or volume adjusted using mercury displacement. Equilibration achieved at atmospheric pressure by hand shaking apparatus, and incrementally adding gas to contacting chamber. Solubility measured by obtaining total uptake of gas by known volume of solvent.

**SOURCE AND PURITY OF MATERIALS:**

1. Prepared by reaction of ethyl iodide with zinc-copper. Purity not measured.

2. Source not given; purity specified as 99 per cent.

**ESTIMATED ERROR:**

$\delta L/L$ = -0.20
(estimated by compiler; see note [4] above)

**REFERENCES:**

**COMPONENTS:**

(1) Ethane; $C_2H_6$; [74-84-0]

(2) Aminocyclohexane (Cyclohexylamine); $C_6H_{13}N$; [108-91-8]

**ORIGINAL MEASUREMENTS:**

Keevil, T.A.; Taylor, D.R.; Streitwieser, A. Jr.

*J. Chem. Eng. Data* 1978, *23*, 237-239.

**VARIABLES:**

$T$/K: 298.35 - 312.85
$P$/kPa: 101.325 (1 atm)

**PREPARED BY:**

W. Hayduk

**EXPERIMENTAL VALUES:**

| $t$/°C | $T$/K | Mole fraction[1] / $10^4 x_1$ | Ostwald coefficient[2] $L$/cm$^3$ cm$^{-3}$ |
|---|---|---|---|
| 25.2 | 298.35 | 145 (146)[3] | 2.99 |
| 30.0 | 303.15 | 139 (139) | 2.90 |
| 35.0 | 308.15 | 130 (130) | 2.75 |
| 39.7 | 312.85 | 124 (124) | 2.38 |

[1] Original data.

[2] Ostwald coefficient calculated by compiler.

[3] From equation of smoothed data calculated by compiler for temperature range from 298.35 to 312.85 K:

$$\ln x_1 = 1032.2/T - 7.6890$$

Correlation coefficient = 0.9966

**AUXILIARY INFORMATION**

**METHOD/APPARATUS/PROCEDURE:**

Removable glass gas-solvent contactor rotated in constant temperature bath for equilibration. Provision for evacuating apparatus and separately charging dry gas and degassed solvent. Volume of solvent charged initially measured. Initial gas volume and initial and final gas pressures measured manometrically over mercury. Incremental addition of gas possible.

**SOURCE AND PURITY OF MATERIALS:**

1. No information given.
2. Dried over lithium cyclohexylamine and degassed.

**ESTIMATED ERROR:**

$\delta x_1/x_1 = 0.05$
(estimated by compiler)

**REFERENCES:**

**COMPONENTS:**

(1) Ethane; $C_2H_6$; [74-84-0]

(2) Benzene, 1,1′-methylenebis, (Diphenylmethane); $C_{13}H_{12}$; [101-81-5]

**ORIGINAL MEASUREMENTS:**

Cukor, P.M.; Prausnitz, J.M.

*J. Phys. Chem.* 1972, *76*, 598-601

**VARIABLES:**

$T$/K: 300-475

**PREPARED BY:**

C.L. Young

**EXPERIMENTAL VALUES:**

| $T$/K | Henry's Constant[a] /atm | Mole fraction[b] of ethane in liquid, $/10^4 x_{C_2H_6}$ |
|---|---|---|
| 300 | 81.8 | 122 (118)[c] |
| 325 | 105 | 95.2(94.5) |
| 350 | 130 | 76.9(78.2) |
| 375 | 155 | 64.5(66.3) |
| 400 | 180 | 55.6(57.3) |
| 425 | 203 | 49.3(50.5) |
| 450 | 221 | 45.2(45.1) |
| 475 | 232 | 43.1(40.7) |

[a] Quoted in supplementary material for original paper

[b] Calculated by compiler for a partial pressure of 1 atmosphere

[c] From equation of smoothed data by compiler:

$$\ln x_1 = 867.63/T - 7.3303$$

Correlation coefficient = 0.9965

**AUXILIARY INFORMATION**

**METHOD/APPARATUS/PROCEDURE:**

Volumetric apparatus similar to that described by Dymond and Hildebrand (1). Pressure measured with a null detector and precision gauge. Details in ref. (2).

**SOURCE AND PURITY OF MATERIALS:**

No details given

**ESTIMATED ERROR:**

$\delta T$/K = ±0.05; $\delta x_{C_2H_6}$ = ±2%

**REFERENCES:**

1. Dymond, J.; Hildebrand, J.H. *Ind. Eng. Chem. Fundam.* 1967, *6*, 130.

2. Cukor, P.M.; Prausnitz, J.M. *Ind. Eng. Chem. Fundam.* 1971, *10*, 638.

**COMPONENTS:**

(1) Ethane; $C_2H_6$; [74-84-0]

(2) Acetic acid, methyl ester or methyl acetate; $C_3H_6O_2$; [79-20-9]

**ORIGINAL MEASUREMENTS:**

Horiuti, J.

*Sci. Pap. Inst. Phys. Chem. Res. (Jpn)* 1931/32, *17*, 125 - 256.

**VARIABLES:**

$T$/K: 273.15 - 313.15

$p_1$/kPa: 101.325 (1 atm)

**PREPARED BY:**

M. E. Derrick
H. L. Clever

**EXPERIMENTAL VALUES:**

| $T$/K | Mol Fraction $10^2 x_1$ | Bunsen Coefficient $\alpha/cm^3(STP)cm^{-3}atm^{-1}$ | Ostwald Coefficient $L/cm^3 cm^{-3}$ |
|---|---|---|---|
| 273.15 | 1.446 | 4.195 | 4.195 |
| 278.15 | 1.337 | 3.907 | 3.979 |
| 283.15 | 1.258 | 3.647 | 3.780 |
| 288.15 | 1.179 | 3.394 | 3.580 |
| 293.15 | 1.113 | 3.181 | 3.414 |
| 298.15 | 1.048 | 2.974 | 3.246 |
| 303.15 | 0.9941 | 2.799 | 3.106 |
| 308.15 | 0.9406 | 2.629 | 2.966 |
| 313.15 | 0.8881 | 2.464 | 2.825 |

The mole fraction and Bunsen coefficient values were calculated by the compiler with the assumption the gas is ideal and that Henry's law is obeyed.

Smoothed Data: For use between 273.15 and 313.15 K, by compilers:

$$\ln x_1 = -7.9987 + 10.2610/(T/100K)$$

The standard error about the regression line is $4.55 \times 10^{-5}$.

| $T$/K | Mol Fraction $10^2 x_1$ |
|---|---|
| 273.15 | 1.438 |
| 283.15 | 1.259 |
| 288.15 | 1.182 |
| 293.15 | 1.113 |
| 298.15 | 1.049 |
| 303.15 | 0.991 |
| 313.15 | 0.890 |

**AUXILIARY INFORMATION**

**METHOD/APPARATUS/PROCEDURE:**

The apparatus consists of a gas buret, a solvent reservoir, and an absorption pipet. The volume of the pipet is determined at various meniscus heights by weighing a quantity of water. The meniscus height is read with a cathetometer.

The dry gas is introduced into the degassed solvent. The gas and solvent are mixed with a magnetic stirrer until saturation. Care is taken to prevent solvent vapor from mixing with the solute gas in the gas buret. The volume of gas is determined from the gas buret readings, the volume of solvent is determined from the meniscus height in the absorption pipet.

**SOURCE AND PURITY OF MATERIALS:**

(1) Ethane. A 60% solution of potassium acetate was electrolyzed between Pt and Cu electrodes. The gas was passed through several wash solutions, dried, and fractionated from liquid air. Boiling point (766 mmHg) -88.3°C.

(2) Methyl acetate. Merck. Extra pure grade. Dried with $P_2O_5$. Distilled several times. Boiling point (760 mmHg) 57.12°C.

**ESTIMATED ERROR:**

$\delta T/K = 0.05$
$\delta x_1/x_1 = 0.01$

**REFERENCES:**

**COMPONENTS:**

(1) Ethane; $C_2H_6$; [74-84-0]

(2) 2-Ethoxyethanol (Ethylcellosolve); $C_4H_{10}O_2$; [110-80-5]

**ORIGINAL MEASUREMENTS:**

Ezheleva, A.E.; Zorin, A.D.

*Tr. Khim. Khim. Tech. (Gorkii)* 1961, *1*, 37-40.

**VARIABLES:**

$T$/K: 303.15 - 343.15

$P$/kPa: 101.325 and above

**PREPARED BY:**

W. Hayduk

**EXPERIMENTAL VALUES:**

| $t$/°C | $T$/K | Mole fraction[1] / $x_1$ | Henry's constant[2] $H$/atm |
|---|---|---|---|
| 30 | 303.15 | 0.0270 (0.0262)[3] | 37.04 |
| 40 | 313.15 | 0.0197 (0.0206) | 50.76 |
| 50 | 323.15 | 0.0164 (0.0165) | 60.98 |
| 60 | 333.15 | 0.0138 (0.0133) | 72.46 |
| 70 | 343.15 | 0.0108 (0.0109) | 92.59 |

[1] Original data, given as the inverse of Henry's constant which is equivalent to mole fraction at a gas partial pressure of 101.325 kPa.

[2] Henry's constant calculated by compiler.

[3] From equation of smoothed data calculated by compiler for temperature range from 303.15 to 343.15 K:

$$\ln x_1 = 2280.0/T - 11.161$$

Correlation coefficient = 0.9955

**AUXILIARY INFORMATION**

**METHOD/APPARATUS/PROCEDURE:**

The apparatus consists of a two-chamber, rocking contacting device with separate gas and liquid chambers joined by two tubes and micro valves. Gas chamber equipped with a pressure gauge. After evacuation gas and deaerated solvent separately charged, and then contacted by opening the micro valves and by rocking. Solubility calculated from knowledge of volume of system, volume of solvent charged, and initial and final gas pressures.

Solvents considered non-volatile; gas partial pressure considered to be total pressure.

**SOURCE AND PURITY OF MATERIALS:**

1. Stated as chromatography pure.

2. Vacuum fractionated.

Purities not specified.

**ESTIMATED ERROR:**

$\delta T$/K = 0.05

$\delta H/H$ = 0.05

(estimated by compiler)

**REFERENCES:**

**COMPONENTS:**

(1) Ethane; $C_2H_6$; [74-84-0]

(2) 2-Propanone or acetone; $C_3H_6O$; [67-64-1]

**ORIGINAL MEASUREMENTS:**

Horiuti, J.

*Sci. Pap. Inst. Phys. Chem. Res. (Jpn)* 1931/32, *17*, 125 - 256.

**VARIABLES:**

$T$/K: 273.15 - 313.15
$p_1$/kPa: 101.325 (1 atm)

**PREPARED BY:**

M. E. Derrick
H. L. Clever

**EXPERIMENTAL VALUES:**

| $T$/K | Mol Fraction $10^2 x_1$ | Bunsen Coefficient $\alpha/cm^3 (STP) cm^{-3} atm^{-1}$ | Ostwald Coefficient $L/cm^3 cm^{-3}$ |
|---|---|---|---|
| 273.15 | 1.322 | 4.202 | 4.202 |
| 278.15 | 1.233 | 3.890 | 3.961 |
| 283.15 | 1.158 | 3.628 | 3.761 |
| 288.15 | 1.085 | 3.374 | 3.559 |
| 293.15 | 1.023 | 3.158 | 3.389 |
| 298.15 | 0.9720 | 2.977 | 3.225 |
| 303.15 | 0.9090 | 2.763 | 3.067 |
| 308.15 | 0.8552 | 2.580 | 2.911 |
| 313.15 | 0.8128 | 2.434 | 2.790 |

The mole fraction and Bunsen coefficient values were calculated by the compiler with the assumption the gas is ideal and that Henry's law is obeyed.

Smoothed Data: For use between 273.15 and 313.15 K, by compilers:

The 298.15 K value was omitted from the linear regression.

$$\ln x_1 = -8.1377 + 10.4138/(T/100K)$$

The standard error about the regression line is $2.2 \times 10^{-5}$.

| $T$/K | Mol Fraction $10^2 x_1$ |
|---|---|
| 273.15 | 1.323 |
| 283.15 | 1.156 |
| 288.15 | 1.085 |
| 293.15 | 1.020 |
| 298.15 | 0.961 |
| 303.15 | 0.907 |
| 313.15 | 0.813 |

**AUXILIARY INFORMATION**

**METHOD/APPARATUS/PROCEDURE:**

The apparatus consists of a gas buret, a solvent reservoir, and an absorption pipet. The volume of the pipet is determined at various meniscus heights by weighing a quantity of water. The meniscus height is read with a cathetometer.

The dry gas is introduced into the degassed solvent. The gas and solvent are mixed with a magnetic stirrer until saturation. Care is taken to prevent solvent vapor from mixing with the solute gas in the gas buret. The volume of gas is determined from the gas buret readings, the volume of solvent is determined from the meniscus height in the absorption pipet.

**SOURCE AND PURITY OF MATERIALS:**

(1) Ethane. A 60% solution of potassium acetate was electrolyzed between Pt and Cu electrodes. The gas was passed through several wash solutions, dried, and fractionated from liquid air. Boiling point (766 mmHg) -88.3°C.

(2) Acetone. Nippon Pure Chemical Co. or Merck. Extra pure grade. Recrystallized with sodium sulfite and stored over calcium chloride. Fractionated, boiling point (760 mmHg) 56.09°C.

**ESTIMATED ERROR:**

$\delta T$/K = 0.05
$\delta x_1/x_1$ = 0.01

**REFERENCES:**

**COMPONENTS:**

(1) Ethane; $C_2H_6$; [74-84-0]

(2) 1,4-Dioxane; $C_4H_8O_2$; [123-91-1]

**ORIGINAL MEASUREMENTS:**

Ben-Naim, A.; Yaacobi, M.

*J. Phys. Chem.* 1974, *78*, 175-8

**VARIABLES:**

$T$/K: 283.15-303.15

**PREPARED BY:**

C.L.Young

**EXPERIMENTAL VALUES:**

| $T$/K | Ostwald coefficient,* $L$ | Mole fraction[+] at partial pressure of 101.3kPa, $x_{C_2H_6}$ |
|---|---|---|
| 283.15 | 2.590 | 0.00931 (0.00947)** |
| 288.15 | 2.548 | 0.00905 (0.00897) |
| 293.15 | 2.465 | 0.00866 (0.00851) |
| 298.15 | 2.347 | 0.00816 (0.00809) |
| 303.15 | 2.201 | 0.00757 (0.00771) |

* Smoothed values obtained from the equation

$kT \ln L = -15{,}942.3 + 115.36\ (T/K) - 0.20190\ (T/K)^2$ cal mol$^{-1}$
where k is in units of cal mol$^{-1}$ K$^{-1}$

\+ calculated by compiler assuming the ideal gas law for ethane.

** From alternate equation of smoothed data:

$$\ln x_1 = 884.70/T - 7.7842$$

Correlation coefficient = 0.9808

**AUXILIARY INFORMATION**

**METHOD/APPARATUS/PROCEDURE:**

The apparatus was similar to that described by Ben-Naim and Baer (1) and Wen and Hung (2). It consists of three main parts, a dissolution cell of 300 to 600 cm$^3$ capacity, a gas volume measuring column, and a manometer. The solvent is degassed in the dissolution cell, the gas is introduced and dissolved while the liquid is kept stirred by a magnetic stirrer immersed in the water bath. Dissolution of the gas results in the change in the height of a column of mercury which is measured by a cathetometer.

**SOURCE AND PURITY OF MATERIALS:**

1. Matheson sample, purity 99.9 mol per cent.

2. AR grade

**ESTIMATED ERROR:**

$\delta T/K = \pm 0.1$; $\delta x_{C_2H_6} = \pm 2\%$ (estimated by compiler).

**REFERENCES:**


1. Ben-Naim, A.; Baer, S. *Trans. Faraday Soc.* 1963, *59*, 2735.

2. Wen, W.-Y.; Hung, J.H. *J. Phys. Chem.* 1970, *74*, 170.

**COMPONENTS:**

(1) Ethane; $C_2H_6$; [74-84-0]

(2) Acetic acid; $C_2H_4O_2$; [64-19-7]

**ORIGINAL MEASUREMENTS:**

Barton, J.R.; Hsu, C.C.

*Chem. Eng. Sci.* 1972, *27*, 1315-1323.

**VARIABLES:**

$T$/K = 295.15
$P$/kPa = 101.325

**PREPARED BY:**

W. Hayduk

**EXPERIMENTAL VALUES:**

| $T$/°C | $T$/K | Ostwald coefficient[1] $L$/cm$^3$ cm$^{-1}$ | Bunsen coefficient[2] $\alpha$/cm$^3$(STP)cm$^{-3}$atm$^{-1}$ | Mole fraction[3] $10^4 x_1$ |
|---|---|---|---|---|
| 22.0 | 293.15 | 2.013 | 1.876 | 48.5 (47.8)[1] |

[1] Calculated by compiler; assuming ideal gas behavior.

[2] From reference (1) below.

[3] Original data.

**AUXILIARY INFORMATION**

**METHOD/APPARATUS/PROCEDURE:**

The solubility apparatus consisted of two glass bulbs of accurately measured volume, mounted together and immersed in a bath. The bulbs could be separately charged, interconnected, as well as agitated when required. Vapor-saturated gas was charged to one bulb while deaerated solvent was charged to the other, completely filling the bulb in each case. Saturated gas was stored above mercury in one leg of a manometer which served as a pressure measuring device as well as a variable volume reservoir. Precision tubing was used in the manometer permitting accurate determinations of gas volume Deaeration was by distillation at total reflux.

Details in reference (1).

**SOURCE AND PURITY OF MATERIALS:**

1. Matheson Chemically pure grade. Specified purity 99.0 per cent.

2. Baker. Reagent grade. Specified purity 99.9 per cent.

**ESTIMATED ERROR:**

$\delta T$/K = 0.05

$\delta\alpha/\alpha$ = 0.005 (authors)

**REFERENCES:**

(1) Barton, J.R.
*Ph.D. Thesis, Chem. Eng.* 1970, Queen's Univ., Kingston, Ont., Canada

**COMPONENTS:**

(1) Ethane; $C_2H_6$; [74-84-0]

(2) 1,1'-Oxybis, 2-Chloroethane (Chlorex); $C_4H_8Cl_2O$; [111-44-4]

**ORIGINAL MEASUREMENTS:**

Ezheleva, A.E.; Zorin, A.D.

*Tr. Khim. Khim. Tech. (Gorkii)* 1961, *1*, 37-40.

**VARIABLES:**

$T$/K: 303.15 - 343.15
$P$/kPa: 101.325 and above

**PREPARED BY:**

W. Hayduk

**EXPERIMENTAL VALUES:**

| $t$/C | $T$/K | Mole fraction[1] / $x_1$ | Henry's constant[2] $H$/atm |
|---|---|---|---|
| 30 | 303.15 | 0.0235 (0.0244)[3] | 42.55 |
| 40 | 313.15 | 0.0221 (0.0219) | 45.25 |
| 50 | 323.15 | 0.0203 (0.0198) | 49.26 |
| 60 | 333.15 | 0.0195 (0.0180) | 51.28 |
| 70 | 343.15 | 0.0153 (0.0165) | 65.36 |

[1] Original data, given as the inverse of Henry's constant which is equivalent to mole fraction at a gas partial pressure of 101.325 kPa.

[2] Actual Henry's constant calculated by compiler.

[3] From equation of smoothed data calculated by compiler for temperature range from 303.15 to 343.15 K:

$$\ln x_1 = 1012.3/T - 7.0539$$

Correlation coefficient = 0.9341

**AUXILIARY INFORMATION**

**METHOD/APPARATUS/PROCEDURE:**

The apparatus consists of a two-chamber, rocking contacting device with separate gas and liquid chambers joined by two tubes and micro valves. Gas chamber equipped with a pressure gauge. After evacuation gas and deaerated solvent separately charged, and then contacted by opening the micro valves and by rocking. Solubility calculated from knowledge of volume of system, volume of solvent charged, and initial and final gas pressures.

Solvents considered non-volatile; gas partial pressure considered to be total pressure.

**SOURCE AND PURITY OF MATERIALS:**

1. Stated as chromatography pure.

2. Vacuum fractionated.

Purities not specified.

**ESTIMATED ERROR:**

$\delta T$/K = 0.05
$\delta H/H$ = 0.05
(estimated by compiler)

**REFERENCES:**

**COMPONENTS:**

(1) Ethane; $C_2H_6$; [74-84-0]

(2) Nitrobenzene; $C_6H_5NO_2$; [98-95-3]

**ORIGINAL MEASUREMENTS:**

Ezheleva, A.E.; Zorin, A.D.;

*Tr. Khim. Khim. Tech. (Gorkii)* 1961, *1*, 37-40.

**VARIABLES:**

$T$/K: 303.15 - 343.15
$P$/kPa: 101.325 and above

**PREPARED BY:**

W. Hayduk

**EXPERIMENTAL VALUES:**

| $t$/ °C | $T$/K | Mole fraction[1] / $x_1$ | Henry's constant[2] $H$/atm |
|---|---|---|---|
| 30 | 303.15 | 0.0070 (0.0067)[3] | 142.9 |
| 40 | 313.15 | 0.0055 (0.0056) | 181.8 |
| 50 | 323.15 | 0.0044 (0.0048) | 227.3 |
| 60 | 333.15 | 0.0042 (0.0041) | 238.1 |
| 70 | 343.15 | 0.0036 (0.0035) | 277.8 |

[1] Original data, given as the inverse of Henry's constant which is equivalent to mole fraction at a gas partial pressure of 101.325 kPa.

[2] Actual Henry's constant calculated by compiler.

[3] From equation of smoothed data:

$$\ln x_1 = 1674.2/T - 10.529$$

Correlation coefficient = 0.9813

**AUXILIARY INFORMATION**

**METHOD/APPARATUS/PROCEDURE:**

The apparatus consists of a two-chamber, rocking contacting device with separate gas and liquid chambers joined by two tubes and micro valves. Gas chamber equipped with a pressure gauge. After evacuation gas and deaerated solvent separately charged, and then contacted by opening the micro valves and by rocking. Solubility calculated from knowledge of volume of system, volume of solvent charged and initial and final gas pressures.

Solvents considered non-volatile; gas partial pressure considered to be total pressure.

**SOURCE AND PURITY OF MATERIALS:**

1. Stated as chromatography pure.

2. Vacuum fractionated.

Purities not specified.

**ESTIMATED ERROR:**

$\delta T$/K = 0.05
$\delta H/H$ = 0.05
(estimated by compiler)

**REFERENCES:**

COMPONENTS:

(1) Ethane; $C_2H_6$; [74-84-0]

(2) Nitrobenzene; $C_6H_5NO_2$; [98-95-3]

ORIGINAL MEASUREMENTS:

Lenoir, J-Y.; Renault, P.; Renon, H.

*J. Chem. Eng. Data* 1971, *16*, 340-2.

VARIABLES:

$T$/K: 298.2

PREPARED BY:

C. L. Young

EXPERIMENTAL VALUES:

| $T$/K | Henry's constant $H_{C_2H_6}$/atm | Mole fraction at 1 atm* $x_{C_2H_6}$ |
|---|---|---|
| 298.2 | 128 | 0.00781 |

* Calculated by compiler assuming a linear function of $p_{C_2H_6}$ vs $x_{C_2H_6}$, i.e., $x_{C_2H_6}$ (1 atm) = $1/H_{C_2H_6}$

AUXILIARY INFORMATION

METHOD/APPARATUS/PROCEDURE:

A conventional gas-liquid chromatographic unit fitted with a thermal conductivity detector was used. The carrier gas was helium. The value of Henry's law constant was calculated from the retention time. The value applies to very low partial pressures of gas and there may be a substantial difference from that measured at 1 atm. pressure. There is also considerable uncertainty in the value of Henry's constant since surface adsorption was not allowed for although its possible existence was noted.

SOURCE AND PURITY OF MATERIALS:

(1) L'Air Liquide sample, minimum purity 99.9 mole per cent.

(2) Touzart and Matignon or Serlabo sample, purity 99 mole per cent.

ESTIMATED ERROR:

$\delta T$/K = $\pm 0.1$; $\delta H$/atm = $\pm 6\%$ (estimated by compiler).

REFERENCES:

COMPONENTS:

(1) Ethane; $C_2H_6$; [74-84-0]

(2) Benzenamine (Aniline); $C_6H_7N$; [62-53-3]

ORIGINAL MEASUREMENTS:

Lenoir, J-Y.; Renault, P.; Renon, H.

*J. Chem. Eng. Data* 1971, *16*, 340-3.

VARIABLES:

$T$/K: 298.2

PREPARED BY:

C. L. Young

EXPERIMENTAL VALUES:

| $T$/K | Henry's constant $H_{C_2H_6}$/atm | Mole fraction at 1 atm* $x_{C_2H_6}$ |
|---|---|---|
| 298.2 | 183 | 0.00546 |

* Calculated by compiler assuming a linear function of $P_{C_2H_6}$ vs $x_{C_2H_6}$, i.e., $x_{C_2H_6}$(1 atm) = $1/H_{C_2H_6}$.

AUXILIARY INFORMATION

METHOD/APPARATUS/PROCEDURE:

A conventional gas-liquid chromatographic unit fitted with a thermal conductivity detector was used. The carrier gas was helium. The value of Henry's law constant was calculated from the retention time. The value applies to very low partial pressures of gas and there may be a substantial difference from that measured at 1 atm. pressure. There is also considerable uncertainty in the value of Henry's constant since surface adsorption was not allowed for although its possible existence was noted.

SOURCE AND PURITY OF MATERIALS:

(1) L'Air Liquide sample, minimum purity 99.9 mole per cent.

(2) Touzart and Matignon or Serlabo sample, purity 99 mole per cent.

ESTIMATED ERROR:

$\delta T$/K = $\pm 0.1$; $\delta H$/atm = $\pm 6\%$ (estimated by compiler).

REFERENCES:

**COMPONENTS:**

(1) Ethane; $C_2H_6$; [74-84-0]

(2) 1,2-Epoxyethane (Ethylene oxide); $C_2H_2O$; [75-21-8]

**ORIGINAL MEASUREMENTS:**

Olson, J.D.

*J. Chem. Eng. Data* 1977, *22*, 326-329.

**VARIABLES:**

$T$/K: 273.15 - 323.15
$P$/kPa: 203 - 840

**PREPARED BY:**

W. Hayduk

**EXPERIMENTAL VALUES:**

| $t$/°C | $T$/K | Henry's constant[1] $H$/atm | Mole fraction[2] / $x_1$ | |
|---|---|---|---|---|
| 0 | 273.15 | 84.3 | 0.01186 | (0.01176)[3] |
| 25 | 298.15 | 109 | 0.00917 | (0.00933) |
| 50 | 323.15 | 129 | 0.00775 | (0.00768) |

[1] Original data; Henry's constants extrapolated to zero gas partial pressure.

[2] Mole fraction calculated by compiler assuming constant $H$ and gas partial pressure of 101.325 kPa. It is noted that solvent normal boiling point is 286.7 K.

[3] From equation of smoothed data developed by compiler:

$$\ln x_1 = 753.75/T - 7.2022$$

Correlation coefficient = 0.9974

**AUXILIARY INFORMATION**

**METHOD/APPARATUS/PROCEDURE:**

The apparatus consisted of an accurate gravimetric method for determining masses of solvent and gas charged into stainless steel bomb of predetermined volume. Gas introduced at pressures of up to 840 kPa measured by bourdon gauge. Equilibration by shaking for 2 to 4 h aided by several loose balls in bomb. Pressure measurements along with known volumes and masses of gas and solvent permitted calculation of Henry's constant. Detailed volume change corrections made for both phases.

**SOURCE AND PURITY OF MATERIALS:**

1. Matheson research grade. Purity 99.96 per cent.

2. UCC commercial grade. GC analysis indicated volatile impurities less than 100 ppm.

**ESTIMATED ERROR:**

$\delta T$/K = 0.10
$\delta H/H$ = 0.03 (author)

**REFERENCES:**

**COMPONENTS:**

(1) Ethane; $C_2H_6$; [74-84-0]

(2) 1,2-Epoxyethane (Ethylene oxide); $C_2H_2O$; [75-21-8]

**ORIGINAL MEASUREMENTS:**

Hess, L.G.; Tilton, V.V.

*Ind. Eng. Chem.* 1950, *42*, 1251-1258.

**VARIABLES:**

$T$/K: 303.15 - 318.15
$P$/kPa: 308.1 - 583.8

**PREPARED BY:**

W. Hayduk

**EXPERIMENTAL VALUES:**

| $t^1$/°C | $T^2$/K | Total pressure[1], pounds per square inch gage / psig | Mass percent[1] ethane in solution | Henry's constant[2] $H$/atm | Mole fraction[2] / $x_1$ |
|---|---|---|---|---|---|
| 30 | 303.15 | 30 | 0.7 | 96.8 | 0.0103 |
| 30 | 303.15 | 40 | 1.2 | 97.0 | |
| 30 | 303.15 | 50 | 1.7 | | |
| 45 | 318.15 | 50 | 0.7 | 116.1 | 0.00861 |
| 45 | 318.15 | 60 | 1.1 | 116.0 | |
| 45 | 318.15 | 70 | 1.5 | | |

[1] Original data.

[2] Calculated by compiler. Original data obeys Henry's law hence Henry's law constant and mole fraction corresponding to gas partial pressure of 101.325 kPa calculated. It is noted that the solvent normal boiling point is 286./K.

Equation through data points:

$$\ln x_1 = 1152.4/T - 8.3769$$

AUXILIARY INFORMATION

**METHOD/APPARATUS/PROCEDURE:**

Experiments were performed with a high pressure, steel flow apparatus consisting of two presaturators for the gas and an equilibrium vessel containing a stirrer operated by a solenoid. The gas is supersaturated in the first saturator at a temperature 10°K above the equilibration temperature. A steady flow of gas is made for at least 2 h after which liquid and vapor samples are withdrawn for analysis at 1-h intervals. Equilibrium indicated by constant consecutive compositions of both phases.

Details in reference (1).

**SOURCE AND PURITY OF MATERIALS:**

Source and purities not available.

**ESTIMATED ERROR:**

$\delta T$/K = 0.1

$\delta x_1/x_1 = \delta H/H = 0.10$

(estimated by compiler)

**REFERENCES:**

1. Wan, S.-W.; Dodge, B.F. *Ind. Eng. Chem.* 1940, *32*, 95.

**COMPONENTS:**

(1) Ethane; $C_2H_6$; [74-84-0]

(2) N-Methylacetamide; $C_3H_7NO$; [79-16-3]

**ORIGINAL MEASUREMENTS:**

Wood, R. H.; DeLaney, D. E.

*J. Phys. Chem.* 1968, *72*, 4651 - 4654.

**VARIABLES:**

$T$/K: 308.15 - 343.15
$p$/kPa: 101.325 (1 atm)

**PREPARED BY:**

P. L. Long
H. L. Clever

**EXPERIMENTAL VALUES:**

The experimental data were not included in the paper. They are available in a thesis (1). The authors obtained the equation

$$\ln x_1 = 6.704 + 352.7/(T/\mathrm{K}) - 2.32 \ln (T/\mathrm{K})$$

by a linear regression of their experimental data.

The smoothed ethane mole fraction solubilities at 101.325 kPa ethane pressure were given in the paper at five degree intervals from 308.15 to 343.15 K. The Bunsen and Ostwald coefficients were calculated by the compiler assuming ideal gas behavior.

Smoothed Data:

| $T$/K | Mol Fraction $10^3 x_1$ | Bunsen Coefficient $\alpha/\mathrm{cm^3(STP)\,cm^{-3}atm^{-1}}$ | Ostwald Coefficient $L/\mathrm{cm^3 cm^{-3}}$ |
|---|---|---|---|
| 308.15 | 4.191 | 1.221 | 1.377 |
| 313.15 | 3.963 | 1.149 | 1.317 |
| 318.15 | 3.752 | 1.083 | 1.261 |
| 323.15 | 3.557 | 1.022 | 1.209 |
| 328.15 | 3.376 | 0.966 | 1.160 |
| 333.15 | 3.207 | 0.913 | 1.114 |
| 338.15 | 3.050 | 0.864 | 1.070 |
| 343.15 | 2.903 | 0.819 | 1.029 |

AUXILIARY INFORMATION

**METHOD/APPARATUS/PROCEDURE:**

The apparatus is described in the thesis (1). A gas buret is connected to a solvent buret through a three-way capillary stopcock. A measured volume of gas is contacted to a known volume of degassed solvent; when equilibrium is reached the total pressure and volume of the system is measured (1).

The apparatus and procedure were checked by measuring the solubility of argon in water at 298.15 K. The Bunsen coefficient of 0.03105 checked well with the literature (2).

**SOURCE AND PURITY OF MATERIALS:**

(1) Ethane. Source not given. Chemically pure grade stated to be 99.0 per cent pure.

(2) N-Methylacetamide. Source not given. Recrystallized three times in a dry box. Typically had a water content of 0.04 mol per cent after a solubility run.

**ESTIMATED ERROR:**

Duplicate runs checked to within 0.1 per cent (authors).

**REFERENCES:**

1. DeLaney, D. E.
M. S. Thesis, 1968
University of Delaware

2. Ben-Naim, A.; Baer, S.
*Trans. Faraday Soc.* 1963, *59*,2735;
*Ibid.* 1964 , *60*, 1736.

**COMPONENTS:**

(1) Ethane; $C_2H_6$; [74-84-0]

(2) 1-Methyl -2-pyrrolidinone; $C_5H_9NO$: [872-50-4]

**ORIGINAL MEASUREMENTS:**

Rivas, O.R.; Prausnitz, J.M.

*Am. Inst. Chem. Eng. J.* 1979, *25*, 975-984.

**VARIABLES:**

$T$/K: 263.15-373.15

**PREPARED BY:**

C.L. Young

**EXPERIMENTAL VALUES:**

| $T$/K | Henry's constant, $H$ /MPa | Mole fraction of ethane in liquid, $x_{C_2H_6}$ + |
|---|---|---|
| 263.15 | 11.99 | 0.008451 (0.00818)* |
| 298.15 | 19.00 | 0.005333 (0.00553) |
| 323.15 | 23.64 | 0.004286 (0.00441) |
| 348.15 | 27.99 | 0.003620 (0.00362) |
| 373.15 | 32.04 | 0.003162 (0.00306) |

+ at a partial pressure of 101.3 kPa calculated by compiler assuming Henry's law applies at that pressure.

* From equation of smoothed data:

$$\ln x_1 = 877.89/T - 8.1418$$

Correlation coefficient = 0.9965

AUXILIARY INFORMATION

**METHOD/APPARATUS/PROCEDURE:**

Volumetric apparatus with a fused quartz precision bourdon pressure gauge. Solubility apparatus carefully thermostatted. Solvent degassed *in situ*. Apparatus described in ref. (1) and modifications given in source.

**SOURCE AND PURITY OF MATERIALS:**

1. and 2. Purity at least 99 mole per cent.

**ESTIMATED ERROR:**

$\delta T$/K = ±0.05; $\delta x_{C_2H_6}$ = ±1%.

**REFERENCES:**

1. Cukor, P.M.; Prausnitz, J.M. *Ind. Eng. Chem. Fundam.* 1971, *10*, 638.

**COMPONENTS:**

(1) Ethane; $C_2H_6$; [74-84-0]

(2) 1-Methyl-2-pyrrolidinone; $C_5H_9NO$; [872-50-4]

**ORIGINAL MEASUREMENTS:**

Lenoir, J-Y.; Renault, P.; Renon, H.

*J. Chem. Eng. Data* 1971, *16*, 340-2

**VARIABLES:**

$T$/K: 298.15

**PREPARED BY:**

C.L. Young

**EXPERIMENTAL VALUES:**

| $T$/K | Henry's Constant $H_{C_2H_6}$/atm | Mole fraction at 1 atm* $x_{C_2H_6}$ |
|---|---|---|
| 298.15 | 149 | 0.00671 |

* Calculated by compiler assuming a linear function of $p_{C_2H_6}$ vs $x_{C_2H_6}$, i.e., $x_{C_2H_6}$ (1 atm) = $1/H_{C_2H_6}$

**AUXILIARY INFORMATION**

**METHOD/APPARATUS/PROCEDURE:**

A conventional gas-liquid chromatographic unit fitted with a thermal conductivity detector was used. The carrier gas was helium. The value of Henry's law constant was calculated from the retention time. The value applies to very low partial pressures of gas and there may be a substantial difference from that measured at 1 atm. pressure. There is also considerable uncertainty in the value of Henry's constant since surface adsorption was not allowed for although its possible existence was noted.

**SOURCE AND PURITY OF MATERIALS:**

(1) L'Air Liquide sample, minimum purity 99.9 mole per cent.

(2) Touzart and Matignon or Serlabo sample, purity 99 mole per cent.

**ESTIMATED ERROR:**

$\delta T$/K = $\pm 0.1$; $\delta H$/atm = $\pm 6\%$ (estimated by compiler).

**REFERENCES:**

COMPONENTS:

(1) Ethane; $C_2H_6$; [74-84-0]

(2) Hexamethylphosphoric triamide; $C_6H_{18}N_3PO$; [680-31-9]

ORIGINAL MEASUREMENTS:

Lenoir, J-Y.; Renault, P.; Renon, H.

*J. Chem. Eng. Data* 1971, *16*, 340-2.

VARIABLES:

$T$/K: 298.2

PREPARED BY:

C. L. Young

EXPERIMENTAL VALUES:

| $T$/K | Henry's constant $H_{C_2H_6}$/atm | Mole fraction at 1 atm* $x_{C_2H_6}$ |
|---|---|---|
| 298.2 | 57.6 | 0.0174 |

* Calculated by compiler assuming a linear function of $p_{C_2H_6}$ vs $x_{C_2H_6}$, i.e., $x_{C_2H_6}$(1 atm) = $1/H_{C_2H_6}$

AUXILIARY INFORMATION

METHOD/APPARATUS/PROCEDURE:

A conventional gas-liquid chromatographic unit fitted with a thermal conductivity detector was used. The carrier gas was helium. The value of Henry's law constant was calculated from the retention time. The value applies to very low partial pressures of gas and there may be a substantial difference from that measured at 1 atm. pressure. There is also considerable uncertainty in the value of Henry's constant since surface adsorption was not allowed for although its possible existence was noted.

SOURCE AND PURITY OF MATERIALS:

(1) L'Air Liquide sample, minimum purity 99.9 mole per cent.

(2) Touzart and Matignon or Serlabo sample, purity 99 mole per cent.

ESTIMATED ERROR:

$\delta T$/K = $\pm 0.1$; $\delta H$/atm = $\pm 6\%$ (estimated by compiler).

REFERENCES:

**COMPONENTS:**

(1) Ethane; $C_2H_6$; [74-84-0]

(2) 2-Furancarboxaldehyde (furfural); $C_5H_4O_2$; [98-01-1]

**ORIGINAL MEASUREMENTS:**

Ezheleva, A.E.; Zorin, A.D.

*Tr. Khim. Khim. Tech. (Gorkii)* 1961, *1*, 37-40.

**VARIABLES:**

$T$/K: 303.15 - 343.15
$P$/kPa: 101.325 and above

**PREPARED BY:**

W. Hayduk

**EXPERIMENTAL VALUES:**

| $t$/°C | $T$/K | Mole fraction[1] / $x_1$ | Henry's constant[2] $H$/atm |
|---|---|---|---|
| 30 | 303.15 | 0.0081 (0.0082)[3] | 123.5 |
| 40 | 313.15 | 0.0075 (0.0074) | 133.3 |
| 50 | 323.15 | 0.0066 (0.0067) | 151.5 |
| 60 | 333.15 | 0.0062 (0.0061) | 161.3 |
| 70 | 343.15 | 0.0056 (0.0056) | 178.6 |

[1] Original data, given as the inverse of Henry's constant which is equivalent to mole fraction at a gas partial pressure of 101.325 kPa.

[2] Actual Henry's constant calculated by compiler.

[3] From equation of smoothed data calculated by compiler for temperature range from 303.15 to 343.15 K:

$$\ln x_1 = 965.68/T - 7.9936$$

Correlation coefficient = 0.9956

**AUXILIARY INFORMATION**

**METHOD/APPARATUS/PROCEDURE:**

Two chamber rocking contacting devices. Separate gas and liquid chambers joined by two tubes and micro valves. Gas chamber equipped with a pressure gauge. After evacuation gas and deaerated solvent separately charged, and then contacted by opening the micro valves and by rocking. Solubility calculated from knowledge of volume of system, volume of solvent charged, and initial and final gas pressures.

Solvents considered non-volatile; gas partial pressure considered to be total pressure.

**SOURCE AND PURITY OF MATERIALS:**

1. Stated as chromatography pure.

2. Vacuum fractionated.

Purities not specified.

**ESTIMATED ERROR:**

$\delta T$/K = 0.05
$\delta H/H$ = 0.05
(estimated by compiler)

**REFERENCES:**

**COMPONENTS:**

(1) Ethane; $C_2H_6$; [74-84-0]

(2) N,N-Dimethyl formic acid (dimethyl formamide); $C_3H_7NO$; [68-12-2]

**ORIGINAL MEASUREMENTS:**

Howard, W.B.; Schoch, E.P.; Mayforth, F.R.

*Petrol. Refiner* 1954, *33*, 143-146.

**VARIABLES:**

$T$/K: 273.15 - 298.15

**PREPARED BY:**

W. Hayduk

**EXPERIMENTAL VALUES:**

| $T$/K | Bunsen Coefficient[1] $\alpha$/cm$^3$(STP) cm$^{-3}$atm$^{-1}$ | Ostwald Coefficient[2] $L$/cm$^3$cm$^{-3}$ | Mole fraction[2] $10^4 x_1$ |
|---|---|---|---|
| 273.15 | 2.0 | 2.00 | 67.7 |
| 298.15 | 1.5 | 1.64 | 51.4 |

[1] Data as listed in paper; original source indicated as Technical literature from Grasselli Chemicals Department, E.I. du Pont de Nemours.

[2] Ostwald coefficient and mole fraction calculated by compiler assuming ideal gas behavior.

AUXILIARY INFORMATION

**METHOD/APPARATUS/PROCEDURE:**

Description of apparatus and method not available.

**SOURCE AND PURITY OF MATERIALS:**

Source, purities, not available.

**ESTIMATED ERROR:**

$\delta\alpha/\alpha = 0.10$

(estimated by compiler)

**REFERENCES:**

**COMPONENTS:**

(1) Ethane; $C_2H_6$; [74-84-0]

(2) Sulfinylbismethane or dimethyl sulfoxide; $C_2H_6OS$ ($CH_3SOCH_3$); [67-68-5]

**ORIGINAL MEASUREMENTS:**

Dymond, J. H.

*J. Phys. Chem.* 1967, *71*, 1829-1831.

**VARIABLES:**

$T$/K: 298.15
$p$/kPa: 101.325 (1 atm)

**PREPARED BY:**

M. E. Derrick
H. L. Clever

**EXPERIMENTAL VALUES:**

| $T$/K | Mol Fraction $10^3 x_1$ | Bunsen Coefficient $\alpha/cm^3(STP)cm^{-3}atm^{-1}$ | Ostwald Coefficient $L/cm^3cm^{-3}$ |
|---|---|---|---|
| 298.15 | 1.78 | 0.560 | 0.611 |

The Bunsen and Ostwald coefficients were calculated by the compiler.

AUXILIARY INFORMATION

**METHOD/APPARATUS/PROCEDURE:**

The liquid is saturated with the gas at a gas partial pressure of 1 atm.

The apparatus is that described by Dymond and Hildebrand (1). The apparatus uses an all-glass pumping system to spray slugs of degassed solvent into the gas. The amount of gas dissolved is calculated from the initial and final gas pressure.

**SOURCE AND PURITY OF MATERIALS:**

(1) Ethane. Phillips Petroleum Co. Research grade. Dried.

(2) Dimethylsulfoxide. Matheson, Coleman and Bell Co. Spectro-quality. Dried and fractionally frozen. m.p. 18.37°C.

**ESTIMATED ERROR:**

**REFERENCES:**

1. Dymond, J.; Hildebrand, J. H. *Ind. Eng. Chem. Fundam.* 1967, *6*, 130.

**COMPONENTS:**

(1) Ethane; $C_2H_6$; [74-84-0]

(2) Sulfinylbismethane, (Dimethylsulfoxide); $C_2H_6SO$; [67-68-5]

**ORIGINAL MEASUREMENTS:**

Lenoir, J-Y.; Renault, P.; Renon, H.

*J. Chem. Eng. Data* 1971, *16*, 340-342

**VARIABLES:**

$T$/K: 298.15

**PREPARED BY:**

C.L. Young

**EXPERIMENTAL VALUES:**

| $T$/K | Henry's Constant $H_{C_2H_6}$/atm | Mole fraction at 1 atm* $x_{C_2H_6}$ |
|---|---|---|
| 298.15 | 386 | 0.00259 |

* Calculated by compiler assuming a linear function of $p_{C_2H_6}$ vs $x_{C_2H_6}$, i.e., $x_{C_2H_6}$ (1 atm) = $1/H_{C_2H_6}$

**AUXILIARY INFORMATION**

**METHOD/APPARATUS/PROCEDURE:**

A conventional gas-liquid chromatographic unit fitted with a thermal conductivity detector was used. The carrier gas was helium. The value of Henry's law constant was calculated from the retention time. The value applies to very low partial pressures of gas and there may be a substantial difference from that measured at 1 atm. pressure. There is also considerable uncertainty in the value of Henry's constant since surface adsorption was not allowed for although its possible existence was noted.

**SOURCE AND PURITY OF MATERIALS:**

(1) L'Air Liquide sample, minimum purity 99.9 mole per cent.

(2) Touzart and Matignon or Serlabo sample, purity 99 mole per cent.

**ESTIMATED ERROR:**

$\delta T$/K = ±0.1; $\delta H$/atm = ±6% (estimated by compiler).

**REFERENCES:**

COMPONENTS:

(1) Ethane; $C_2H_6$; [74-84-0]
(2) 4-Methyl- 1,3-Dioxolan-2-one, (Propylene carbonate); $C_4H_6O_3$; [108-32-7]

ORIGINAL MEASUREMENTS:

Rivas, O.R.; Prausnitz, J.M.

*Am. Inst. Chem. Eng. J.* 1979, *25*, 975-984.

VARIABLES:

$T$/K: 263.15-373.15

PREPARED BY:

C.L. Young

EXPERIMENTAL VALUES:

| $T$/K | Henry's constant, $H$ /MPa | Mole fraction of ethane in liquid, $x_{C_2H_6}$ + |
|---|---|---|
| 263.15 | 22.88 | 0.004429 (0.00430)* |
| 298.15 | 33.81 | 0.002997 (0.00309) |
| 323.15 | 40.76 | 0.002486 (0.00255) |
| 348.15 | 46.96 | 0.002158 (0.00216) |
| 373.15 | 52.42 | 0.001933 (0.00188) |

+ at a partial pressure of 101.3 kPa calculated by compiler assuming Henry's law applies at that pressure.

* From equation of smoothed data:

$$\ln x_1 = 741.72/T - 8.2668$$

Correlation coefficient = 0.9960

AUXILIARY INFORMATION

METHOD/APPARATUS/PROCEDURE:

Volumetric apparatus with a fused quartz precision bourdon pressure gauge. Solubility apparatus carefully thermostatted. Solvent degassed *in situ*. Apparatus described in ref. (1) and modifications given in source.

SOURCE AND PURITY OF MATERIALS:

1. and 2. Purity at least 99 mole per cent.

ESTIMATED ERROR:

$\delta T/K = \pm 0.05$; $\delta x_{C_2H_6} = \pm 1\%$.

REFERENCES:

1. Cukor, P.M.; Prausnitz, J.M. *Ind. Eng. Chem. Fundam.* 1971, *10*, 638.

COMPONENTS:

(1) Ethane; $C_2H_6$; [74-84-0]

(2) 4-Methyl-1,3-dioxolan-2-one, (Propylene Carbonate); $C_4H_6O_3$; [108-32-7]

ORIGINAL MEASUREMENTS:

Lenoir, J-Y.; Renault, P.; Renon, H.

*J. Chem. Eng. Data*, 1971, *16*, 340-2.

VARIABLES:

$T$/K: 298.2-343.2

PREPARED BY:

C. L. Young

EXPERIMENTAL VALUES:

| $T$/K | Henry's constant $H_{C_2H_6}$/atm | Mole fraction at 1 atm* $x_{C_2H_6}$ |
|---|---|---|
| 298.2 | 232 | 0.00431 (0.00430)† |
| 323.2 | 280 | 0.00357 (0.00359) |
| 343.2 | 314 | 0.00318 (0.00317) |

* Calculated by compiler assuming a linear function of $p_{C_2H_6}$ vs $x_{C_2H_6}$, i.e., $x_{C_2H_6}$(1 atm) = $1/H_{C_2H_6}$

† From equation of smoothed data:

$$\ln x_1 = 693.94/T - 7.7763$$

Correlation coefficient = 0.9994

AUXILIARY INFORMATION

METHOD/APPARATUS/PROCEDURE:

A conventional gas-liquid chromatographic unit fitted with a thermal conductivity detector was used. The carrier gas was helium. The value of Henry's law constant was calculated from the retention time. The value applies to very low partial pressures of gas and there may be a substantial difference from that measured at 1 atm. pressure. There is also considerable uncertainty in the value of Henry's constant since surface adsorption was not allowed for although its possible existence was noted.

SOURCE AND PURITY OF MATERIALS:

(1) L'Air Liquide sample, minimum purity 99.9 mole per cent.

(2) Touzart and Matignon or Serlabo sample, purity 99 mole per cent.

ESTIMATED ERROR:

$\delta T$/K = $\pm 0.1$; $\delta H$/atm = $\pm 6\%$ (estimated by compiler).

REFERENCES:

COMPONENTS:

(1) Ethane; $C_2H_6$; [74-84-0]

(2) 2-(2-Aminoethoxy)-ethanol, (Diglycolamine); $C_4H_{11}NO_2$; [929-06-6]

ORIGINAL MEASUREMENTS:

Rivas, O.R. Prausnitz, J.M.

*Am. Inst. Chem. Eng. J.* 1979, *25*, 975-984.

VARIABLES:

$T$/K: 298.15-373.15

PREPARED BY:

C.L. Young

EXPERIMENTAL VALUES:

| $T$/K | Henry's constant, $H$ /MPa | Mole fraction of ethane,[+] in liquid, $x_{C_2H_6}$ |
|---|---|---|
| 298.15 | 47.13 | 0.002150 (0.00211)* |
| 323.15 | 57.05 | 0.001776 (0.00182) |
| 348.15 | 64.14 | 0.001580 (0.00161) |
| 373.15 | 68.46 | 0.001480 (0.00145) |

\+ at a partial pressure of 101.3 kPa calculated by compiler assuming Henry's law applies at that pressure.

* From equation of smoothed data for temperatures between 298.15 and 373.15 K:

$$\ln x_1 = 557.10/T - 8.0318$$

Correlation coefficient = 0.9876

AUXILIARY INFORMATION

METHOD/APPARATUS/PROCEDURE:

Volumetric apparatus with a fused quartz precision bourdon pressure gauge. Solubility apparatus carefully thermostatted. Solvent degassed *in situ*. Apparatus described in ref. (1). and modifications given in source.

SOURCE AND PURITY OF MATERIALS:

1. Purity at least 99 mole per cent.

2. Purity at least 97 mole per cent.

ESTIMATED ERROR:

$\delta T$/K = ±0.05; $\delta x_{C_2H_6}$ = ±1%.

REFERENCES:

1. Cukor, P.M.; Prausnitz, J.M. *Ind. Eng. Chem. Fundam.* 1971, *10*, 638.

COMPONENTS:

(1) Ethane; $C_2H_6$; [74-84-0]

(2) 2-Aminoethanol, (Monoethanolamine); $C_2H_7NO$: [141-43-5]

ORIGINAL MEASUREMENTS:

Rivas, O.R.; Prausnitz, J.M.

*Am. Inst. Chem. Eng. J.* 1979, *25*, 975-984.

VARIABLES:

$T$/K: 298.15-373.15

PREPARED BY:

C.L. Young

EXPERIMENTAL VALUES:

| $T$/K | Henry's constant, $H$ /MPa | Mole fraction of[+] ethane in liquid, $x_{C_2H_6}$ |
|---|---|---|
| 298.15 | 126.1 | 0.0008035 (0.000792)* |
| 323.15 | 142.1 | 0.0007131 (0.000703) |
| 348.15 | 152.5 | 0.0006644 (0.000672) |
| 373.15 | 158.5 | 0.0006393 (0.000630) |

\+ at a partial pressure of 101.3 kPa calculated by compiler assuming Henry's law applies at that pressure.

* from equation of smoothed data:

$$\ln x_1 = 340.93/T - 8.2839$$

Correlation coefficient = 0.9858

AUXILIARY INFORMATION

METHOD/APPARATUS/PROCEDURE:

Volumetric apparatus with a fused quartz precision bourdon pressure gauge. Solubility apparatus carefully thermostatted. Solvent degassed *in situ*. Apparatus described in ref. (1) and modifications given in source.

SOURCE AND PURITY OF MATERIALS:

1. and 2. Purity at least 99 mole per cent.

ESTIMATED ERROR:

$\delta T/K = \pm 0.05$; $\delta x_{C_2H_6} = \pm 1\%$.

REFERENCES:

1. Cukor, P.M., Prausnitz, J.M. *Ind. Eng. Chem. Fundam.* 1971, *10*, 638.

COMPONENTS:

(1) Ethane; $C_2H_6$; [74-84-0]

(2) Octamethylcyclotetrasiloxane $C_8H_{24}O_4Si_4$; [556-67-2]

ORIGINAL MEASUREMENTS:

Chappelow, C.C.; Prausnitz, J.M.

*Am. Inst. Chem. Eng. J.* 1974, *20*, 1097-1104.

VARIABLES:

$T$/K: 300-425

PREPARED BY:

C.L. Young

EXPERIMENTAL VALUES:

| $T$/K | Henry's Constant[a] /atm | Mole fraction[b] of ethane at 1 atm. partial pressure, $x_{C_2H_6}$ |
|---|---|---|
| 300 | 19.4 | 0.0515 (0.0497)[c] |
| 325 | 26.7 | 0.0375 (0.0375) |
| 350 | 35.6 | 0.0281 (0.0294) |
| 375 | 43.5 | 0.0230 (0.0238) |
| 400 | 50.4 | 0.0198 (0.0198) |
| 425 | 56.6 | 0.0177 (0.0169) |

[a] Authors stated measurements were made at several pressures and values of solubility used were all within the Henry's Law region.

[b] Calculated by compiler assuming linear relationship between mole fraction and pressure.

[c] From equation of smoothed data:

$$\ln x_1 = 1102.8/T - 6.6773$$

Correlation coefficient = 0.9958

AUXILIARY INFORMATION

METHOD/APPARATUS/PROCEDURE:

Volumetric apparatus similar to that described by Dymond and Hildebrand (1). Pressure measured with a null detector and precision gauge. Details in ref. (2).

SOURCE AND PURITY OF MATERIALS:

Solvent degassed, no other details given.

ESTIMATED ERROR:

$\delta T$/K = ±0.1; $\delta x_{C_2H_6}$ = ±1%

REFERENCES:

1. Dymond, J.; Hildebrand, J.H. *Ind. Eng. Chem. Fundam.* 1967, *6*, 130.
2. Cukor, P.M.; Prausnitz, J.M. *Ind. Eng. Chem. Fundam.* 1971, *10*, 638.

**COMPONENTS:**

(1) Ethane; $C_2H_6$; [74-84-0]

(2) Thiophene, tetrahydro-1, 1-dioxide, (Sulfolane); $C_4H_8O_2S$; [126-33-0]

**ORIGINAL MEASUREMENTS:**

Rivas, O.R.; Prausnitz, J.M.

*Am. Inst. Chem. Eng. J.* 1979, *25*, 975-984.

**VARIABLES:**

$T$/K: 303.15-373.15

**PREPARED BY:**

C.L. Young

**EXPERIMENTAL VALUES:**

| $T$/K | Henry's constant, /MPa | Mole fraction of[+] ethane in liquid, $x_{C_2H_6}$ | |
|---|---|---|---|
| 303.15 | 48.69 | 0.002081 | (0.00203)* |
| 323.15 | 58.30 | 0.001738 | (0.00178) |
| 348.15 | 67.32 | 0.001505 | (0.00154) |
| 373.15 | 73.00 | 0.001388 | (0.00136) |

+ at a partial pressure of 101.3 kPa calculated by compiler assuming Henry's law applies at that pressure.

* from equation of smoothed data between 303.15 and 373.15 K:

$$\ln x_1 = 654.93/T - 8.3581$$

Correlation coefficient = 0.9889

AUXILIARY INFORMATION

**METHOD/APPARATUS/PROCEDURE:**

Volumetric apparatus with a fused quartz precision bourdon pressure gauge. Solubility apparatus carefully thermostatted. Solvent degassed *in situ*. Apparatus described in ref. (1) and modifications given in source.

**SOURCE AND PURITY OF MATERIALS:**

1. and 2. Purity at least 99 mole per cent.

**ESTIMATED ERROR:**

$\delta T$/K = ±0.05; $\delta x_{C_2H_6}$ = ±1%.

**REFERENCES:**

1. Cukor, P.M.; Prausnitz, J.M. *Ind. Eng. Chem. Fundam.* 1971, *10*, 638.

**COMPONENTS:**

(1) Ethane; $C_2H_6$; [74-84-0]

(2) Deuterium oxide (heavy water); $D_2O$; [7789-20-0]

**ORIGINAL MEASUREMENTS:**

Ben-Naim, A.; Wilf, J.;Yaacobi, M.

*J. Phys. Chem.* 1973, *77*, 95-102.

**VARIABLES:**

$T$/K: 278.15-298.15

$P$/kPa: 101.325 (1 atm)

**PREPARED BY:**

W. Hayduk

**EXPERIMENTAL VALUES:**

| $t$/°C | $T$/K | Ostwald Coefficient[1] $L$/cm$^3$cm$^{-3}$ | Bunsen Coefficient[2] $\alpha$/cm$^3$(STP)cm$^{-3}$atm$^{-1}$ | Mole Fraction[2] /$10^4 x_1$ |
|---|---|---|---|---|
| 5 | 278.15 | 0.0889 | 0.0873 | 0.706 (0.692)[3] |
| 10 | 283.15 | 0.0733 | 0.0707 | 0.571 (0.578) |
| 15 | 288.15 | 0.0621 | 0.0589 | 0.476 (0.485) |
| 20 | 293.15 | 0.0539 | 0.0502 | 0.406 (0.410) |
| 25 | 298.15 | 0.0479 | 0.0439 | 0.355 (0.348) |

[1] From original data

[2] Bunsen coefficient and mole fraction calculated by compiler assuming ideal gas behavior.

[3] From equation of smoothed data:

$$\ln x_1 = 2850.7/T - 19.827$$

Correlation coefficient = 0.9977

**AUXILIARY INFORMATION**

**METHOD/APPARATUS/PROCEDURE:**

The method is volumetric utilizing an all-glass apparatus consisting of a dissolution cell of 300 to 600 cm$^3$ capacity, a gas volume measuring column, and a manometer. The solvent is degassed in the dissolution cell. Gas dissolves while the liquid is stirred using a magnetic stirrer. The volume of gas confined over mercury is read initially and after equilibration, by means of a cathetometer.

The apparatus is described by Ben-Naim and Baer (1) but it includes the modification introduced by Wen and Hung (2) of replacing the stopcocks with Teflon needle valves.

**SOURCE AND PURITY OF MATERIALS:**

1. Matheson; minimum specified purity 99.9 mole per cent.

2. Darmstadt; specified purity 99.75 per cent

**ESTIMATED ERROR:**

$\delta L/L = 0.005$

Estimated by compiler

**REFERENCES:**

1. Ben-Naim, A.; Baer, S. *Trans. Faraday Soc.* 1963, *59*, 2735.

2. Wen, W.-Y.; Hung, J.H. *J.Phys. Chem.* 1970, *74*, 170.

**COMPONENTS:**

(1) Ethane; $C_2H_6$; [74-84-0]

(2) Esters of phosphoric acid

**ORIGINAL MEASUREMENTS:**

Lenoir, J-Y.; Renault, P.; Renon, H.

*J. Chem. Eng. Data* 1971, *16*, 340-342

**VARIABLES:**

$T$/K = 298.2 - 343.2

**PREPARED BY:**

C. L. Young

**EXPERIMENTAL VALUES:**

| $T$/K | Henry's constant $H_{C_2H_6}$/atm | Mole fraction at 1 atm* $x_{C_2H_6}$ |
|---|---|---|
| Phosphoric acid, trimethyl ester; $C_3H_9O_4P$; [512-56-1] | | |
| 325.2 | 250 | 0.00400 |
| Phosphoric acid, triethyl ester; $C_6H_{15}O_4P$; [78-40-0] | | |
| 325.2 | 92.4 | 0.0108 |
| Phosphoric acid, tripropyl ester; $C_9H_{21}O_4P$; [513-08-6] | | |
| 298.2 | 39.0 | 0.0256 |
| 323.2 | 51.1 | 0.0196 |
| 343.2 | 68.0 | 0.0147 |
| Phosphoric acid, tributyl ester; $C_{12}H_{27}O_4P$; [126-73-8] | | |
| 325.2 | 46.4 | 0.0216 |
| Phosphoric acid, tris(2-methyl propyl) ester; $C_{12}H_{27}O_4P$; [126-71-6] | | |
| 325.2 | 42.5 | 0.0235 |

*Calculated by compiler assuming a linear function of $p_{C_2H_6}$ vs $x_{C_2H_6}$

i.e., $x_{C_2H_6}$(1 atm) = $1/H_{C_2H_6}$

AUXILIARY INFORMATION

**METHOD/APPARATUS/PROCEDURE:**

A conventional gas-liquid chromatographic unit fitted with a thermal conductivity detector was used. The carrier gas was helium. The value of Henry's law constant was calculated from the retention time. The value applies to very low partial pressures of gas and there may be a substantial difference from that measured at 1 atm. pressure. There is also considerable uncertainty in the value of Henry's constant since surface adsorption was not allowed for although its possible existence was noted.

**SOURCE AND PURITY OF MATERIALS:**

(1) L'Air Liquide sample, minimum purity 99.9 mole per cent.

(2) Touzart and Matignon or Serlabo sample, purity 99 mole per cent.

**ESTIMATED ERROR:**

$\delta T$/K = $\pm 0.1$; $\delta H$/atm = $\pm 6\%$ (estimated by compiler).

**REFERENCES:**

| COMPONENTS: | ORIGINAL MEASUREMENTS: |
|---|---|
| (1) Ethane; $C_2H_6$; [74-84-0]<br>(2) Dog blood and lung tissue. | Young, I.H.; Wagner, P.D.<br>*J. Appl. Physiol.* 1979, *46*, 1207-10. |
| VARIABLES: | PREPARED BY: |
| $T$/K = 310.15 | C.L. Young |

EXPERIMENTAL VALUES:

| $T$/K | Component 2 | Solubility, $S^+$ | No. of Observations | Bunsen[#] coefficient $\alpha$ |
|---|---|---|---|---|
| 310.15 | Lung tissue | 0.0136 ± 0.0015 | 25 | 0.0910 |
| | Blood | 0.0161 ± 0.0014 | 18 | 0.108 |

$$S^+ = 10^2 \times \frac{\text{Volume of gas dissolved converted to at 1 atm pressure}}{\text{Volume of liquid} \times 760}$$

# calculated by compiler assuming Henry's law holds up to 1 atm pressure.

AUXILIARY INFORMATION

METHOD/APPARATUS/PROCEDURE:

Approximately 10 $cm^3$ of each sample (Lung tissue homogenate or blood) was introduced into a 50 $cm^3$ syringe with 30 $cm^3$ of a gas mixture containing 0.0031 mole per cent of ethane. After equilibriation gas phase analysed by gas chromatography. All gas expelled from syringe and 15 $cm^3$ of nitrogen added and re-equilibriated. Samples of gas analysed by gas chromatography. Details in source.

SOURCE AND PURITY OF MATERIALS:

1. No details given.
2. Lung tissue obtained from eight mongrel dogs. Each animal being heparinized before being killed. Lung tissue allowed to drain and then portions with few major vessels and cartilage homogenised. Blood obtained from some animals.

ESTIMATED ERROR:

$\delta T$/K = ±0.1

REFERENCES:

COMPONENTS:

(1) Ethane; $C_2H_6$; [74-84-0]

(2) Various organic solvents and hydrogen sulfide at elevated pressures

EVALUATOR:

Colin L. Young
School of Chemistry
University of Melbourne
Parkville, Victoria 3052
Australia

CRITICAL EVALUATION:

There appears to be no solubility data at elevated pressures for the solvents 1,1-oxybisethane (diethyl ether), 1-propene, hydrocarbon oil, and hydrogen sulfide, with which to compare the data of Ohgaki *et al.* (1), McKay *et al.* (2), Sage *et al.* (3) and Robinson *et al.* (4), respectively. These data are classified as tentative.

Ohgaki *et al.* (1) have used an apparatus of proven design and their data for ethane in benzene are consistent both with the low pressure data of Jadot (5), Horiuti (6) and Armitage *et al.* (7) and the high pressure data of Kay and Nevens (8). The data of Ohgaki *et al.* (1) are in fair agreement with the low pressure data of Horiuti (6) for the solvent methyl acetate and slightly lower than Horiuti's data for 2-propanone (acetone). Therefore the data of Ohgaki *et al.* (1) for the solubility of ethane in various solvents are classified as tentative. Kay and Nevens (8) determined the bubble point and dew point of mixtures of known composition and while this method is suitable for the purpose of the original work, the results obtained are not very suitable to give a detailed account of the solubility as a function of pressure.

The high pressure solubility data in methanol of Ohgaki *et al.* (1) and of Ma and Kohn (9) are completely consistent with one another as well as with the low pressure data of Ben-Naim and Yaacobbi (10), Boyer and Bircher (11) and Gjaldbaek and Niemann (12). The data in methanol are also classified as tentative.

Ethane solubilities at high pressure have been determined in certain military fuels by Findl *et al.* (13). These data are classified as tentative.

References

1. Ohgaki, K.; Sano, F.; Katayama, T. *J. Chem. Eng. Data* 1976, *21*, 55-58.

2. McKay, R.A.; Reamer, H.H.; Sage, B.H.; Lacey, W.N. *Ind. Eng. Chem.* 1951, *43*, 2112-2117.

3. Sage, B.H.; Davies, J.A.; Sherbourn, J.E.; Lacey, W.N. *Ind. Eng. Chem.* 1936, *28*, 1328-1333.

4. Robinson, D.B.; Kalra, H.; Krishnan, T.; Miranda, R.D. *Proc. Annu. Conv. Gas Process. Assoc. Tech. Pap.* 1975, *54*, 25-31.

5. Jadot, J.J. *Chim. Phys.* 1972, *69*, 1036-1040.

6. Horiuti, J. *Sci. Pap. Inst. Phys. Chem. Res. (Jpn)* 1931, *17*, 126-256.

7. Armitage, D.A.; Linford, R.G.; Thornhill, D.G.T. *Ind. Eng. Chem. Fundam.* 1978, *17*, 362-364.

8. Kay, W.B.; Nevens, T.D. *Chem. Eng. Prog. Pymp. Ser. no. 3* 1952, *48*, 108-114.

9. Ma, Y.H.; Kohn, J.P. *J. Chem. Eng. Data* 1964, *9*, 3-5.

| COMPONENTS: | EVALUATOR: |
|---|---|
| (1) Ethane; $C_2H_6$; [74-84-0]<br>(2) Various organic solvents and hydrogen sulfide at elevated pressures | Colin L. Young<br>School of Chemistry<br>University of Melbourne<br>Parkville, Victoria 3052<br>Australia |

CRITICAL EVALUATION:

...continued


10. Ben-Naim, A.; Yaacobi, M. *J. Phys. Chem.* 1974, *78*, 175-178.

11. Boyer, F.L.; Bircher, L.J. *J. Phys. Chem.* 1960, *64*, 1330-1331.

12. Gjaldbaek, J.C.; Niemann, H. *Acta Chem. Scand.* 1958, *12*, 1015-1023.

13. Findl, E.; Brande, H.; Edwards, H. *U.S. Dept. Comm., Office Tech. Ser. Report No. AD 274 623*, 1960, 216pp.

COMPONENTS:

(1) Ethane; $C_2H_6$; [74-84-0]

(2) 1,1'-Oxybisethane, (Diethyl-ether); $C_4H_{10}O$; [60-29-7]

ORIGINAL MEASUREMENTS:

Ohgaki, K.; Sano, F.; Katayama, T.

*J. Chem. Eng. Data* 1976, *21*, 55-8

VARIABLES:

$T$/K = 298.15

$P$/MPa = 9.5-38.6

PREPARED BY:

C.L. Young

EXPERIMENTAL VALUES:

| $T$/K | $P/10^5$Pa | Mole fraction of ethane in liquid, $x_{C_2H_6}$ | in vapor, $y_{C_2H_6}$ |
|---|---|---|---|
| 298.15 | 9.559 | 0.2529 | 0.9244 |
| | 15.077 | 0.4010 | 0.9545 |
| | 17.733 | 0.4732 | 0.9630 |
| | 30.312 | 0.7840 | 0.9782 |
| | 35.221 | 0.8879 | 0.9788 |
| | 38.567 | 0.9588 | 0.9798 |

AUXILIARY INFORMATION

METHOD/APPARATUS/PROCEDURE:

Static equilibrium cell fitted with windows and magnetic stirrer. Temperature of thermostatic liquid measured with platinum resistance thermometer. Pressure measured using dead weight gauge and differential pressure transducer. Samples of vapor and liquid analysed by gas chromatography. Details in source and ref. (1).

SOURCE AND PURITY OF MATERIALS:

1. Takachiho Kagakukogyo Co. sample purity better than 99.7 mole per cent.
2. Merck Co. sample purity about 99.993 mole per cent.

ESTIMATED ERROR: $\delta T$/K $=\pm 0.01$; $\delta P/10^5$Pa $= \pm 0.01$; $\delta x_{C_2H_6}$ (for $x_{C_2H_6} < 0.5$) $= \pm 1\%$. $\delta(1-x_{C_2H_6})$ (for $x_{C_2H_6} > 0.5$) $= \pm 1\%$. (similarly for vapor composition $y$).

REFERENCES:

1. Ohgaki, K.; Katayama, T.

*J. Chem. Eng. Data* 1975, *20*, 264

| COMPONENTS: | ORIGINAL MEASUREMENTS: |
|---|---|
| (1) Ethane; $C_2H_6$; [74-84-0]<br>(2) 1-Propene; $C_3H_6$; [115-07-1] | McKay, R. A.; Reamer, H. H.; Sage, B. H.; Lacey, W. N. *Ind. Eng. Chem.* 1951, *43*, 2112-2117. |
| VARIABLES: | PREPARED BY: |
| $T$/K = 260.9-310.9<br>$P$/MPa = 0.689-2.76 | C. L. Young |

EXPERIMENTAL VALUES:

| $T$/K | $P$/MPa | Mole fraction of ethane in liquid, $x_{C_2H_6}$ | Mole fraction of ethane in gas, $y_{C_2H_6}$ | $T$/K | $P$/MPa | Mole fraction of ethane in liquid, $x_{C_2H_6}$ | Mole fraction of ethane in gas, $y_{C_2H_6}$ |
|---|---|---|---|---|---|---|---|
| 260.9 | 0.689 | 0.228 | 0.513 | 310.9 | 3.10 | 0.461 | 0.626 |
| | 1.03 | 0.494 | 0.761 | | 3.45 | 0.554 | 0.697 |
| | 1.38 | 0.745 | 0.894 | | 3.79 | 0.643 | 0.759 |
| | 1.72 | 0.977 | 0.991 | | 4.14 | 0.727 | 0.813 |
| 277.6 | 0.689 | 0.014 | 0.045 | | 4.48 | 0.809 | 0.863 |
| | 1.03 | 0.209 | 0.452 | | 4.83 | 0.894 | 0.912 |
| | 1.38 | 0.390 | 0.646 | | 4.98 | 0.930 | 0.930 |
| | 1.72 | 0.568 | 0.775 | 344.3 | 3.45 | 0.062 | 0.111 |
| | 2.07 | 0.739 | 0.873 | | 3.79 | 0.130 | 0.205 |
| | 2.41 | 0.896 | 0.951 | | 4.14 | 0.199 | 0.277 |
| 310.9 | 1.72 | 0.048 | 0.118 | | 4.48 | 0.269 | 0.330 |
| | 2.07 | 0.157 | 0.317 | | 4.83 | 0.338 | 0.357 |
| | 2.41 | 0.260 | 0.447 | | 4.86 | 0.350 | 0.350 |
| | 2.76 | 0.361 | 0.543 | | | | |

AUXILIARY INFORMATION

METHOD/APPARATUS/PROCEDURE:

General description of cell given in ref. (1). The cell was fitted with port which enabled isobaric, isothermal sampling. Quantity of propene determined by catalytic hydrogenation as in ref. (2) and (3).

SOURCE AND PURITY OF MATERIALS:

1. Crude sample fractionated twice.
2. Prepared by dehydration of alcohol over aluminium oxide. Fractionated.

ESTIMATED ERROR:

$\delta T$/K = ±0.1; $\delta P$/MPa = ±0.007; $\delta x_{C_2H_6}$, $\delta y_{C_2H_6}$ = ±0.003.

REFERENCES:


1. Sage, B.H.; Lacey, W.N. *Trans. Am. Inst. Mining Met. Eng.* 1940, *136*, 136.
2. McMillan, W.A.; Cole, H.A.; Richie, A.V. *Ind.Eng.Chem.Anal.Ed.* 1936, *8*, 2658.
3. Sage, B.H.; Lacey, W.N. *Ind. Eng. Chem.* 1948 *40*, 1299.

COMPONENTS:

(1) Ethane; $C_2H_6$; [74-84-0]

(2) Hydrocarbon oil

ORIGINAL MEASUREMENTS:

Sage, B. H.; Davies, J. A.; Sherborne, J. E.; Lacey, W. N.

*Ind. Eng. Chem.*

1936, *28*, 1328-1333.

VARIABLES:

$T$/K = 294.3-361.0

$P$/MPa = 0.95-16.35

PREPARED BY:

C. L. Young

EXPERIMENTAL VALUES:

| $T$/°F | $T$/K[a] | $P$/psia | $P$/MPa[a] | Solubility, $S$ /wt-% |
|---|---|---|---|---|
| 70.0 | 294.3 | 139 | 0.95 | 3.19 |
| | | 204 | 1.41 | 5.79 |
| | | 412 | 2.84 | 16.43 |
| | | 506 | 3.49 | 26.32 |
| | | 562 | 3.87 | 49.82 |
| 100.0 | 311.0 | 172 | 1.19 | 3.19 |
| | | 264 | 1.82 | 5.79 |
| | | 568 | 3.92 | 16.43 |
| | | 716 | 4.94 | 26.32 |
| | | 1141 | 7.87 | 49.82 |
| 130.0 | 327.6 | 207 | 1.43 | 3.19 |
| | | 330 | 2.28 | 5.79 |
| | | 736 | 5.07 | 16.43 |
| | | 988 | 6.81 | 26.32 |
| | | 1608 | 11.09 | 49.82 |
| 160.0 | 344.3 | 245 | 1.69 | 3.19 |
| | | 400 | 2.76 | 5.79 |
| | | 933 | 6.43 | 16.43 |
| | | 1310 | 9.03 | 26.32 |
| | | 2021 | 13.93 | 49.82 |
| 190.0 | 361.0 | 285 | 1.97 | 3.19 |
| | | 476 | 3.28 | 5.79 |
| | | 1140 | 7.86 | 16.43 |
| | | 1626 | 11.21 | 26.32 |
| | | 2372 | 16.35 | 49.82 |

(cont.)

AUXILIARY INFORMATION

METHOD/APPARATUS/PROCEDURE:

Contents of variable volume cell brought to equilibrium at desired temperature and pressure and volume determined. Volume varied by admission or removal of mercury. Bubble point determined from change in slope of pressure-volume curve.

Details given in ref. (1).

SOURCE AND PURITY OF MATERIALS:

1. Sample from Carbide and Carbon Chemicals Corp. Fractionated.
2. Non-waxy asphalt crude oil with molecular weight of between 335 & 340 (by freezing point depression).

ESTIMATED ERROR:

$\delta T$/K = ±0.13; $\delta P$/psia = ±1;

$\delta S/S$ = ±0.001.

REFERENCES:

1. Sage, B. H.; Backus, H. S.; Lacey, W. N. *Ind. Eng. Chem.* 1935, *27*, 686.

| COMPONENTS: | ORIGINAL MEASUREMENTS: |
|---|---|
| (1) Ethane; $C_2H_6$; [74-84-0] | Sage, B. H., Davies, J. A.; |
| (2) Hydrocarbon oil | Sherborne, J. E.; Lacey, W. N. |
| | *Ind. Eng. Chem.* |
| | 1936, *28*, 1328-1333. |

EXPERIMENTAL VALUES:

| $T$/°F | $T$/K[a] | $P$/psia | $P$/MPa[a] | Solubility, $S$ /wt-% |
|---|---|---|---|---|
| 220.0 | 377.6 | 329 | 2.27 | 3.19 |
| | | 560 | 3.86 | 5.79 |
| | | 1350 | 9.31 | 16.43 |
| | | 1920 | 13.24 | 26.32 |
| | | 2690 | 18.55 | 49.82 |

Specific volume data given in source.

[a]Calculated by compiler.

COMPONENTS:

(1) Ethane; $C_2H_6$; [74-84-0]

(2) Benzene; $C_6H_6$; [71-43-2]

ORIGINAL MEASUREMENTS:

Ohgaki, K.; Sano, F. Katayama, T.

*J. Chem. Eng. Data* 1976, *21*, 55-8.

VARIABLES:

$T/K = 298.15$

$P/MPa = 7.7-38.0$

PREPARED BY:

C.L. Young

EXPERIMENTAL VALUES:

| $T/K$ | $P/10^5 Pa$ | Mole fraction of ethane in liquid, $x_{C_2H_6}$ | in vapor, $y_{C_2H_6}$ |
|---|---|---|---|
| 298.15 | 7.759 | 0.1201 | 0.9801 |
| | 13.490 | 0.2202 | 0.9876 |
| | 20.383 | 0.3747 | 0.9910 |
| | 25.243 | 0.5355 | 0.9920 |
| | 28.795 | 0.6494 | 0.9927 |
| | 34.411 | 0.8602 | 0.9933 |
| | 38.007 | 0.9299 | 0.9937 |

AUXILIARY INFORMATION

METHOD/APPARATUS/PROCEDURE:

Static equilibrium cell fitted with windows and magnetic stirrer. Temperature of thermostatic liquid measured with platinum resistance thermometer. Pressure measured using dead weight gauge and differential pressure transducer. Samples of vapor and liquid analysed by gas chromatography. Details in source and ref. (1).

SOURCE AND PURITY OF MATERIALS:

1. Takachiho Kagakukogyo Co. sample purity better than 99.7 mole per cent.

2. Baker Chem. Co. sample purity about 99.993 mole per cent.

ESTIMATED ERROR: $\delta T/K = \pm 0.01$; $\delta P/10^5 Pa = \pm 0.01$; $\delta x_{C_2H_6}$ (for $x_{C_2H_6} < 0.5$) = ±1%; $\delta(1-x_{C_2H_6})$ (for $x_{C_2H_6} > 0.5$) = ±1%. (similarly for vapor composition $y$).

REFERENCES:

1. Ohgaki, K.; Katayama, T.

*J. Chem. Eng. Data* 1975, *20*, 264.

| COMPONENTS: | ORIGINAL MEASUREMENTS: |
|---|---|
| (1) Ethane; $C_2H_6$; [74-84-0]<br>(2) Benzene; $C_6H_6$; [71-43-2] | Kay, W. B.; Nevens, T. D.<br>*Chem. Eng. Prog. Symp. Ser. no. 3* 1952, *48*, 108-114. |
| **VARIABLES:** | **PREPARED BY:** |
| $T$/K: 269-504 $P$/MPa: 0.7-10.0 | C. L. Young |

EXPERIMENTAL VALUES:

| t/°C | $T$/K[a] | $P$/psia | $P$/MPa[a] | Mole fraction of ethane in liquid, $x_{C_2H_6}$ | Mole fraction of ethane in vapor, $y_{C_2H_6}$ |
|---|---|---|---|---|---|
| 4.5 | 277.7 | 100 | 0.689 | 0.1520 | - |
| 32.2 | 305.4 | 150 | 1.03 | 0.1520 | - |
| 55.9 | 329.1 | 200 | 1.38 | 0.1520 | - |
| 77.3 | 350.5 | 250 | 1.72 | 0.1520 | - |
| 97.0 | 370.2 | 300 | 2.07 | 0.1520 | - |
| 115.3 | 388.5 | 350 | 2.41 | 0.1520 | - |
| 132.0 | 405.2 | 400 | 2.76 | 0.1520 | - |
| 147.3 | 420.5 | 450 | 3.10 | 0.1520 | - |
| 161.7 | 434.9 | 500 | 3.45 | 0.1520 | - |
| 175.0 | 448.2 | 550 | 3.79 | 0.1520 | - |
| 187.3 | 460.5 | 600 | 4.14 | 0.1520 | - |
| 199.2 | 472.4 | 650 | 4.48 | 0.1520 | - |
| 210.4 | 483.6 | 700 | 4.83 | 0.1520 | - |
| 221.4 | 494.6 | 750 | 5.17 | 0.1520 | - |
| 232.1 | 505.3 | 800 | 5.52 | 0.1520 | - |
| 242.7 | 515.9 | 850 | 5.86 | 0.1520 | - |
| 254.5 | 527.7 | 900 | 6.21 | 0.1520 | - |
| 209.8 | 483.0 | 300 | 2.07 | - | 0.1520 |
| 219.7 | 492.9 | 350 | 2.41 | - | 0.1520 |
| 228.2 | 501.4 | 400 | 2.76 | - | 0.1520 |
| 235.7 | 508.9 | 450 | 3.10 | - | 0.1520 |
| 242.4 | 515.6 | 500 | 3.45 | - | 0.1520 |

(cont.)

AUXILIARY INFORMATION

METHOD/APPARATUS/PROCEDURE:

Samples of known composition confined in thick-walled glass tube over mercury. Temperature measured with thermocouple and pressure with Bourdon gauge. Dew and bubble points determined.

SOURCE AND PURITY OF MATERIALS:

1. Phillips Petroleum Co. sample; purity 99.9 mole per cent. Dried with phosphorus pentoxide.
2. Analytical grade reagent, fractionally distilled and degassed.

ESTIMATED ERROR:

$\delta T$/K = ±0.05; $\delta P$/MPa = ±0.007; $\delta x_{C_2H_6}$, $\delta y_{C_2H_6}$ = ±0.0002.

REFERENCES:

COMPONENTS:

(1) Ethane; $C_2H_6$; [74-84-0]

(2) Benzene; $C_6H_6$; [71-43-2]

ORIGINAL MEASUREMENTS:

Kay, W. B.; Nevens, T. D.

*Chem. Eng. Prog. Symp. Ser. no. 3* 1952, *48*, 108-114.

EXPERIMENTAL VALUES:

| t/°C | $T$/K[a] | $P$/psia | $P$/MPa[a] | Mole fraction of ethane in liquid, $x_{C_2H_6}$ | Mole fraction of ethane in vapor, $y_{C_2H_6}$ |
|---|---|---|---|---|---|
| 248.6 | 521.8 | 550 | 3.79 | - | 0.1520 |
| 254.4 | 527.6 | 600 | 4.14 | - | 0.1520 |
| 259.4 | 532.6 | 650 | 4.48 | - | 0.1520 |
| 263.9 | 537.1 | 700 | 4.83 | - | 0.1520 |
| 267.7 | 540.9 | 750 | 5.17 | - | 0.1520 |
| 270.7 | 543.9 | 800 | 5.52 | - | 0.1520 |
| 271.8 | 545.0 | 850 | 5.86 | - | 0.1520 |
| 269.7 | 542.9 | 900 | 6.21 | - | 0.1520 |
| 1.4 | 274.6 | 200 | 1.38 | 0.3922 | - |
| 13.7 | 286.9 | 250 | 1.72 | 0.3922 | - |
| 24.6 | 297.8 | 300 | 2.07 | 0.3922 | - |
| 35.1 | 308.3 | 350 | 2.41 | 0.3922 | - |
| 44.7 | 317.9 | 400 | 2.76 | 0.3922 | - |
| 53.8 | 327.0 | 450 | 3.10 | 0.3922 | - |
| 62.5 | 335.7 | 500 | 3.45 | 0.3922 | - |
| 71.1 | 344.3 | 550 | 3.79 | 0.3922 | - |
| 74.3 | 347.5 | 600 | 4.14 | 0.3922 | - |
| 87.5 | 360.7 | 650 | 4.48 | 0.3922 | - |
| 95.7 | 368.9 | 700 | 4.83 | 0.3922 | - |
| 104.0 | 377.2 | 750 | 5.17 | 0.3922 | - |
| 112.4 | 385.6 | 800 | 5.52 | 0.3922 | - |
| 120.6 | 393.8 | 850 | 5.86 | 0.3922 | - |
| 128.7 | 401.9 | 900 | 6.21 | 0.3922 | - |
| 136.8 | 410.0 | 950 | 6.55 | 0.3922 | - |
| 145.2 | 418.4 | 1000 | 6.89 | 0.3922 | - |
| 154.2 | 427.4 | 1050 | 7.24 | 0.3922 | - |
| 163.0 | 436.2 | 1100 | 7.58 | 0.3922 | - |
| 172.3 | 445.5 | 1150 | 7.93 | 0.3922 | - |
| 183.1 | 456.3 | 1200 | 8.27 | 0.3922 | - |
| 196.3 | 469.5 | 1250 | 8.62 | 0.3922 | - |
| 186.0 | 459.2 | 300 | 2.07 | - | 0.3922 |
| 193.7 | 466.9 | 350 | 2.41 | - | 0.3922 |
| 200.3 | 473.5 | 400 | 2.76 | - | 0.3922 |
| 206.3 | 479.5 | 450 | 3.10 | - | 0.3922 |
| 211.8 | 485.0 | 500 | 3.45 | - | 0.3922 |
| 216.9 | 490.1 | 550 | 3.79 | - | 0.3922 |
| 221.6 | 494.8 | 600 | 4.14 | - | 0.3922 |
| 225.5 | 498.7 | 650 | 4.48 | - | 0.3922 |
| 229.2 | 502.4 | 700 | 4.83 | - | 0.3922 |
| 232.3 | 505.5 | 750 | 5.17 | - | 0.3922 |
| 234.9 | 508.1 | 800 | 5.52 | - | 0.3922 |
| 237.1 | 510.3 | 850 | 5.86 | - | 0.3922 |
| 238.8 | 512.0 | 900 | 6.21 | - | 0.3922 |
| 240.0 | 513.2 | 950 | 6.55 | - | 0.3922 |
| 240.5 | 513.7 | 1000 | 6.89 | - | 0.3922 |
| 240.2 | 513.4 | 1050 | 7.24 | - | 0.3922 |
| 239.2 | 512.4 | 1100 | 7.58 | - | 0.3922 |
| 236.4 | 509.6 | 1150 | 7.93 | - | 0.3922 |
| 230.8 | 504.0 | 1200 | 8.27 | - | 0.3922 |
| 222.3 | 495.5 | 1250 | 8.62 | - | 0.3922 |
| 5.7 | 278.9 | 250 | 1.72 | 0.5023 | - |
| 15.2 | 288.4 | 300 | 2.07 | 0.5023 | - |
| 24.1 | 297.3 | 350 | 2.41 | 0.5023 | - |
| 32.5 | 305.7 | 400 | 2.76 | 0.5023 | - |
| 40.3 | 313.5 | 450 | 3.10 | 0.5023 | - |
| 47.6 | 320.8 | 500 | 3.45 | 0.5023 | - |
| 54.8 | 328.0 | 550 | 3.79 | 0.5023 | - |
| 61.8 | 335.0 | 600 | 4.14 | 0.5023 | - |

(cont.)

| COMPONENTS: | ORIGINAL MEASUREMENTS: |
|---|---|
| (1) Ethane; $C_2H_6$; [74-84-0] | Kay, W. B.; Nevens, T. D. |
| (2) Benzene; $C_6H_6$; [71-43-2] | *Chem. Eng. Prog. Symp. Ser. no. 3* |
| | 1952, *48*, 108-114. |

EXPERIMENTAL VALUES:

| t/°C | $T$/K [a] | $P$/psia | $P$/MPa [a] | Mole fraction of ethane in liquid, $x_{C_2H_6}$ | in vapor, $y_{C_2H_6}$ |
|---|---|---|---|---|---|
| 68.6 | 341.8 | 650 | 4.48 | 0.5023 | - |
| 75.2 | 348.4 | 700 | 4.83 | 0.5023 | - |
| 81.8 | 355.0 | 750 | 5.17 | 0.5023 | - |
| 88.3 | 361.5 | 800 | 5.52 | 0.5023 | - |
| 94.8 | 368.0 | 850 | 5.86 | 0.5023 | - |
| 101.4 | 374.6 | 900 | 6.21 | 0.5023 | - |
| 107.9 | 381.1 | 950 | 6.55 | 0.5023 | - |
| 113.7 | 386.9 | 1000 | 6.89 | 0.5023 | - |
| 121.2 | 394.4 | 1050 | 7.24 | 0.5023 | - |
| 128.3 | 401.5 | 1100 | 7.58 | 0.5023 | - |
| 135.7 | 408.9 | 1150 | 7.93 | 0.5023 | - |
| 143.3 | 416.5 | 1200 | 8.27 | 0.5023 | - |
| 151.3 | 424.5 | 1250 | 8.62 | 0.5023 | - |
| 160.2 | 433.4 | 1300 | 8.96 | 0.5023 | - |
| 171.8 | 445.0 | 1350 | 9.31 | 0.5023 | - |
| 172.5 | 445.7 | 300 | 2.07 | - | 0.5023 |
| 179.2 | 452.4 | 350 | 2.41 | - | 0.5023 |
| 185.6 | 458.8 | 400 | 2.76 | - | 0.5023 |
| 191.2 | 464.4 | 450 | 3.10 | - | 0.5023 |
| 196.4 | 469.6 | 500 | 3.45 | - | 0.5023 |
| 201.2 | 474.4 | 550 | 3.79 | - | 0.5023 |
| 205.5 | 478.7 | 600 | 4.14 | - | 0.5023 |
| 209.1 | 482.3 | 650 | 4.48 | - | 0.5023 |
| 212.2 | 485.4 | 700 | 4.83 | - | 0.5023 |
| 215.0 | 488.2 | 750 | 5.17 | - | 0.5023 |
| 217.4 | 490.6 | 800 | 5.52 | - | 0.5023 |
| 219.6 | 492.8 | 850 | 5.86 | - | 0.5023 |
| 221.3 | 494.5 | 900 | 6.21 | - | 0.5023 |
| 222.5 | 495.7 | 950 | 6.55 | - | 0.5023 |
| 223.4 | 496.6 | 1000 | 6.89 | - | 0.5023 |
| 223.6 | 496.8 | 1050 | 7.24 | - | 0.5023 |
| 223.4 | 496.6 | 1100 | 7.58 | - | 0.5023 |
| 222.5 | 495.7 | 1150 | 7.93 | - | 0.5023 |
| 220.3 | 493.5 | 1200 | 8.27 | - | 0.5023 |
| 217.5 | 490.7 | 1250 | 8.62 | - | 0.5023 |
| 211.7 | 484.9 | 1300 | 8.96 | - | 0.5023 |
| 201.4 | 474.6 | 1350 | 9.31 | - | 0.5023 |
| 5.3 | 278.5 | 300 | 2.07 | 0.7231 | - |
| 12.7 | 285.9 | 350 | 2.41 | 0.7231 | - |
| 19.4 | 292.6 | 400 | 2.76 | 0.7231 | - |
| 25.7 | 298.9 | 450 | 3.10 | 0.7231 | - |
| 31.8 | 305.0 | 500 | 3.45 | 0.7231 | - |
| 37.6 | 310.8 | 550 | 3.79 | 0.7231 | - |
| 43.1 | 316.3 | 600 | 4.14 | 0.7231 | - |
| 48.3 | 321.5 | 650 | 4.48 | 0.7231 | - |
| 53.4 | 326.6 | 700 | 4.83 | 0.7231 | - |
| 58.2 | 331.4 | 750 | 5.17 | 0.7231 | - |
| 63.0 | 336.2 | 800 | 5.52 | 0.7231 | - |
| 67.8 | 341.0 | 850 | 5.86 | 0.7231 | - |
| 72.6 | 345.8 | 900 | 6.21 | 0.7231 | - |
| 77.3 | 350.5 | 950 | 6.55 | 0.7231 | - |
| 81.8 | 355.0 | 1000 | 6.89 | 0.7231 | - |
| 86.7 | 359.2 | 1050 | 7.24 | 0.7231 | - |
| 91.9 | 365.1 | 1100 | 7.58 | 0.7231 | - |
| 97.1 | 370.3 | 1150 | 7.93 | 0.7231 | - |
| 102.6 | 375.8 | 1200 | 8.27 | 0.7231 | - |
| 108.5 | 381.7 | 1250 | 8.62 | 0.7231 | - |
| 115.5 | 388.7 | 1300 | 8.96 | 0.7231 | - |
| 124.4 | 397.6 | 1350 | 9.31 | 0.7231 | - |

(cont.)

| COMPONENTS: | ORIGINAL MEASUREMENTS: |
|---|---|
| (1) Ethane; $C_2H_6$; [74-84-0] | Kay, W. B.; Nevens, T. D. |
| (2) Benzene; $C_6H_6$; [71-43-2] | *Chem. Eng. Prog. Symp. Ser. no. 3.* 1952, *48*, 108-114. |

EXPERIMENTAL VALUES:

| t/°C | $T$/K [a] | $P$/psia | $P$/MPa [a] | Mole fraction of ethane in liquid, $x_{C_2H_6}$ | Mole fraction of ethane in vapor, $y_{C_2H_6}$ |
|---|---|---|---|---|---|
| 140.0 | 413.2 | 300 | 2.07 | - | 0.7231 |
| 145.3 | 418.5 | 350 | 2.41 | - | 0.7231 |
| 150.2 | 423.4 | 400 | 2.76 | - | 0.7231 |
| 154.5 | 427.7 | 450 | 3.10 | - | 0.7231 |
| 158.2 | 431.4 | 500 | 3.45 | - | 0.7231 |
| 161.9 | 435.1 | 550 | 3.79 | - | 0.7231 |
| 165.0 | 438.2 | 600 | 4.14 | - | 0.7231 |
| 167.9 | 441.1 | 650 | 4.48 | - | 0.7231 |
| 170.3 | 443.5 | 700 | 4.83 | - | 0.7231 |
| 172.2 | 445.4 | 750 | 5.17 | - | 0.7231 |
| 173.8 | 447.0 | 800 | 5.52 | - | 0.7231 |
| 175.1 | 448.3 | 850 | 5.86 | - | 0.7231 |
| 176.3 | 449.5 | 900 | 6.21 | - | 0.7231 |
| 177.1 | 450.3 | 950 | 6.55 | - | 0.7231 |
| 177.5 | 450.7 | 1000 | 6.89 | - | 0.7231 |
| 177.5 | 450.7 | 1050 | 7.24 | - | 0.7231 |
| 177.4 | 450.6 | 1100 | 7.58 | - | 0.7231 |
| 176.7 | 449.9 | 1150 | 7.93 | - | 0.7231 |
| 175.6 | 448.8 | 1200 | 8.27 | - | 0.7231 |
| 173.6 | 446.6 | 1250 | 8.62 | - | 0.7231 |
| 170.8 | 444.0 | 1300 | 8.92 | - | 0.7231 |
| 166.2 | 439.4 | 1350 | 9.31 | - | 0.7231 |
| 157.2 | 430.4 | 1400 | 9.65 | - | 0.7231 |
| 134.9 | 408.1 | 1450 | 10.00 | - | 0.7231 |
| 0.6 | 273.8 | 300 | 2.07 | 0.8523 | - |
| 7.7 | 280.9 | 350 | 2.41 | 0.8523 | - |
| 14.0 | 287.2 | 400 | 2.76 | 0.8523 | - |
| 19.8 | 293.0 | 450 | 3.10 | 0.8523 | - |
| 25.2 | 298.4 | 500 | 3.45 | 0.8523 | - |
| 30.2 | 303.4 | 550 | 3.79 | 0.8523 | - |
| 35.0 | 308.2 | 600 | 4.14 | 0.8523 | - |
| 39.7 | 312.9 | 650 | 4.48 | 0.8523 | - |
| 44.1 | 317.3 | 700 | 4.83 | 0.8523 | - |
| 48.6 | 321.8 | 750 | 5.17 | 0.8523 | - |
| 53.0 | 326.2 | 800 | 5.52 | 0.8523 | - |
| 57.4 | 330.6 | 850 | 5.86 | 0.8523 | - |
| 61.6 | 334.8 | 900 | 6.21 | 0.8523 | - |
| 66.2 | 339.4 | 950 | 6.55 | 0.8523 | - |
| 71.0 | 344.2 | 1000 | 6.89 | 0.8523 | - |
| 76.0 | 349.2 | 1050 | 7.24 | 0.8523 | - |
| 81.8 | 355.0 | 1100 | 7.58 | 0.8523 | - |
| 102.5 | 375.7 | 250 | 1.72 | - | 0.8523 |
| 108.4 | 381.6 | 300 | 2.07 | - | 0.8523 |
| 113.4 | 386.6 | 350 | 2.41 | - | 0.8523 |
| 117.2 | 390.4 | 400 | 2.76 | - | 0.8523 |
| 120.3 | 393.5 | 450 | 3.10 | - | 0.8523 |
| 123.3 | 396.5 | 500 | 3.45 | - | 0.8523 |
| 126.1 | 399.3 | 550 | 3.79 | - | 0.8523 |
| 128.3 | 401.5 | 600 | 4.14 | - | 0.8523 |
| 130.0 | 403.2 | 650 | 4.48 | - | 0.8523 |
| 131.5 | 404.7 | 700 | 4.83 | - | 0.8523 |
| 132.6 | 405.8 | 750 | 5.17 | - | 0.8523 |
| 133.4 | 406.6 | 800 | 5.52 | - | 0.8523 |
| 133.9 | 407.1 | 850 | 5.86 | - | 0.8523 |
| 134.2 | 407.4 | 900 | 6.21 | - | 0.8523 |
| 134.0 | 407.2 | 950 | 6.55 | - | 0.8523 |
| 133.4 | 406.6 | 1000 | 6.89 | - | 0.8523 |
| 132.7 | 405.9 | 1050 | 7.24 | - | 0.8523 |
| 131.2 | 404.4 | 1100 | 7.58 | - | 0.8523 |

(cont.)

COMPONENTS:

(1) Ethane; $C_2H_6$; [74-84-0]

(2) Benzene; $C_6H_6$; [71-43-2]

ORIGINAL MEASUREMENTS:

Kay, W. B.; Nevens, T. D.

*Chem. Eng. Prog. Symp. Ser. no. 3*

1952, *48*, 108-114.

EXPERIMENTAL VALUES:

| t/°C | $T$/K [a] | $P$/psia | $P$/MPa [a] | Mole fraction of ethane in liquid, $x_{C_2H_6}$ | Mole fraction of ethane in vapor, $y_{C_2H_6}$ |
|---|---|---|---|---|---|
| 129.0 | 402.2 | 1150 | 7.93 | - | 0.8523 |
| 125.2 | 398.4 | 1200 | 8.27 | - | 0.8523 |
| 115.9 | 389.1 | 1250 | 8.62 | - | 0.8523 |
| 108.5 | 381.7 | 1250 | 8.62 | - | 0.8523 |
| 96.3 | 369.5 | 1200 | 8.27 | - | 0.8523 |
| 85.2 | 358.4 | 1150 | 7.93 | - | 0.8523 |
| -3.5 | 269.7 | 300 | 2.07 | 0.9481 | - |
| 3.7 | 276.9 | 350 | 2.41 | 0.9481 | - |
| 9.9 | 283.1 | 400 | 2.76 | 0.9481 | - |
| 15.3 | 288.5 | 450 | 3.10 | 0.9481 | - |
| 20.3 | 293.5 | 500 | 3.45 | 0.9481 | - |
| 23.0 | 296.2 | 550 | 3.79 | 0.9481 | - |
| 29.3 | 302.5 | 600 | 4.14 | 0.9481 | - |
| 33.5 | 306.7 | 650 | 4.48 | 0.9481 | - |
| 37.4 | 310.6 | 700 | 4.83 | 0.9481 | - |
| 41.2 | 314.4 | 750 | 5.17 | 0.9481 | - |
| 45.6 | 318.8 | 800 | 5.52 | 0.9481 | - |
| 50.6 | 323.8 | 850 | 5.86 | 0.9481 | - |
| 77.1 | 350.3 | 400 | 2.76 | - | 0.9481 |
| 79.6 | 352.8 | 450 | 3.10 | - | 0.9481 |
| 81.3 | 354.5 | 500 | 3.45 | - | 0.9481 |
| 82.7 | 355.9 | 550 | 3.79 | - | 0.9481 |
| 83.3 | 356.5 | 600 | 4.14 | - | 0.9481 |
| 83.7 | 356.9 | 650 | 4.48 | - | 0.9481 |
| 83.8 | 357.0 | 700 | 4.83 | - | 0.9481 |
| 83.5 | 356.7 | 750 | 5.17 | - | 0.9481 |
| 83.1 | 356.3 | 800 | 5.52 | - | 0.9481 |
| 81.9 | 355.1 | 850 | 5.86 | - | 0.9481 |
| 79.6 | 352.8 | 900 | 6.21 | - | 0.9481 |
| 74.2 | 347.4 | 950 | 6.55 | - | 0.9481 |
| 64.9 | 338.1 | 1000 | 6.89 | - | 0.9481 |
| 56.7 | 329.9 | 1050 | 7.24 | - | 0.9481 |

[a] Calculated by compiler.

COMPONENTS:

(1) Ethane; $C_2H_6$; [74-84-0]

(2) Hydrogen sulfide; $H_2S$; [7783-06-4]

ORIGINAL MEASUREMENTS:

Robinson, D. B.; Kalra, H.; Krishnan, T.; Miranda, R. D.

*Proc. Annu. Conv. Gas Process. Assoc. Tech. Pap.* 1975, *54*, 25-31.

VARIABLES:

$T$/K = 283.15-255.32

$P$/MPa = 0.5-2.7

PREPARED BY:

C. L. Young

EXPERIMENTAL VALUES:

| $T$/K | $P$/MPa | Mole fraction of ethane in liquid, $x_{C_2H_6}$ | Mole fraction of ethane in vapor, $y_{C_2H_6}$ | $T$/K | $P$/MPa | Mole fraction of ethane in liquid, $x_{C_2H_6}$ | Mole fraction of ethane in vapor, $y_{C_2H_6}$ |
|---|---|---|---|---|---|---|---|
| 283.15 | 1.575 | 0.0241 | 0.1174 | 255.93 | 1.292 | 0.3316 | 0.6360 |
| | 1.727 | 0.0498 | 0.2013 | 227.93 | 0.256 | 0.0175* | 0.2203 |
| | 1.917 | 0.0849 | 0.2848 | | 0.359 | 0.0600 | 0.4592 |
| | 2.073 | 0.1314 | 0.3716 | | 0.439 | 0.1119 | 0.5830 |
| | 2.348 | 0.2264 | 0.4791 | | 0.522 | 0.2114 | 0.6619 |
| | 2.527 | 0.3126 | 0.5334 | | 0.587 | 0.4065 | 0.7203 |
| | 2.644 | 0.3938 | 0.5784 | 199.93 | 0.0652 | 0.0085* | 0.2602 |
| 255.32 | 0.643 | 0.0095* | 0.0888 | | 0.099 | 0.0291 | 0.5272 |
| | 0.774 | 0.0375 | 0.2509 | | 0.136 | 0.0627 | 0.6614 |
| | 1.016 | 0.1169 | 0.4442 | | 0.168 | 0.1313 | 0.7441 |
| | 1.211 | 0.2330 | 0.5597 | | 0.188 | 0.2189 | 0.7774 |

Additional vapour-liquid equilibrium data in source.

* smoothed value.

AUXILIARY INFORMATION

METHOD/APPARATUS/PROCEDURE:

Cell fitted with two movable pistons which enabled cell contents to be circulated in external line. Fitted with optical system which allowed measurement of refractive index. Temperature measured with iron-constantan thermocouple and pressure with strain gauge transducer. Details in ref. (1). Components charged into cell, mixed by piston movement. Samples withdrawn and analysed by G.C. Details in ref. (1).

SOURCE AND PURITY OF MATERIALS:

1. Phillips Research; purity 99.9 mole per cent or better.

2. Matheson C.P. grade sample; distilled final purity 99.8 mole per cent.

ESTIMATED ERROR:

$\delta T$/K = ±0.05; $\delta P$/MPa = ±0.02; $\delta x_{C_2H_6}$, $\delta y_{C_2H_6}$ = ±0.003.

REFERENCES:

1. Besserer, G. J.; Robinson, D. B. *Can. J. Chem. Eng.* 1971, *49*, 651.

COMPONENTS:

(1) Ethane; $C_2H_6$; [74-84-0]

(2) Acetic acid, methyl ester; (Methyl acetate); $C_3H_6O_2$; [79-20-9]

ORIGINAL MEASUREMENTS:

Ohgaki, K.; Sano, F.; Katayama, T.

*J. Chem. Eng. Data* 1976, *21*, 55-8.

VARIABLES:

$T$/K = 298.15

$P$/MPa = 4.8-38.5

PREPARED BY:

C.L. Young

EXPERIMENTAL VALUES:

| $T$/K | $P/10^5$Pa | Mole fraction of ethane in liquid, $x_{C_2H_6}$ | in vapor, $y_{C_2H_6}$ |
|---|---|---|---|
| 298.15 | 4.805 | 0.0487 | 0.9320 |
| | 9.934 | 0.1072 | 0.9648 |
| | 16.877 | 0.1895 | 0.9762 |
| | 25.283 | 0.3047 | 0.9816 |
| | 32.222 | 0.4489 | 0.9832 |
| | 33.279 | 0.4601 | 0.9834 |
| | 34.108 | 0.5214 | 0.9837 |
| | 34.468 | 0.5784 | 0.9838 |
| | 36.510 | 0.7921 | 0.9840 |
| | 38.495 | 0.9085 | 0.9846 |

AUXILIARY INFORMATION

METHOD/APPARATUS/PROCEDURE:

Static equilibrium cell fitted with windows and magnetic stirrer. Temperature of thermostatic liquid measured with platinum resistance thermometer. Pressure measured using dead weight gauge and differential pressure transducer. Samples of vapor and liquid analysed by gas chromatography. Details in source and ref. (1).

SOURCE AND PURITY OF MATERIALS:

1. Phillips Petroleum Co., sample purity at least 99.96 mole per cent.

2. Merck Co., sample purity about 99.99 mole per cent.

ESTIMATED ERROR: $\delta T$/K = ±0.01; $\delta P/10^5$Pa = ±0.01; $\delta x_{C_2H_6}$ (for $x_{C_2H_6} < 0.5$) = ±1%; $\delta(1-x)_{C_2H_6}$ (for $x_{C_2H_6} > 0.5$) = ±1%. (Similarly for vapor compositions).

REFERENCES:

1. Ohgaki, K.; Katayama, T. *J. Chem. Eng. Data* 1975, *20*, 264.

COMPONENTS:

(1) Ethane; $C_2H_6$; [74-84-0]

(2) 2-Propanone, (Acetone); $C_3H_6O$; [67-64-1]

ORIGINAL MEASUREMENTS:

Ohgaki, K.; Sano, F.; Katayama, T.

*J. Chem. Eng. Data* 1976, *21*, 55-8.

VARIABLES:

$T$/K = 298.15

$P$/MPa = 4.8-39.4

PREPARED BY:

C.L. Young

EXPERIMENTAL VALUES:

| $T$/K | $P/10^5$Pa | Mole fraction of ethane in liquid, $x_{C_2H_6}$ | in vapor, $y_{C_2H_6}$ |
|---|---|---|---|
| 298.15 | 4.804 | 0.0427 | 0.9371 |
| | 9.838 | 0.0916 | 0.9648 |
| | 17.696 | 0.1721 | 0.9769 |
| | 26.389 | 0.2826 | 0.9809 |
| | 33.619 | 0.4485 | 0.9819 |
| | 35.622 | 0.5770 | 0.9817 |
| | 36.242 | 0.6919 | 0.9821 |
| | 39.365 | 0.9268 | 0.9841 |

AUXILIARY INFORMATION

METHOD/APPARATUS/PROCEDURE:

Static equilibrium cell fitted with windows and magnetic stirrer. Temperature of thermostatic liquid measured with platinum resistance thermometer. Pressure measured using dead weight gauge and differential pressure transducer. Samples of vapor and liquid analysed by gas chromatography. Details in source and ref. (1).

SOURCE AND PURITY OF MATERIALS:

1. Phillips Petroleum Co. sample purity at least 99.96 mole per cent.
2. Wako Pure Chemical Co. sample purity about 99.99 mole per cent.

ESTIMATED ERROR: $\delta T$/K = ±0.01; $\delta P/10^5$Pa= ±0.01; $\delta x_{C_2H_6}$ (for $x_{C_2H_6} < 0.5$) = ±1%; $\delta(1-x_{C_2H_6})$ (for $x_{C_2H_6} > 0.5$) = ±1%. (similarly for vapor composition, $y$).

REFERENCES:

1. Ohgaki, K.; Katayama, T. *J. Chem. Eng. Data* 1975, *20*, 264.

COMPONENTS:

(1) Ethane; $C_2H_6$; [74-84-0]

(2) Methanol; $CH_4O$; [67-56-1]

ORIGINAL MEASUREMENTS:

Ohgaki, K.; Sano, F.; Katayama, T.

*J. Chem. Eng. Data* 1976, *21*, 55-8.

VARIABLES:

$T$/K = 298.15

$P$/MPa = 10.9-41.3

PREPARED BY:

C.L. Young

EXPERIMENTAL VALUES:

| $T$/K | $P/10^5$Pa | Mole fraction of ethane in liquid, $x_{C_2H_6}$ | Mole fraction of ethane in vapor, $y_{C_2H_6}$ |
|---|---|---|---|
| 298.15 | 10.938 | 0.0403 | 0.9822 |
| | 21.021 | 0.0880 | 0.9887 |
| | 31.432 | 0.1755 | 0.9914 |
| | 38.563 | 0.2728 | 0.9921 |
| | 41.244 | 0.3511 | 0.9922 |

AUXILIARY INFORMATION

METHOD/APPARATUS/PROCEDURE:

Static equilibrium cell fitted with windows and magnetic stirrer. Temperature of thermostatic liquid measured with platinum resistance thermometer. Pressure measured using dead weight gauge and differential pressure transducer. Samples of vapor and liquid analysed by gas chromatography. Details in source and ref. (1).

SOURCE AND PURITY OF MATERIALS:

1. Phillips Petroleum Co. sample purity at least 99.96 mole per cent.

2. Wako Pure Chemical Co. sample purity about 99.99 mole per cent.

ESTIMATED ERROR: $\delta T$/K = ±0.01; $\delta P/10^5$Pa = ±0.01; $\delta x_{C_2H_6}$ (for $x_{C_2H_6} < 0.5$) = ±1%; $\delta(1-x_{C_2H_6})$ (for $x_{C_2H_6} > 0.5$) = ±1%. (similarly for vapor composition $y$).

REFERENCES:

1. Ohgaki, K.; Katayama, T.

*J. Chem. Eng. Data* 1975, *20*, 264.

COMPONENTS:

(1) Ethane; $C_2H_6$; [74-84-0]

(2) Methanol; $CH_3OH$; [67-56-1]

ORIGINAL MEASUREMENTS:

Ma, Y. H.; Kohn, J. P.
*J. Chem. Eng. Data*
1964, *9*, 3-5.

VARIABLES:

$T$/K = 323.15-373.15
$P$/MPa= 1.01-6.08

PREPARED BY:

C. L. Young

EXPERIMENTAL VALUES:

| $T$/K | $P$/MPa | Mole fraction of ethane in liquid, $x_{C_2H_6}$ | Mole fraction of ethane in vapor, $y_{C_2H_6}$ | $T$/K | $P$/MPa | Mole fraction of ethane in liquid, $x_{C_2H_6}$ | Mole fraction of ethane in vapor, $y_{C_2H_6}$ |
|---|---|---|---|---|---|---|---|
| 373.15 | 1.01 | 0.0085 | - | 323.15 | 3.04 | 0.0990 | 0.943 |
| | 2.03 | 0.0284 | 0.687 | | 4.05 | 0.1454 | 0.960 |
| | 3.04 | 0.0510 | 0.756 | | 5.07 | 0.2045 | 0.970 |
| | 4.05 | 0.0719 | 0.808 | | 6.08 | 0.2753 | 0.978 |
| | 5.07 | 0.0967 | 0.842 | 298.15 | 1.01 | 0.0370 | - |
| | 6.08 | 0.1290 | 0.864 | | 2.03 | 0.0871 | - |
| 348.15 | 1.01 | 0.0181 | - | | 3.04 | 0.1665 | - |
| | 2.03 | 0.0433 | 0.855 | | 4.05 | 0.3210 | - |
| | 3.04 | 0.0718 | 0.897 | | 4.128 | 0.3528 | - |
| | 4.05 | 0.1015 | 0.926 | 273.15 | 1.01 | 0.0764 | - |
| | 5.07 | 0.1332 | 0.944 | | 2.03 | 0.2225 | - |
| | 6.08 | 0.1738 | 0.949 | | 2.34 | 0.4068 | - |
| 323.15 | 1.01 | 0.0270 | - | 248.15 | 1.01 | 0.2085 | - |
| | 2.03 | 0.0619 | 0.907 | | 1.21 | 0.3850 | - |

AUXILIARY INFORMATION

METHOD/APPARATUS/PROCEDURE:

Pyrex glass cell. Temperature measured with platinum resistance thermometer and pressure with Bourdon gauge. Bubble and dew points of mixtures of known composition determined. Experimental data quoted obtained by smoothing. Details of apparatus and procedure in ref. (1).

SOURCE AND PURITY OF MATERIALS:

1. Pure grade Matheson sample, vented at 233 K, until ⅓ removed; final purity at least 99.4 mole per cent.
2. J. T. Baker sample, purity 99.8 mole per cent, less than 0.09 mole per water.

ESTIMATED ERROR:

$\delta T$/K = ±0.1; $\delta P$/MPa = ±0.0007; $\delta x_{C_2H_6}$ = ±0.004; $\delta y_{C_2H_6}$ = ±0.006.

REFERENCES:

1. Kohn, J. P.
*Am. Inst. Chem. Eng. J.*
1961, *7*, 514.

**COMPONENTS:**

(1) Ethane; $C_2H_6$; [74-84-0]

(2) Military fuels JP-1 and JP-4

**ORIGINAL MEASUREMENTS:**

Findl, E.; Brande, H.; Edwards, H.

U.S. Dept. Comm., Office Tech. Ser. Report No. AD 274 623, 1960, 216 pp.

*Chem. Abstr.* 1963, *58*, 6628c.

**VARIABLES:**

$T$/K = 310.9 - 533.2

$p_1$/kPa = 172.4 - 2744.

**PREPARED BY:**

H.L. Clever

**METHOD/APPARATUS/PROCEDURE:**

The apparatus was constructed of two concentric glass tubes, the inner of which was of constant bore and which contained a glass bead slightly smaller than the tube diameter. The glass bead served to mix the solvent and also to measure the solution viscosity. Constant temperature mineral oil was circulated through the annular space.

The air content of the solvent, initially air-saturated, was determined by Orsat analysis. The solvent was charged into the cell until a vapor/liquid ratio of about 0.15 was attained. After the cell was allowed to reach the desired temperature, ethane gas was introduced from a high pressure cylinder held at constant temperature (150 ± 1°F). The cell was rocked until constant pressure was attained. The quantity of gas dissolved was calculated taking into account the pressure difference in the ethane supply cylinder, pressure in the equilibration cell, gas compressibility, initial dissolved air content of the solvent and the solvent vapor pressure.

Solubilities were reported graphically as standard volume per unit mass, and Ostwald coefficient. Values were read from the graphs by the compiler.

Experimental details reported earlier (1).

## AUXILIARY INFORMATION

**METHOD/APPARATUS/PROCEDURE:**

See above.

**SOURCE AND PURITY OF MATERIALS:**

(1) Stated to be of high purity.

(2) The fuels were specified as MIL-F-5624c Grade JP-4 and MIL-F-2558 Gradc JP-1 formerly designated as Shell UMF, Grade C.

A table of fuel characteristics was included in the report.

**ESTIMATED ERROR:**

$\delta T$/°F = ± 1

$\delta p$/lb in$^{-2}$ = ± 1

$\delta L/L$ = ± 0.02 (minimum), max.≈0.05

**REFERENCES:**

1. Schlagel, L.A.; Findl, E.; Edwards, H.

Ing Er. Rept. 183, Thompson Products, Inc. Inglewood Lab., Inglewood, CA, USA, Aug 19, 1955.

| COMPONENTS: | ORIGINAL MEASUREMENTS: |
|---|---|
| (1) Ethane; $C_2H_6$; [74-84-0] | Findl, E.; Brande, H.; Edwards, H. |
| (2) Military fuel JP-1 | U. S. Dept. Comm., Office Tech. Ser. Report No. *AD 274 623*, 1960, 216 pp. |
| | *Chem. Abstr.* 1963, *58*, 6628c. |

EXPERIMENTAL VALUES:

| Temperature $t/^0C$ | $T/K$ | Ethane Pressure $p_1$/psia | Solubility $/cm^3(STP)\ g^{-1}$ | Ostwald Coefficient $L/cm^3\ cm^{-3}$ |
|---|---|---|---|---|
| 100 | 310.9 | 45 | 8.3 | 2.86 |
| | | 96 | 20.8 | 3.35 |
| | | 190 | 52.0 | 3.21 |
| | | 195 | 49.0 | 2.91 |
| | | 292.5 | 83.0 | 3.09 |
| | | 292.5 | 88.0 | 2.89 |
| | | 390 | - | 2.90 |
| | | 390 | - | 2.94 |
| | | | | [3.36($p$→0) - 2.89 ($p$=400 psia)][1] |
| 200 | 366.5 | 42.7 | 6.0 | 2.17,2.25 |
| | | 92.0 | 13.0 | 2.14 |
| | | 93.0 | 13.5 | 2.19 |
| | | 194 | 33.0 | 2.05 |
| | | 291 | 42.0 | 1.95 |
| | | 293 | 43.0 | 2.05 |
| | | 393 | 58.0 | 1.87 |
| | | 393 | 59.0 | 1.92 |
| | | | | [2.25($p$→0) - 1.90 ($p$=400 psia)][1] |
| 300 | 422.0 | 42 | 2.5 | - |
| | | 94 | 5.5 | - |
| | | 193 | 12.8 | 1.15 |
| | | 193 | 13.5 | 1.15 |
| | | 291 | 21.5 | 1.17 |
| | | 293 | 22.5 | 1.20 |
| | | 391 | 30.5 | 1.22 |
| | | 393 | 30.5 | 1.17 |
| | | | | [1.15($p$→0) - 1.22 ($p$=400 psia)][1] |
| 400 | 477.6 | 80 | 3.1 | -- |
| | | 82.5 | 3.5 | -- |
| | | 180 | 10.2 | 0.95 |
| | | 182.5 | 9.3 | 1.04 |
| | | 185 | 11.0 | - |
| | | 187.5 | 10.0 | - |
| | | 279 | 18.0 | 1.14 |
| | | 285 | 16.0 | 0.98 |
| | | 285 | 17.2 | - |
| | | 291 | 16.1 | 1.10 |
| | | 379 | 25.5 | 1.13 |
| | | 382.5 | 24.0 | 1.06 |
| | | 390 | 24.8 | 0.99 |
| | | 397.5 | 22.5 | - |
| | | | | [1.00($p$→0) - 1.09 ($p$=400 psia)][1] |
| 500 | 533.2 | 65 | 2.3 | 0.77 |
| | | 166 | 6.5 | 0.82 |
| | | 168 | 6.5 | 0.89 |
| | | 269 | 12.5 | 0.90 |
| | | 270 | 13.2 | 0.91 |
| | | 372 | 18.0 | 0.92 |
| | | 379 | 20.0 | - |
| | | | | [0.79($p$→0) - 0.95 ($p$=400 psia)][1] |

[1] From authors' line through the Ostwald coefficient values.

psia ≡ pounds per square inch absolute. One lb $in^{-2}$ ≡ 6.89476 kPa.

COMPONENTS:

(1) Ethane; $C_2H_6$; [74-84-0]

(2) Military fuel JP-4

ORIGINAL MEASUREMENTS:

Findl, E.; Brande, H.; Edwards, H.

U. S. Dept. Comm., Office Tech. Ser. Report No. *AD 274 623*, 1960, 216 pp.

*Chem. Abstr.* 1963, *58*, 6628c.

EXPERIMENTAL VALUES:

| Temperature $t/^{0}C$ | $T/K$ | Ethane Pressure $p_1$/psia | Solubility /$cm^3$(STP) $g^{-1}$ | Ostwald Coefficient $L/cm^3\ cm^{-3}$ |
|---|---|---|---|---|
| 100 | 310.9 | 95 | 31.0 | 3.76 |
| | | 96 | 33.0 | 3.97 |
| | | 140 | - | 3.95 |
| | | 196 | 75.0 | 3.91 |
| | | 199 | - | 3.88 |
| | | 287.5 | - | 3.87 |
| | | 293 | - | 3.80 |
| | | 296 | - | 3.80 |
| | | | | [4.05($p$→0)-3.72($p$=400psia] |
| 200 | 366.5 | 87 | 13.0 | 2.05 |
| | | 187.5 | 31.0 | 2.13 |
| | | 188 | - | 2.24 |
| | | 281.5 | 50.0 | 2.13 |
| | | 284 | 53.5 | 2.22 |
| | | 376 | - | 2.13 |
| | | 389 | 77.0 | 2.20 |
| | | | | [2.27($p$→0) - 2.17 ($p$=400 psia)][1] |
| 300 | 422.0 | 60 | 5.0 | - |
| | | 95 | 8.1 | 1.45 |
| | | 160 | 16.2 | 1.56 |
| | | 161 | 17.4 | 1.61 |
| | | 204 | 22.5 | 1.60 |
| | | 262 | 30.5 | 1.58 |
| | | 264 | - | 1.64 |
| | | 305 | 35.0 | - |
| | | 355 | 44.0 | 1.68 |
| | | 356 | 42.0 | 1.59 |
| | | | | [1.60($p$→0 to $p$=400 psia)][1] |
| 400 | 477.6 | 63 | - | 1.30 |
| | | 112.5 | 7.9 | 1.25 |
| | | 120 | 8.4 | 1.20 |
| | | 212.5 | 18.0 | 1.29 |
| | | 215 | 18.0 | 1.31 |
| | | 243 | 21.0 | 1.27 |
| | | 307.5 | 27.0 | 1.32 |
| | | 310 | 27.0 | 1.34 |
| | | | | [1.29($p$→0) - 1.34 ($p$=400 psia)][1] |
| 500 | 533.2 | 25 | - | 1.18 |
| | | 67 | - | 1.13 |
| | | 70 | - | 1.12 |
| | | 87.5 | 4.0 | - |
| | | 121 | 6.5 | 1.05 |
| | | 121 | 7.0 | 1.11 |
| | | 149 | 9.0 | - |
| | | 173 | 11.5 | 1.08 |
| | | 210 | - | 1.11 |
| | | 221 | 15.5 | 1.10 |
| | | | | [1.09($p$→0) - 1.15 ($p$=400 psia)][1] |

[1] From authors' line through the Ostwald coefficient values.

# SYSTEM INDEX

Underlined numbers refer to evaluations. Other numbers refer to compiled tables.


A

Acetamide, *N*-methyl- 195 - 199, 215
Acetic acid 195 - 199, 208
Acetic acid, ethyl ester 195 - 199, 201
methyl ester 195 - 199, 204
232, 233, 245
pentyl ester 195 - 199, 200
Acetone see 2-propanone
Aminocyclohexane see cyclohexanamine
2-Aminoethanol see ethanol, 2-amino-
Ammonium bromide (aqueous) 29, 30, 36
Ammonium chloride (aqueous) 29, 30, 35
Amyl acetate see acetic acid, pentyl ester
Aniline see benzenamine

B

Benzenamine 195 - 199, 212
Benzene 138 - 141,
155 - 157
232, 233,
238 - 243
Benzene (ternary) 163 - 165
Benzenemethanol 166, 167, 194
Benzene, chloro- 138 - 141,
158, 159
Benzene, methyl- 138 - 141, 154
Benzene, 1,1'-methylenebis- 195 - 199, 203
Benzene, nitro- 195 - 199,
210, 211
Benzyl alcohol see benzenemethanol
1,1-Bicyclohexyl 138 - 141, 150
Blood (dog) 195 - 199, 231
1-Butanamine, 1,1,2,2,3,3,4,4,4-nonafluoro-
*N*,*N*-bis(nonafluorobutyl)- 138 - 141, 142, 143
Butanamine, *N*,*N*,*N*-tributyl-, bromide (aqueous) 29, 30, 40
Butane 110 - 112, 116
Butane (ternary) 137
*iso*-Butane see propane, 2-methyl-
Butane, 2,2'-dimethyl- 77, 78, 105
1-Butanol 166, 167,
179 - 181

C

Calcium chloride (aqueous) 30, 43
Carbon disulfide 138 - 141,
161, 162
Carbon tetrachloride see methane, tetrachloro-
Cesium chloride (aqueous) 33, 59
Chorex see ethane, 1,1'-oxybis[2-chloro-
Chlorobenzene see benzene, chloro-
Cyclohexanamine 195 - 199, 202
Cyclohexane 138 - 141,
148, 149
Cyclohexanol 166, 167, 189
Cyclohexylamine see cyclohexanamine
Cyclotetrasiloxane, octamethyl- 195 - 199, 227

D

Decahydronaphthalene see naphthalene (decahydro-
Decalin see naphthalene (decahydro-

| | | |
|---|---|---|
| Decane | | 77, 78, 93, 94, 110 - 112, 129 - 131 |
| Diethanolamine | see ethanol, 2,2'-iminobis- | |
| Diethylether | see ethane, 1,1'-oxybis- | |
| Diglycolamine | see ethanol, 2-(2-aminoethoxy)- | |
| 2,2'-Dimethylbutane | see butane, 2,2'-dimethyl- | |
| *N*,*N*-Dimethylformamide | see formamide, *N*,*N*-dimethyl- | |
| *N*,*N*-Dimethylglycine, potassium salt | see glycine, *N*,*N*-dimethyl-, potassium salt | |
| Dimethylsulfoxide | see methane, sulfinylbis- | |
| 1,4-Dioxane | | 195 - 199, 207 |
| 1,4-Dioxane (aqueous) | | 64, 66 |
| 1,3-Dioxolane-2-one, 4-methyl- | | 195 - 199, 223, 224 |
| Diphenylmethane | see benzene, 1,1'-methylenebis- | |
| Dipropylene glycol | see propanol, oxybis- | |
| DMSO | see methane, sulfinylbis- | |
| Docosane | | 77, 78, 104 |
| Dodecane | | 77, 78, 95, 96, 110 - 112, 132 - 135 |

E

| | | |
|---|---|---|
| Eicosane | | 77, 78, 102, 103, 110 - 112, 136 |
| 1,2-Epoxyethylene | see oxirane | |
| Ethanamine, *N*,*N*,*N*-triethyl-, bromide (aqueous) | | 29, 30, 38 |
| Ethanamine, 2-hydroxy-, *N*,*N*,*N*-tris(2-hydroxyethyl)-, bromide (aqueous) | | 29, 30, 42 |
| 1,2-Ethanediol | | 166, 167, 190, 192 |
| Ethane, 1,1'-oxybis- | | 232, 233, 234 |
| 1,1'-oxybis-[2-chloro- | | 195 - 199, 209 |
| 1,1,2-trichloro-1,2,2-trifluoro- | | 138 - 141, 146, 147 |
| 1,1,2-trichloro-1,2,2-trifluoro- (ternary) | | 163 - 165 |
| Ethanol | | 166, 167, 172 - 175 |
| Ethanol (aqueous) | | 64, 69, 70 |
| 2-amino- | | 195 - 199, 226 |
| 2-amino- (aqueous) | | 74, 75 |
| 2-(2-aminoethoxy)- | | 195 - 199, 225 |
| 2-ethoxy- | | 195 - 199, 205 |
| 2,2'-iminobis- (aqueous) | | 74, 76 |
| Ethene (ternary) | | 23, 24 - 26 |
| Ethyl acetate | see acetic acid, ethyl ester | |
| Ethyl cellosolve | see ethanol, 2-ethoxy- | |
| 2-Ethoxyethanol | see ethanol, 2-ethoxy- | |
| Ethylene glycol | see 1,2-ethanediol | |
| Ethylene oxide | see oxirane | |

F

| | | |
|---|---|---|
| Formamide, *N*,*N*-dimethyl- | | 195 - 199, 220 |
| Freon 113 | see ethane, 1,1,2-trichloro-1,2,2-trifluoro- | |
| Fuel (military) | | 232, 233, 249 - 251 |
| 2-Furancarboxaldehyde | | 195 - 199, 219 |
| Furfuryl | see 2-furancarboxaldehyde | |

G

| | | |
|---|---|---|
| Glycine, *N*,*N*-dimethyl-, potassium salt | | 60 |
| α,*D*-Glucopyranoside-β-*D*-fructofuranosyl- | | 64, 68 |

Guanidine monohydrochloride (aqueous) 29, 30, 41
Guanidinium chloride see guanidine monohydrochloride

H

Heavy water see water-*D2*
Heptadecane 77, 78, 98
2,2,4,4,6,8,8-Heptamethylnonane see nonane, 2,2,4,4,6,8,8-heptamethyl-
Heptane 77, 78, 85 - 89, 110 - 112, 125, 126
Heptane, hexadecafluoro- 138 - 141, 144, 145
1-Heptanol 166, 167, 187
Hexadecafluoroheptane see heptane, hexadecafluoro-
Hexadecane 77, 78, 97 - 100
Hexadecane (ternary) 109
Hexamethylphosphoric triamide see phosphoric triamide, hexamethyl-
Hexane 77, 78, 79 - 84, 110 - 112, 121 - 124
Hexane (ternary) 109
*neo*-Hexane see butane, 2,2'-dimethyl-
1-Hexanol 166, 167, 185, 186
Hydrocarbon oil 232, 233, 236, 237
Hydrogen sulfide 232, 233, 244
2-Hydroxy-*N*,*N*,*N*-tris(2-hydroxyethyl)-ethanaminium bromide) see ethanamine, 2-hydroxy-, *N*,*N*,*N*-tris(2-hydroxyethyl)-, bromide

J

JP-1 see fuel
JP-4 see fuel

L

Lithium chloride (aqueous) 30, 31, 44, 45
Lung tissue (dog) 195 - 199, 231

M

Methanamine, *N*,*N*,*N*-trimethyl-, bromide 29, 30, 37
Methane, sulfinylbis- 195 - 199, 221, 222
Methane, sulfinylbis- (aqueous) 64, 67
Methane, tetrachloro- 138 - 141, 152, 153
Methanol 166, 167, 168 - 171, 232, 233, 247, 248
*N*-methylacetamide see acetamide, *N*-methyl-
1-Methylnaphthalene see naphthalene, 1-methyl-
2-Methylpropane see propane, 2-methyl-
4-methyl-1,3-dioxolane-2-one see 1,3-dioxolan-2-one, 4-methyl-

1-Methyl-2-pyrrolidinone see 2-pyrrolidinone, 1-methyl-
Methylacetate see acetic acid, methyl ester
Monoethanolamine see ethanol, 2-amino-

N

Naphthalene, decahydro- 138 - 141, 151
Naphthalene, 1-methyl- 138 - 141, 160
Nitrobenzene see benzene, nitro-
Nonane 77, 78, 92, 93
Nonane, 2,2,4,4,6,8,8-heptamethyl- 77, 78, 107

O

Octadecane 77, 78, 99, 101
Octamethylcyclotetrasiloxane see cyclotetrasiloxane, octamethyl-
Octane 77, 78, 89 - 91, 110 - 112, 127, 128
*iso*-Octane see pentane, 2,2,4-trimethyl-
1-Octanol 166, 167, 188
Oil see hydrocarbon oil
Oxirane 195 - 199, 213, 214
1,1'-Oxybis-2-chloroethane see ethane, 1,1'-oxybis[2-chloro-

P

Pentane 77, 78, 79, 110 - 112, 119, 120
Pentane (ternary) 137
Pentane, 2,2,4-trimethyl- 77, 78, 106
1-Pentanol 166, 167, 182 - 184
1-Pentanol (multicomponent) 61, 62, 63
Pentyl acetate see acetic acid, pentyl ester
Perfluoroheptane see heptane, hexadecafluoro-
Perfluorotributylamine see 1-butanamine,1,1,2,2,3,3,4,4,4-nonafluoro-*N*,*N*-bis(nonafluorobutyl)-
Phenol 166, 167, 191
Phosphoric acid, tributyl ester 195 - 199, 230
Phosphoric acid, triethyl ester 195 - 199, 230
Phosphoric acid, trimethyl ester 195 - 199, 230
Phosphoric acid, tripropyl ester 195 - 199, 230
Phosphoric acid, tris (2-methyl propyl) ester 195 - 199, 230
Phosphoric triamide, hexamethyl- 195 - 199, 218
Potassium chloride (aqueous) 32, 57
Potassium iodide (aqueous) 32, 33, 58
Propanamine, *N*,*N*,*N*-tripropyl-, bromide (aqueous) 29, 30, 39
Propane 110 - 112, 113 - 115
Propane, 2-methyl- 110 - 112, 117, 118
1-Propanol 166, 167, 176 - 178
1-Propanol (aqueous) 64, 65
Propanol, oxybis- 166, 167, 193
2-Propanone 195 - 199, 206 232, 233, 246
1-Propene 232, 233, 235

Propylene see 1-propene
Propylene carbonate see 1,3-dioxolan-2-one, 4-methyl-
2-Pyrrolidinone, 1-methyl- 195 - 199, 216, 217

S

SDS see sulfuric acid, monododecyl ester sodium salt
Sodium bromide (aqueous) 32, 51
chloride (aqueous) 31, 32, 46 - 50, 55, 56
dodecyl sulfate see sulfuric acid, monododecyl ester sodium salt
iodide (aqueous) 32, 52
sulfate see sulfuric acid, sodium salt
Squalane see tetracosane, 2,6,10,15,19,23-hexamethyl-
Sucrose see α-*D*-glucopyranoside-β-*D*-fructofuranosyl
Sulfolane see thiophene, tetrahydro-1,1-dioxide
Sulfuric acid 29, 32, 34, 53
Sulfuric acid, monododecyl ester, sodium salt (aqueous and multicomponent) 32, 54 - 56, 61, 62, 63
sodium salt (aqueous) 32, 53

T

Tetracosane, 2,6,10,15,19,23-hexamethyl- 77, 78, 108
Tetraethanolammonium bromide see ethanamine, 2-hydroxy-, *N*,*N*,*N*-tris(2-hydroxyethyl)-, bromide
Tetraethylammonium bromide see ethanamine, *N*,*N*,*N*-triethyl-, bromide
Tetramethylammonium bromide see methanamine, *N*,*N*,*N*-trimethyl-, bromide
Tetrapropylammonium bromide see propanamine, *N*,*N*,*N*-tripropyl-, bromide
Tetrabutylammonium bromide see butanamine, *N*,*N*,*N*-tributyl-, bromide
Thiophene, tetrahydro-1,1-dioxide 195 - 199, 226
Toluene see benzene, methyl-
1,1,2-Trichloro-1,2,2-trifluoroethane see ethane, 1,1,2-trichloro-1,2,2-trifluoro-
*N*,*N*,*N*-Tributylbutanaminium bromide see butanamine, *N*,*N*,*N*-tributyl-, bromide
*N*,*N*,*N*-Triethylethanaminium bromide see ethanamine, *N*,*N*,*N*-triethyl-, bromide
*N*,*N*,*N*-Trimethylmethanaminium bromide see methanamine, *N*,*N*,*N*-trimethyl-, bromide
2,2,4-Trimethylpentane see pentane, 2,2,4-trimethyl-
*N*,*N*,*N*-Tripropylpropanaminium bromide see propanamine, *N*,*N*,*N*-tripropyl-, bromide

U

Urea (aqueous) 64, 71 - 73

W

Water-*D2* 195 - 199, 229
Water 1, 2, 3 - 15, 16, 17 - 22
Water (+ ethane + ethene) 23, 24 - 26

# REGISTRY NUMBER INDEX

Underlined numbers refer to evaluations. Other numbers refer to compiled tables.


| Registry No. | Pages |
|---|---|
| 50-01-1 | 29, 30, 41 |
| 56-23-5 | 138-141, 152, 153 |
| 57-13-6 | 64, 71-73 |
| 57-59-1 | 64, 68 |
| 60-29-7 | 232, 233, 234 |
| 63-53-3 | 195-199, 212 |
| 64-17-5 | 64, 69, 70, 166, 167, 172-175 |
| 64-19-7 | 195-199, 208 |
| 64-20-0 | 29, 30, 37 |
| 67-56-1 | 166, 167, 168-171, 232, 233, 247, 248 |
| 67-64-1 | 195-199, 206, 232, 233, 246 |
| 67-68-5 | 64, 67, 195-199, 221, 222 |
| 68-12-2 | 195-199, 220 |
| 71-23-8 | 64, 65, 166, 167, 176-178 |
| 71-36-3 | 166, 167, 179-181 |
| 71-41-0 | 61, 62, 63, 166, 167, 182-184 |
| 71-43-2 | 138-141, 155-157, 163-165, 232, 233, 238-243 |
| 71-91-0 | 29, 30, 38 |
| 74-85-1 | 23, 24-26 |
| 74-98-6 | 110-112, 113-115 |
| 75-15-0 | 138-141, 161, 162 |
| 75-21-8 | 195-199, 213, 214 |
| 75-28-5 | 110-112, 117, 118 |
| 75-83-2 | 105 |
| 76-13-1 | 138-141, 146, 147, 163-165 |
| 78-40-0 | 195-199, 230 |
| 79-16-3 | 195-199, 215 |
| 79-20-9 | 195-199, 204, 232, 233, 245 |
| 91-17-8 | 138-141, 151 |
| 92-51-3 | 138-141, 150 |
| 98-01-1 | 195-199, 219 |
| 98-95-3 | 195-199, 210, 211 |
| 100-51-6 | 166, 167, 194 |
| 100-81-5 | 195-199, 203 |
| 106-97-8 | 110-112, 116, 137 |
| 107-21-1 | 166, 167, 190, 192 |
| 108-32-7 | 195-199, 223, 224 |
| 108-88-3 | 138-141, 154 |
| 108-90-7 | 138-141, 158, 159 |
| 108-91-8 | 195-199, 202 |
| 108-93-0 | 166, 167, 189 |
| 108-95-2 | 166, 167, 191 |
| 109-66-0 | 77, 78, 79, 110-112, 119, 120, 137 |
| 110-54-3 | 77, 78, 79-84, 109, 110-112, 121-124 |
| 110-80-5 | 195-199, 205 |
| 110-82-7 | 138-141, 148, 149 |
| 110-98-5 | 166, 167, 193 |
| 111-01-3 | 77, 78, 108 |
| 111-27-3 | 166, 167, 185, 186 |
| 111-42-2 | 74, 76 |

| | |
|---|---|
| 111-44-4 | 195-199, 209 |
| 111-65-9 | 77, 78, 89-93, 110-112, 127, 128 |
| 111-70-6 | 166, 167, 187 |
| 111-84-2 | 77, 78, 91-93 |
| 111-87-5 | 166, 167, 186 |
| 112-40-3 | 77, 78, 95, 96, 110-112, 132-135 |
| 112-95-8 | 77, 78, 102, 103, 110-112, 136 |
| 115-07-1 | 232, 233, 235 |
| 123-91-1 | 64, 66, 195-199, 207 |
| 124-18-5 | 77, 78, 93, 94, 110-112, 129-131 |
| 126-33-0 | 195-199, 228 |
| 126-71-6 | 195-199, 230 |
| 126-73-8 | 195-199, 230 |
| 141-43-5 | 74, 75, 195-199, 226 |
| 141-78-6 | 195-199, 201 |
| 142-82-5 | 77, 78, 85-89, 110-112, 125, 126 |
| 151-21-3 | 32, 54-56, 61, 62, 63 |
| 311-89-7 | 138-141, 142, 143 |
| 335-57-9 | 138-141, 144, 145 |
| 512-56-1 | 195-199, 230 |
| 513-08-6 | 195-199, 230 |
| 540-84-1 | 106 |
| 544-76-3 | 77, 78, 97-100, 109 |
| 556-67-2 | 195-199, 227 |
| 593-45-3 | 77, 78, 99, 101 |
| 628-63-7 | 195-199, 200 |
| 629-78-7 | 77, 78, 98 |
| 629-97-0 | 77, 78, 104 |
| 680-31-9 | 195-199, 218 |
| 872-50-4 | 195-199, 216, 217 |
| 929-06-6 | 195-199, 225 |
| 1321-94-4 | 138-141, 160 |
| 1643-19-2 | 29, 30, 40 |
| 1941-30-6 | 29, 30, 39 |
| 4328-04-5 | 29, 30, 42 |
| 4390-04-9 | 107 |
| 7447-40-7 | 32, 57 |
| 7447-41-8 | 30, 31, 44, 45 |
| 7647-14-5 | 31, 32, 46-50 |
| 7647-15-6 | 32, 51 |
| 7647-17-8 | 33, 59 |
| 7664-93-9 | 29, 32, 34, 53 |
| 7681-11-0 | 32, 33, 58 |
| 7681-17-8 | 32, 33 |
| 7681-82-5 | 32, 52 |
| 7732-18-5 | 1, 2, 3-15, 16, 17-22, 23, 24-26, 27-33, 34-60, 61, 62, 63, 64, 65-73, 74, 75, 76 |
| 7757-82-6 | 32, 53 |
| 7783-06-4 | 232, 233, 244 |
| 7789-20-0 | 195-199, 229 |
| 10043-52-4 | 30, 43 |
| 12124-97-9 | 29, 30, 36 |
| 12125-02-9 | 29, 30, 35 |
| 17647-86-8 | 60 |

# AUTHOR INDEX


ANTHONY, R. G. 20, 24-26
ARMITAGE, D. A. 147, 155, 164, 165
ARRIAGA, J. L. 94, 95
AVDEEVA, O. I. 48

BARTON, J. R. 208
BATTINO, R. 3, 4
BEN-NAIM, A. 6-8, 35, 45, 50-52, 57, 59, 65-71, 148, 169, 173, 177, 180, 183, 186, 207, 229
BESSERER, G. J. 117, 118
BILLETT, F. 11, 44, 47, 58
BIRCHER, L. J. 168, 172, 176, 179, 182, 185, 187, 188
BOYER, F. L. 168, 172, 176, 179, 182, 185, 187, 188
BOZHOVAKAYA, T. K. 48
BRANDE, H. 249 - 251

CHAPPELOW, C. C. 103, 108, 160, 227
CHENG, S. C. 83, 85, 91, 96, 97, 109
CLAES, P. 82, 105
CLAUSSEN, W. F. 12
CLEMENTS, H. E. 154
COFFIN, R. L. 13, 41, 73
CUKOR, P. M. 100, 150, 203
CULBERSON, O. L. 17 - 19
CZAPLINSKI, A. 15, 43, 49
CZERSKI, L. 15, 43, 49

DANNEIL, A. 21, 22
DAVIES, J. A. 236, 237
DECANNIÈRE, L. 82, 105
DELANEY, D. E. 215
DYMOND, J. H. 149, 221

EDWARDS, H. 249 - 251
EUCKEN, A. 10, 46
EZHELEVA, A. E. 205, 209, 210, 219

FINDL, E. 249 - 251
FONTAINE, R. 82, 105
FRANCK, E. U. 21, 22

GARST, A. W. 75, 76
GJALDBAEK, J. C. 87, 90, 92, 144, 145, 162,170, 174, 178, 181, 184, 189, 190

HANDA, Y. P. 3, 4
HARRIS, H. G. 101, 102, 104
HAYDUK, W. 83 - 86, 91, 96, 97, 109
HERLIHY, J. C. 137
HERTZBERG, G. 10, 46
HESS, L. G. 214
HILDEBRAND, J. H. 106, 143, 146, 163
HORIUTI, J. 152, 156, 158, 159, 204, 206
HORN, A. B. 17
HOSKINS, J. C. 55, 56, 62, 63
HOWARD, W. B. 220
HSU, C. C. 208
HUNG, J. H. 5, 36 -40, 42, 72

JADOT, R. 79, 89, 93, 153, 157

KAHRE, L. C. 113
KALRA, H. 244
KATAYAMA, T. 124, 234, 238, 245 - 247
KAY, W. B. 239 - 243
KEEVIL, T. A. 202
KENTON, F. H. 53
KING, A. D. 54 -56, 62, 63
KOBATAKE, Y. 106, 143
KOBE, K. A. 53
KOHN, J. P. 127, 128, 134-136, 248

KRISHNAN, T. 244

LACEY, W. N. 119, 120, 235 - 237
LAWSON, J. D. 75, 76
LEE, K. H. 134, 135
LEGRET, D. 133
LENOIR, J.-Y. 98, 151, 191-194, 211, 212, 217, 218, 222, 224, 230
LEUHDDEMANN, R. 60
LHOTAK, V. 116
LINFORD, R. G. 146, 147, 155, 163 - 165
LUTSYK, A. I. 14, 34

MALIK, S. K. 13, 41, 73
MALIK, V. K. 84, 86
MATHESON, I. B. C. 54
MA, Y. H. 248
MAYFORTH, F. R. 220
McCAFFREY, D. S. 127, 128
McDANIEL, A. S. 81, 88, 171, 175, 200, 201
McKAY, R. A. 235
McKETTA, J. J. 17 - 20, 24 - 26
MEHRA, V. S. 125, 126
MESKEL-LESAVRE, M. 132
MIKSOVSKY, I. 114, 115
MIRANDA, R. D. 244
MISHNINA, T. A. 48
MONFORT, J. P. 94, 95
MORRISON, T. J. 11, 44, 47, 58
MORTIMER, G. A. 80, 154

NEVENS, T. D. 239 - 244
NG, S. 101, 102, 104
NIEMANN, H. 145, 162, 170, 174, 178, 181, 184, 189, 190
NODDES, G. 60

OHGAKI, K. 124, 234, 238, 245 - 247
OLSON, J. D. 213

POLGLASE, M. F. 12
POWELL, R. J. 142, 161
PRAUSNITZ, J. M. 100 - 104, 108,150, 160, 203, 216, 223, 225 - 228
PURI, S. 136

REAMER, H. H. 119, 120, 129 - 131, 235
RENAULT, P. 98, 151,191 - 194, 211, 212, 217, 218, 222, 224, 230
RENON, H. 98, 99, 107, 132, 133, 151, 191 - 194, 211, 212, 217, 218, 222, 224, 230
RETTICH, T. R. 3, 4,
RICHON, D. 99, 107, 132, 133
RIVAS, O. R. 216, 223, 225, 226, 228
ROBINSON, D. B. 117, 118, 244
RODRIGUES, A. B. J. 127, 128
RUDAKOV, E. S. 14, 34

SAGE, B. H. 119, 120, 129 - 131,235 - 237
SANO, F. 124, 234, 238, 245 - 247
SCHOCH, E. P. 220
SCHWARZ, H.-G. 60
SHERBORNE, J. E. 236, 237
SILBERBERG, I. H. 121 - 123
STOLLER, L. 13, 41, 73
STREITWIESER, A. 202

TAYLOR, D. R. 202
THODOS, G. 125, 126, 137
THOMSEN, E. S. 87, 90, 92, 144
THORNHILL, D. G. T. 147, 155, 164, 165
TILQUIN, B. 82, 105
TILTON, V. V. 214

TODHEIDE, K. 21, 22

WAGNER, P. D. 231
WATERS, J. A. 80, 145
WEN, W.-Y. 5, 36 - 40, 42, 72
WETLAUFER, D. B. 13, 41, 73
WICHTERLE, I. 114 - 116
WILF, J. 6, 229
WILHELM, E. 3, 4
WINKLER, L. W. 9
WOOD, R. H. 215

YAACOBI, M. 6 - 8, 35, 45, 50 - 52, 57, 59, 65 - 71, 148, 169, 173, 177, 180, 183, 186, 207, 229
YOUNG, I. H. 231

ZAIS, E. J. 121 - 123
ZORIN, A. D. 205, 209, 210, 219